SECRET LIVES OF
CARNIVOROUS MARSUPIALS

ANDREW BAKER AND CHRIS DICKMAN

CSIRO
PUBLISHING

A catalogue record for this book is available from the National Library of Australia.

Published by

CSIRO Publishing
36 Gardiner Road, Clayton VIC 3168
Private Bag 10, Clayton South VIC 3169
Australia

Telephone: +61 3 9545 8400
Email: publishing.sales@csiro.au
Website: www.publish.csiro.au

Front cover: (clockwise from top left) Lesser Hairy-footed Dunnart (*Sminthopsis youngsoni*), Bobby Tamayo; Pallid Fat-tailed Opossum (*Thylamys pallidior*), Dario Podesta; Common Planigale (*Planigale maculata*), Steve Murphy.

Title page: Patagonian Opossum (*Lestodelphys halli*), Dario Podesta.

Back cover: Spotted-tailed Quoll (*Dasyurus maculatus*), David Nelson.

Set in Adobe 10/14 Palatino LT Std and Trajan
Edited by Peter Storer
Cover design by Andrew Weatherill
Typeset by Desktop Concepts Pty Ltd, Melbourne
Index by Bruce Gillespie
Printed by Ingram Lightning Source

FOREWORD

In so many ways, this beautiful book represents both the beginning and the end for me.

In 1965, Basil Marlow and Jacaranda Press published a fat little black-and-white pocket guide called *Marsupials of Australia*. There were line drawings throughout, but only two photos in the whole book: a benign quoll on the front cover and the ferocious-looking author on the back. When I was first introduced to him in 1976, Basil was Curator of Mammals at the Australian Museum (having succeeded Ellis Troughton), and I was a callow Queensland Museum 'cadet' in my first year of apprenticeship to Mike Archer who was Basil's counterpart in Brisbane. Mike, who had encouraged me to study antechinuses, was introducing me to Marlow. For me it was a nervous meeting with a secret hero, Marlow having published the earliest study of reproduction in a local antechinus species and I couldn't wait to talk carnivorous-marsupial-shop with him. But it was an unpromising start. Basil, just about to retire, was announcing theatrically that his passion at the present time was being lavished on 'pinnipeds'. Then he overheard me asking Mike what pinnipeds were. Before Mike could save me, Basil swung around and glared at me hissing, 'pinn-i-peds are creatures that are *feather-footed* … as opposed to those that are *feather-brained*.' And from then on, until the end of our visit to Sydney, he called me *Dick* (a thinly veiled muddle with the name of the comedian Dick Van Dyke). It was lucky that Queensland had no real pinnipeds, so I never had to interact with Basil again, God bless him (and the seals).

Fortunately, things quickly got better because, very soon after that (1979), the Australian Mammal Society held its Annual Scientific Meeting in Brisbane. One of the rising stars that year was presenting a paper on the comparative ecology of two species of antechinus that lived together in the Brindabellas, on the New South Wales/Australian Capital Territory border. It was Chris Dickman: gracious, convivial, extremely bright but unpretentious – a young man who already knew the power of a quiet voice and the capacity to hold back. Chris, not long arrived from the UK, was fascinated by the bizarre reproductive corner that antechinuses had backed themselves into and wanted to crack the code that differentially triggered their annual reproductive surges up and down the east Australian coastline. I was interested in travelling up and down the same coastline too, but more with a mind to sort out the tangle of names people had attached to the antechinuses that occurred there. Suffice to say that, apart from plenty of other reasons, there was enough common ground in the conundrums associated with marsupial mice in 1979 for Chris and I to forge a friendship then that would be life-long.

Chris was so clever he cracked that enigmatic reproductive code and published the results in a few years, but not only did he demonstrate how the rate of change of photoperiod triggered annual antechinus breeding, he also proved the species he had been studying alongside *Antechinus swainsonii* wasn't what everyone thought it was (*A. stuartii*), but a new species he went on to describe as *Antechinus agilis*. That was just the beginning.

On the other side of New South Wales, where that state borders Queensland, I couldn't resolve the problems I had with antechinus taxonomy by conventional morphometrics. It wasn't until almost 35 years later when Andrew Baker came along, that it all finally made sense! Not long after he completed his doctorate, Andrew worked with me in the Vertebrate Section of the Queensland Museum. We made innumerable trips slopping around the mangroves documenting the decline of Water Mice (*Xeromys*) in response to a huge Gold Coast real-estate development. And we cut up, transported and cleaned the bones of more stinking, stranded whales than we now care to think about. Andrew was sharp, tireless, self-deprecating and quick to exceed all that was ever expected of him, while simultaneously outshining all who instructed him. He was also a talented artist, musician and comic. Soon after his short museum stint, he took up a lecturing job at Queensland University of Technology. By that time he understood, by osmosis, all the problems of antechinus taxonomy in Queensland and was ready to lend a hand in solving them.

In the space of a few years, Andrew and his team of incredible students and co-workers systematically tackled all the problems I was clueless to sort out, as well as the plethora of other snarls they unravelled along the way. Using the combination of biochemical and classical morphometric techniques they resolved the once-murky relationships within the genus with fluorescent clarity, revised the taxonomy of all members of the genus individually, and added five new carnivorous marsupial species to the nation's biodiversity tally.

During my professional life I saw some monumental turnarounds and discoveries ... the recovery of Humpback Whales from the brink of extinction, the rediscovery of the Night Parrot and the living embodiment of Longman's Beaked-whale, known for 120 years from only bones. Any one of those episodes in a lifetime might make you feel as if it were all worthwhile. But, on top of that, to see the problems that dogged your career meticulously worked through, resolved and published all in a manner that made you feel proud of the next generation (when it could have left you defeated) is worth the gold of ten-thousand retirement watches.

I always enjoy it much better when I read a book and know that the author was not just a great writer, but a person of great humanity. Well, here is a double-bunger. A covetable book written by two extremely good examples of the best that can be found in mankind. It's such a pleasure to commend them and their book to you.

Steve Van Dyck, September 2017

Contents

ACKNOWLEDGEMENTS

We are indebted to the vignette contributors who most generously shared their research stories, including: Soledad Albanese, Mike Archer, Diana Fisher, Menna Jones, Marissa Parrott, Silvia Pavan and Emerson Vieira.

We are most grateful also to Steve Van Dyck who composed the Foreword.

Many of the incredible photos that grace these pages were donated by researchers who spent hours of toil in the wilds to secure precious images of their study animals. Our heartfelt thanks are extended to: Auscape, British Museum of Natural History, Jorge Brito, Gary Cranitch, Francisco Fonturbel, Marcela Franco, Henk Godthelp, F. R. Gonzalez, Alex Hartshorne, Andrew Henley, John Hornbuckle, Agustin Iriarte, Pablo Jayat, Steve Murphy, David Nelson, Reed Ojala-Barbour, Museum Victoria, Gustavo Scrocchi, Thiago Semedo, Bobby Tamayo, Hugh Tyndale-Biscoe, Emerson Vieira, Leanne White, Steve Wilson, Paul Zborowski and Anders Zimny.

Many Australian carnivorous marsupial images were sourced from Jiri Lochman at Lochman Transparencies. His wonderful photos allowed us to share some of the secret behaviours of these rarely encountered species. Other excellent shots of South American carnivorous marsupials were sourced from Dario Podesta and Chico Sanchez.

Thanks to Jesse Hawley who carefully researched and reconstructed his artistic impressions of extinct carnivorous marsupials.

Lauren Webb and Briana Melideo (CSIRO) patiently provided publication advice, and Peter Storer very carefully copy edited the entire manuscript.

Many early prints used in the book were the result of painstaking effort by wildlife artists and scientists. They did a marvellous job of representing these bizarre and wonderful creatures whose appearance doubtless stunned them hundreds of years ago. We thank them for their diligence.

Finally, we dedicate this book with appreciation to the thousands of scientists and adventurers over the last 500 years who shared their hard-earned and voluminous expert knowledge of carnivorous marsupials in papers, books, reports and letters. In some cases, such riches came at the expense of their lives. We have drawn deeply from their collective wellspring of knowledge in producing this work.

Andrew Baker's Acknowledgements

I dedicate this book with warmth and love to my wonderful partner, Stephanie and children – Ethan, Oliver, Eleanor and Joseph – without whose unconditional support, patience and sacrifice, the task would have proved impossible.

I thank Nina Prasolova, Narelle McCaffrey, Kate Harbison, Christoph Shrank (Queensland University of Technology), Roberto Miguez (British Museum of Natural History), Meg Lloyd (Queensland Museum), Sandy Ingleby, Patricia Egan and Vanessa Finney (Australian Museum), Ellen Alers (Smithsonian Institution) and personnel from archives of: British Museum of Natural History, Smithsonian Institution, Western Australian Museum, Queensland Museum and Australian Museum. They generously provided and/or interpreted archived mammal collector and taxonomist correspondences.

I am indebted to Steve Reeves who miraculously rendered Oldfield Thomas' handwritten scrawl to English with uncanny accuracy.

Steve Van Dyck, Ian Gynther and Harry Hines provided sage and timely academic advice.

Thomas Mutton, Eugene Mason, Emma Gray and Coral Pearce enthusiastically shared results from their postgraduate research projects.

Chris Dickman's Acknowledgements

I am indebted to my wife, Carol, for her wonderful love and inspirational support over many years, and for her many important, but unappreciated, contributions to mammalogy before and during the writing of this book. It was Carol, for example, who came up with the name *'agilis'* for the new *Antechinus* that we discovered, after being inspired by seeing these cheeky acrobats running amok in our tiny flat in Canberra. Our daughter Alice and son in law Matt have been part of the family enterprise, sharing our passion for marsupials and nurturing assorted creatures that have contributed to our collective understanding of them.

My students, colleagues and other laboratory members contributed many insights, much information and technical and academic advice as writing of this book progressed. Glenda Wardle, Bobby Tamayo, Chin-Liang Beh, Aaron Greenville, David Nelson and Tom Newsome shared their expertise at key times, and Carol McKechnie, Madeline de Montford, Bede Moses, Ché Parker and Stephanie Yip provided very helpful comments on early drafts of the manuscript.

Many visitors to my laboratory from South America have greatly increased my knowledge of the American carnivorous marsupials over the years, and I am particularly indebted to Emerson Vieira, Natália Leiner, Soledad Albanese, Jayme Prevedello, Mariana Ferreira and Camila Cassano for sharing their wisdom. Courtesy of Soledad, I was privileged to see my first South American carnivorous marsupial – a Pallid Fat-tailed Opossum – while visiting Mendoza. Many other friends and colleagues in Australasia and the Americas, too numerous to name, have contributed immeasurably to my work and understanding of the secret lives of carnivorous marsupials, and I am very grateful to all.

1

THE CARNIVOROUS MARSUPIALS

Can you imagine a mammal that weighs as little as a penny yet leaps five times its own height to take down prey far larger than itself? If not, can you visualise a rapacious killer that dispatches each victim with a deft bite to the neck before ripping its head off?

How about flesh-loving devils that relentlessly pursue victims to exhaustion then simply tear them to pieces? The hunting pack squabbles over the carcass in a screaming feast all night long, making enough racket to wake the dead. By morning, not a trace of the victim remains. Even its bones are ground to powder by a formidable predator sporting the most powerful jaws for its size of any living mammal.

What of maniacal male mammals that unceremoniously grab females by the neck to mate in frenzied sex sessions lasting up to 14 hours straight? The males invariably mate themselves to death, poisoned by their own raging hormones. Females are nymphs, and their clutch of young will have several fathers. With greedy mouth and grasping arms, their babies race each other to the mother's pouch. Once there, they attach to the first available nipple, suckling its life-giving milk. With all teats taken, late arrivals to the pouch simply wither and die.

What about prehistoric forms that include apex predator equivalents of lions, tigers, bears, wolves and even sabre-toothed, flesh-eating kangaroos?

Collectively, these animals are known as carnivorous marsupials. They include ~136 living species, inhabiting remote parts of Australia, the Americas and New Guinea. These unique creatures are secretive, rarely encountered and little understood.

Many of Earth's species, among them carnivorous marsupials, have succumbed to the pressures imposed by human sprawl – sent to extinction in the last few hundred years. Many more carnivorous marsupials are under serious threat of extermination, amid Earth's sixth and first human-caused, mass extinction.

How can we save the survivors that remain?

More practically perhaps, as a society, why would we wish to save them?

First, humanity must appreciate these unique and secretive animals and decide they *need* to be saved – this requires knowledge and understanding. Without such knowledge, we can, and will do, little or nothing collectively to intervene. Recent decades are testament to this. To ultimately save and conserve our precious carnivorous marsupials, then, we must first and foremost get to know them. That is the aim of this book.

What are marsupials?

Marsupials form one of three groups that together comprise the living mammals. There are more than 5500 living species of mammals; over 330 of these are marsupials, five are monotremes, while the dominant remainder comprises placental mammals. Most female mammals, including all placentals and marsupials, give birth to live young and then suckle them with nutritious milk, but female monotremes, which include the platypus and echidnas, lay eggs. After hatching, the young of these ancient mammals are fed on milk crudely oozed from skin glands on the mother's belly. Despite these differences, all

Fig. 1.1. The Endangered Black-tailed Dusky Antechinus (*Antechinus arktos*) is found only in the highest, wettest ancient Gondwanan rainforests of eastern Australia. Source: Gary Cranitch.

mammals are united by their ability to produce milk and also by their possession of fur. These traits distinguish them from other vertebrates (animals with backbones). Other modern groups of vertebrates include amphibians, reptiles, birds and fishes.

There are numerous differences between placental and marsupial mammals, especially relating to reproduction. Female placental mammals, including humans, nurture their young before birth with a placenta. Offspring come into the world fairly well developed. Marsupials, in contrast – including iconic species such as koala, kangaroos and wombats – are born at a very immature stage of development and look like pink jelly-beans. Although a simple placenta forms in all marsupial species, it nourishes the young for a relatively short part of their overall development; in some species, this may be just a few days. The tiny young crawl up to the mother's 'marsupium' or pouch, attaching to nipples there. The marsupial pouch may be fully developed, as in kangaroos, or it may be simply a naked depression on the belly with some skin folds that only partially protect the young as they grow. The latter is the case in most carnivorous marsupials. In marsupials, most of the reproductive effort occurs during a prolonged suckling period. Male placentals have testicles behind/under the penis. Marsupial males

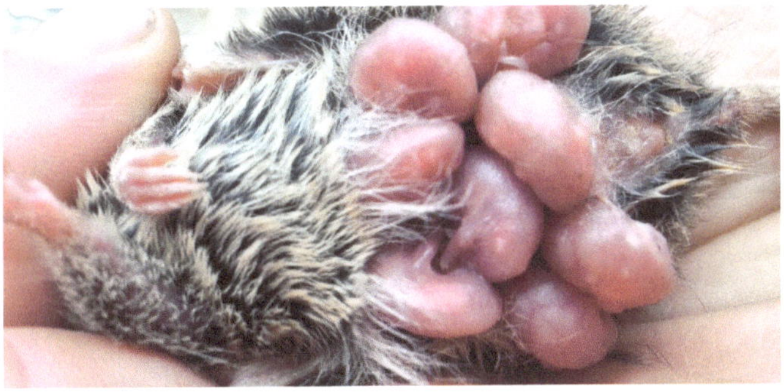

Fig. 1.2. The Silver-headed Antechinus (*Antechinus argentus*) (viewed from the underside) – one of Australia's rarest mammals and likely endangered. This is 'Polly-Jean', a pit-tagged two-year-old mother with eight, 4-week-old young. Antechinuses have no discrete 'pouch', just a depression on the belly which cannot contain the rapidly growing offspring for long. By 6 weeks old, the young are simply too large for the mother to carry around. She leaves them in a nest of leaves while she forages, returning several times during the night to suckle the babies on her nutritious milk. Source: Andrew Baker.

have the opposite arrangement. Male placentals may have non-functioning nipples, while male marsupials typically have no nipples at all, although some at least have mammary cells as pouch young.

What makes a marsupial 'carnivorous'?

Animal dietary strategies have traditionally and simply been divided into three broad categories. Herbivores eat plants, carnivores eat meat and omnivores eat both. However, this classification is too vague to account for the complexity of all feeding strategies. Some animals switch diets at different parts of their life cycle or even between seasons. Others consume mostly flesh and incidentally eat plants when ingesting the guts of their herbivorous prey. Some animals have very specific diets. For example, there are those that consume solely fruit (frugivores) or nectar (nectarivores). Others, such as a few species of bats, eat only blood (sanguivores). There is overlap in many prey categories. For example, species preying mostly on insects are insectivorous, but broadly speaking they may be considered carnivorous because they kill and eat animal prey. Insects are animals too! What's more, finding out an animal's diet is not a trivial exercise, especially for solitary and secretive species that are

Fig. 1.3. Carnivorous marsupial diets vary, but most of the smaller species, such as this Common Planigale (*Planigale maculata*), eat a variety of invertebrates. Source: Steve Murphy.

difficult to find and observe under natural conditions. Such elusive animals must be caught, their scats (i.e. faeces or poo) collected and analysed under a microscope. The scattered hard fragments of pre-digested prey remains in faeces are not easily identified, even by experts. Sometimes, prey items such as soft-bodied worms and slugs cannot be seen in scats at all. In such cases, only an analysis of DNA (see following section) in the faeces may reveal the identity of prey items. In short, this means there is conjecture as to how best to categorise the diet of any animal. For our purposes, it is debatable how many marsupials may be considered carnivorous. In part, this is because not all species have been subjected to study. However, many species that **have** been well studied include variable amounts of fruit or other plant products in their diet. Here, we nominate as carnivorous those marsupials with diets that consist by majority (more than 50%) of vertebrate or invertebrate prey. Even so, some marsupials inevitably fall across the grey borders of 'carnivory' into other dietary categories, and we mention some of these at the end of the next chapter.

Diversity: how many carnivorous marsupials are there?

Earth's creatures are taxonomically positioned using a system invented by Carolus (Carl) Linnaeus in the 1700s to impose a classifying order on relationships. The structuring may be familiar: kingdom, phylum, class, order, family, genus and species. These various groupings describe relationships between organisms and their evolutionary patterns are often represented in diagrams called 'trees of life'. The kingdom is the most all-encompassing category and, for creatures, includes all animal life. Species, by contrast, can be viewed as very fine branches on the tree of life and represent collections of individuals that are similar in appearance, genetic structure and behaviour. Because of the extraordinary diversity of animals – living as well as extinct – these categories are themselves often sub-divided. For example, all mammals are grouped together within the phylum Chordata with the other vertebrate groups mentioned earlier, but are separated from them within their own class, Mammalia. Marsupials are then grouped below the level of class but above the level of order in their own intermediate category, the 'infraclass' Marsupialia. Some scientists have proposed alternatives to this arrangement, but all agree that marsupials form a large and diverse group within the class Mammalia.

Some debate also surrounds the formal definition of species. We return to this issue in more detail in Chapter 4, but provide a brief background here as a primer. For most creatures, a good rule of thumb is that members of the same species can be recognised if they are at least **potentially** able to breed with each other. We emphasise the word 'potentially' here for good reason. Species sometimes have broken and discontinuous distributions due to loss of habitat, changes in climate or other reasons, and these isolated populations may then begin to diverge from each other as they adapt progressively to local conditions. Animals in these isolates could potentially interbreed if they were able to come together, and thus conform to our definition of species, but the isolated populations may be given their own 'subspecies' names if they are different enough to warrant it. How much difference is enough? This remains a vexed question, but increasingly it is being answered by detailed molecular studies of an organism's DNA. DNA (deoxyribonucleic acid), the double-helix molecule that carries the building blocks of life, differs between all living things and, most usefully for biologists, allows us to identify different organisms. It is not usually necessary to describe the entire genetic material of animal species to understand their relationships with each other; comparisons of selected regions of DNA can suffice. In essence, if these selected regions of DNA are compared and show little or no difference between populations, the populations can be considered to belong to the same species. Assuming differences in DNA have gradually accrued over time, increasing differences in the DNA are suggestive of subspecies, and then of species, genera, or even more distant relationships on the basis that they are increasingly less likely to be interbreeding. Here, we note where DNA analyses have been used to identify species and to describe relationships between groups of species, and use the concept of subspecies mostly to describe geographically different populations of the same species.

Genus and species names, which are often of Greek or Latin derivation, are *italicised,* and together form the scientific name of a species. The combination of genus and species (e.g. *Antechinus arktos*) is unique for every organism on Earth. Unlike common names, which can sometimes be the same

Fig. 1.4. An excellent example of living carnivorous marsupial diversity. The extraordinary, Endangered Numbat (*Myrmecobius fasciatus*) is a survivor of an ancient group. DNA and morphological analyses suggest that Numbat ancestors last shared common ancestry with other marsupials ~30–40 million years ago. Source: Jiri Lochman.

for different species, scientific names cannot be confused. To maintain readability we use species' common names throughout the text, but also provide the scientific name at the first mention in each chapter. If we are discussing several species in the same genus, we also use the scientific shorthand of abbreviating the genus name to just its first letter. Thus, if *Antechinus arktos* is mentioned and we then refer to another antechinus species, it will be introduced as *A.* followed by the specific name (e.g. *A. adustus*, *A. stuartii*). We also follow the convention of capitalising the first letters of species' common names, such as Rusty Antechinus (*A. adustus*) and Brown Antechinus (*A. stuartii*). This ensures clarity in referring to the species in question rather than adjectives to describe their colour or behaviour.

The living carnivorous marsupials total ~136 species within 32 genera and six families. Almost half (45%) of these occur only in Australia – some 61 species in 15 genera, three families and two orders. But almost as many carnivorous marsupials (43%) reside only in the Americas, totalling 58 species in 13 genera, three families and three orders. These are found mostly in South America, but also the southern and central reaches of North America, with four species (in as many genera) found in the Central American isthmus and Mexico. Eleven per cent of the world's carnivorous marsupials are found only in New Guinea, including 15 species in six genera, one family and one order. Just two species (1%) of carnivorous marsupial species are shared between northern Australia and New Guinea. These two countries, plus New Zealand, comprise the majority, by landmass, of Australasia. Yet New Zealand has no native marsupials – and just two living native land mammals, both bats (derived from Australia). The Americas share no living marsupials whatsoever with Australasia. Thus, most living carnivorous marsupials are 'endemic' to their landmass – found there and nowhere else. And some species, as we shall discover, have very limited distributions indeed.

So the reader can better appreciate their diversity, we briefly introduce each of the carnivorous marsupial genera in Chapter 2, providing a snapshot of their history of discovery, taxonomy, biology (breeding, behaviour, diet) and conservation. These are themes that we return to in more detail later in the book. We also provide an appendix listing all species with their common and scientific names and family relationships. This will be a handy reference to consult when carnivorous marsupials are named in later chapters.

2

GUIDE TO CARNIVOROUS MARSUPIALS

In the accounts of genera in this chapter, we refer to:

1. 'Size', which equates to the average adult body length, including the head, body and tail, and range of body weights (in brackets).
2. The 'IUCN (International Union for Conservation of Nature) Red List' (see also Chapter 11), which provides a comprehensive and objective global approach for evaluating conservation status including, in increasing level of extinction risk: non-threatened species categories of Least Concern and Near Threatened; threatened species categories of Vulnerable, Endangered and Critically Endangered; and two extinction categories, Extinct in the Wild and Extinct.

Fig. 2.1. The largest living carnivorous marsupial, the Tasmanian Devil (*Sarcophilus harrisii*), can weigh up to an impressive 14 kg – similar to a medium-sized dog. The primary prey are kangaroos, wallabies, possums and wombats, which it kills by ambush or relentlessly running them to death. Once exhausted beyond endurance, the prey are simply torn limb from limb and devoured. Source: Jiri Lochman.

Carnivorous marsupials are classified as part of:

Kingdom Animalia

Phylum Chordata

Class Mammalia

Infraclass Marsupialia

Australian fauna (2 orders, 3 families, 15 genera, 61 endemic species)

Order Dasyuromorphia; Family Dasyuridae (13 genera, 58 species)

Genus *Antechinomys* (Kultarr or Jerboa-marsupial, 1 species)

Distribution: Semi-arid and arid Australia

Habitat: Desert plains, stony and sandy areas where small bushes and grasses predominate, scrub land, clay pans

Size: 17–25 cm (20–30 g)

The Kultarr was named by British naturalist John Gould in 1856. He depicted the animal on a tree branch, but it is in fact active only on the ground. Kultarrs were first considered the marsupial equivalent of the hopping-mouse (jerboa), but in 1867 the Kultarr was given its own genus, *Antechinomys*. The Kultarr's relationships with other Australian marsupials remain uncertain even to the present day. From the 1970s, the Kultarr was recognised as being

Fig. 2.2. The Kultarr (*Antechinomys laniger*). Note the hugely elongated hind feet, which the animal uses in concert with the hands to bound about with astonishing agility. Source: Jiri Lochman.

distinctive morphologically (i.e. in body and skull form), although bearing similarity to the dunnarts (genus *Sminthopsis*). Several DNA studies suggest that the dunnarts are not monophyletic (i.e. more closely related to each other than they are to anything else). In phylogenetic 'trees of life' based on DNA, the Kultarr is nested within the dunnart group. Indeed, one study using both DNA and morphology (body characteristics such as fur colour, measures of external and internal (skull) features), has shown that Kultarrs are most closely related to Long-tailed Dunnarts (*Sminthopsis longicaudata*), excluding the rest of the dunnarts. Two subspecies of Kultarr are recognised, which together possess a wide distribution across arid central Australia.

Kultarrs are small carnivorous marsupials with fawn–sandy brown fur above and white fur on the chest and belly. They have dark fur down the mid-line of the face and around the eye. Their ears are very large; the eyes protrude. They have long, thin hind feet. A long tail ends in a sparse brush of dark hairs. These animals are active at night, spending much time in search of food. They are known to enter torpor (i.e. a period of inactivity; a mini-hibernation) for up to 16 hours. Contrary to popular belief, these animals do not hop, but bound rapidly using hands and feet. This gives the animal great agility and an ability to rapidly change direction, which is useful to evade predators and catch prey. Kultarrs mostly eat invertebrates, including spiders, centipedes, crickets and cockroaches. They have a long breeding season and may mate several times throughout the year. Breeding timing varies depending on location and is likely related to day length. Exact length of gestation (i.e. pregnancy) has not been determined, but is more than 12 days. Females have six to eight teats and there is no discrete pouch, only folds of skin that protect the young during early stages of suckling. Almost nothing is known about their movement patterns. The species is adapted to life in open country and appears to undergo changes in population size throughout the year. Kultarrs are rare and scattered. They are notoriously difficult to catch using standard Elliott (aluminium box) traps, and capture rates in alternative types of trap, such as pitfall traps (where holes are dug in the ground), are usually also very low.

This species is classified as Least Concern on the IUCN Red List. Nevertheless, some populations may be extinct, victim to intensified land use by pastoralists and introduced predators such as the European Red Fox (*Vulpes vulpes*) and feral House Cat (*Felis catus*).

Genus *Antechinus* (antechinuses, 15 species)

Distribution: Eastern, south-western and northern Australia, mostly in coastal or near-coastal regions; one species can be found in semi-arid mulga

Habitat: Open to closed forest and rainforest, but including heaths, sedges and grassland in some species

Size: 12–31 cm (16–178 g)

The first antechinus was captured on Waterhouse Island off the north-east coast of Tasmania. It was named by French naturalist Étienne Geoffroy Saint-Hilaire more than 200 years ago, in 1803. The Frenchman originally classified the beast as a new species of quoll (genus *Dasyurus*). In 1840, another species was discovered in Tasmania and named by the redoubtable George Waterhouse, then Keeper at the British Museum. He nominated the animal as a new species of *Phascogale*. The genus *Antechinus* was formally recognised by William Sharp Macleay when he nominated a third species in this carnivorous marsupial group captured near Sydney Harbour, New South Wales. Macleay was from a family of keen naturalists, which included his father, Alexander and cousin, William. The Macleays' primary focus was on insects, with broader interests that included vertebrates. In 1837, Irish artist and naturalist James Stuart, then working as Colonial Assistant Surgeon, had discovered a little marsupial near Sydney Harbour and passed it on to his friend William Sharp Macleay. Macleay named the new species after Stuart, raising the new genus *Antechinus* at the same time, in 1841. He was just in time for Stuart to recognise the honour: a year later, Stuart was dead after a third bout of typhus contracted during his duties tending the sick on incoming disease-ridden ships. Later, there followed the naming of two further new species of *Antechinus*, by British taxonomist Oldfield Thomas in 1904 and 1923. No other species of *Antechinus* was named until Australian taxonomist Steve Van Dyck's Cinnamon Antechinus (*A. leo*), in 1980. A range of genetic studies in the 1980s–1990s tested relationships among various families and genera of Australian mammals, including dasyurids. This work prompted description of other antechinus species. When Van Dyck conducted his comprehensive review of antechinuses in 2002, he recognised 10 living species. Since 2012, using DNA and morphology, five new species have been discovered, bringing the total to 15 species in the genus (see also Chapter 4, vignette). A further species most likely needs to be split in two, and this taxonomic work is in progress.

Antechinuses are small- to medium-sized carnivorous marsupials that are generally brownish to greyish in colour. The tails are thin and tapering, ranging from somewhat shorter to slightly longer than the head–body length. The fur is dense and generally soft. Species inhabiting colder areas tend to have a thicker, shaggier coat. The heads are conical in shape and ears are small to medium in size. Hands and feet are moderate in size and claws can be long in a couple of species that are adept diggers. Some species have relatively long, narrow snouts that give them a shrew-like appearance. Antechinuses are generalist predators, consuming mostly invertebrates such as beetles, cockroaches, spiders, grasshoppers, bugs, moths/butterflies, millipedes, scorpions, ants, and the like. They will also take a range of vertebrates, including skinks and frogs. The largest species regularly eat small birds and mammals. They are ferocious predators, taking on prey sometimes almost as large as themselves. In captivity, if unwary House Mice (*Mus musculus*) enter their cages or traps, most antechinus species will tear the heads from their hapless prey before quickly devouring the remainder.

Antechinuses are perhaps best known for their semelparous, so-called 'suicidal' reproduction. No male antechinus lives beyond the breeding season, which occurs annually across 1–3 weeks. Breeding is stressful – males and females mate promiscuously in sessions lasting up to 14 hours at a time. The annual male die-off is an inevitable result of surges in male testosterone that ensure a failure in the stress hormone (cortisol) cut-off switch. Excess cortisol causes systemic 'poisoning' and males suffer from internal bleeding. Organ malfunction followed by death results inevitably from immune system failure. This ensures that no male is alive by the time the first babies are born. Gestation is typically about a month long. Females will have four to 14 young, depending on the species. The 3–4 mm young attach themselves to nipples on the mother's pouch-less abdomen. Sunken walls and skin folds afford some protection

Fig. 2.3. The Agile Antechinus (*Antechinus agilis*). Many antechinuses spend as much time in trees as they do on the ground. Some species have been implicated in pollination of various plant species, such as the *Banksia* depicted here. Source: Jiri Lochman.

for the young while the mother moves about. The young remain attached to the nipples for ~6 weeks. They will spend a further 6 weeks in a spherical leaf/grass/twig nest, hidden in a tree trunk, fallen log, crevice or disused burrow. Once weaned, they leave the nest as juveniles after several experimental hunting forays with their mother. The general pattern is for antechinuses to breed in winter and give birth to young in the spring, then disperse in summer. In any given species at a certain location, reproduction is typically strongly synchronised. It will occur at close to the same time every year, depending in large part on day length and the rate at which the day length is increasing. A few species are mostly ground-dwelling, but the rest are as comfortable in trees as they are on the ground. Antechinuses are exceptionally fast-moving, exhibiting short, sharp, staccato/jerky movements. They are extremely agile and can leap a metre from a crouch position. In captivity, most species readily use rat running wheels, but they run on the outside of the wheel, upside down, rather than using it in the traditional manner! Most species are hostile if handled. Larger species will hiss and bite readily, especially males during the breeding season.

Until recently, no antechinus was listed on the IUCN Red List, but two species are now listed (Vulnerable and Near Threatened) and a number are currently under consideration for listing, including three of the five most recently named. Two of these new species apparently have among the smallest distributions of any Australian mammal. Antechinuses generally require forested areas for feeding and nesting, so land clearing, which has been marked and ongoing in eastern and south-western Australia, is a threat. Antechinuses fall prey to a range of feral predators such as cats and foxes. Pigs (*Sus scrofa*), Cattle (*Bos taurus*) and Horses (*Equus caballus*), which are ubiquitous throughout Australia, even in protected areas, trample their habitat. Fires, both planned and accidental, will cause local antechinus numbers to crash. It will take anywhere from 1 to 20 years for populations to re-establish, provided they have been able to take refuge in adjacent unburned areas. Climate change threatens several high-altitude species.

Genus *Dasycercus* (mulgaras, 2 species)

Distribution: Central and Western Australia

Habitat: Arid spinifex grasslands, sand plains, rocky desert (gibber) plains and dune fields

Size: 18–36 cm (60–185 g)

Frederick Waterhouse was an English naturalist whose famous elder brother, George, was Keeper at the British Museum. Frederick sailed for Australia with his wife in the 1850s to make his own mark. In 1856, Frederick's donation of his own valuable insect and bird specimens helped found the South Australian Museum. In 1862, the museum opened and Frederick Waterhouse became Curator. While occupying this position, Waterhouse received a single specimen of an odd carnivorous marsupial from central Australia. Not knowing what to make of it, Waterhouse forwarded the specimen to gifted, but controversial, German-born Australian taxonomist Gerard Krefft, then Director of the Australian Museum. In 1867, Krefft nominated a new genus for the strange mammal, *Chaetocercus*. In the years that followed, Krefft divided his attention between the study of mammals and reptiles. In 1874, Krefft was fired, being famously carried to the museum door and thrown unceremoniously into the street. He had fallen afoul of the powerful Museum Trustees, accusing them of misappropriation of museum funds. They in turn branded Krefft a drunk, falsifier of attendance records and purveyor of dirty postcards! Krefft later sued and won a settlement against the Museum. As it turned out, Krefft's mammal genus *Chaetocercus* had already been coined in 1855 to denote a group of hummingbirds. So, in 1875, Director of the Berlin Zoological Museum Wilhelm Peters nominated the genus *Dasycercus*. A second species, from the Pilbara in Western Australia, was named in 1904 by Assistant Curator at the Australian Museum, Edgar Waite. Waite had received three live specimens from Bernard Woodward, Director of the Western Australian Museum, and named the new species in honour of their collector, A.C. Blyth. DNA studies since 2000 have confirmed the two species are divergent and position them as sister (i.e. most closely related in trees of life, or 'phylogenies') to the Kowari (*Dasyuroides byrnei*).

Fig. 2.4. Mulgaras (*Dasycercus* spp.) are avid diggers. They live in burrows that they excavate in and around sand dunes. Most burrows have multiple entrances or pop-holes that the occupants can emerge from or duck into as need arises. They are not strictly nocturnal and sometimes sunbathe near burrow entrances. Source: Jiri Lochman.

Between 1980 and 2003, a range of ecological studies was conducted on mulgaras, almost all of which were thought at the time to be on the Crest-tailed Mulgara (*Dasycercus cristicauda*). However, it has since been discovered that the study animal was, in most cases, actually the Brush-tailed Mulgara (*D. blythi*), which has a much larger distribution. As a result, there is little ecological information available for the Crest-tailed Mulgara. The two species are arid-dwelling, medium-sized and quite similar in appearance. They are best differentiated by the dorsal crest of dark hairs on the tail of the Crest-tailed Mulgara, which is absent in the bushier-tailed Brush-tailed Mulgara. These animals are pale, sandy brown above and greyish-white below. The hands and feet are pale. The head is solid with a shortish, conical snout. The tail is thick and tapering and furry along its length, and reddish brown at the base with about the last third of the tip black.

Mulgaras are opportunistic and generalist carnivores that consume a range of vertebrates and invertebrates. Vertebrates include mammals (various other dasyurids, rodents), reptiles (dragons, skinks, monitors), frogs and birds. Invertebrates include beetles, bugs, grasshoppers, ants, spiders, centipedes, cockroaches, termites and moths/butterflies. Mulgaras are ground-dwellers and mostly active at night. However, in captivity they are also known to sunbathe: stretched out, heads on each other's backs, at burrow entrances. Basking probably helps with temperature regulation, especially in the depths of winter. When food is plentiful, they store fat in their tails. This can be reabsorbed as an energy source in lean times. Little is known about breeding. Females with up to six young in the pouch have been captured in September and October. Mating has been observed in captive individuals during early winter, with young born after a 5–6 week gestation. The young suckle for 3–4 months and are reproductively mature when 10–11 months old. These animals are probably long lived for small mammals. Some individuals have reproduced each year in captivity for at least 6 years. Mulgaras are generally solitary and share burrows only during mating. As in most mammals, males move more than females. Males move away from where they were born as juveniles, but females often stay closer to their mother.

One species (*D. blythi*) is listed as Least Concern and the other (*D. cristicauda*) as Near Threatened on the IUCN Red List, but neither species could be considered abundant. This is at least partly due to predation by feral cats and Red Foxes. Mulgaras were regularly encountered during the 1900s as travellers moved back and forth along the Canning Stock Route in Western Australia. Today, both species probably still occupy large distributions, but are less plentiful than in times gone by.

Genus *Dasyuroides* (Kowari, 1 species)

Distribution: South-eastern Northern Territory, south-western Queensland and north-eastern South Australia, central Australia

Habitat: Rocky desert (gibber) plains between braided river channels and sand dunes where there is less than 25% cover of shrubs

Size: 24–34 cm (70–175 g)

Baldwin Spencer described the Kowari in 1896, placing it in a unique genus, *Dasyuroides*. Later, the animal was included under the same genus as the mulgaras, but DNA and morphology studies have since shown the Kowari to be distinctive, albeit positioned in phylogenies as sister to the mulgaras. Two subspecies are recognised, but the overall distribution is restricted, limited to a small region of central Australia. These animals are fawn to light brown above and whitish below. The head–body of the Kowari is similar to the mulgaras, but the hind feet and tail are quite different. Kowaris have only four toes on the hind foot, while mulgaras have five. The tail base of the Kowari is never fattened but usually is in mulgaras. The black brush of hair on the Kowari's tail is very long and distinctive. The tails of mulgaras are merely black-tipped or feature a low dorsal crest.

Fig. 2.5. Kowaris (*Dasyuroides byrnei*) are capable killers of various vertebrates. This one is devouring a gecko. Source: Wade Hughes/Lochman LT.

Kowaris are active at night and are highly competent, medium-sized opportunistic predators of both vertebrates and invertebrates. They regularly eat Long-haired Rats (*Rattus villosissimus*) and House Mice, as well as a range of invertebrates. Breeding occurs from May to December, depending on locality. Females carry up to six pouch young. Individuals typically survive only 1–2 years in the wild. The gestation period is ~35 days, and young remain attached to the nipple for ~56 days. Sexual maturity occurs at ~10–11 months, after which the young disperse and are probably solitary until coming together again to mate. Nests are bowl-shaped and constructed of a variety of materials. The female pouch is incomplete and as the young grow they are sometimes bumped and half-dragged across the ground. Indeed, mothers walk in a stiff-legged gait to reduce the amount of contact larger young have with the ground. Even so, young are often bruised and battered by the experience. Although mostly nocturnal, animals will emerge from their burrows and sunbathe on cool winter mornings. When conditions are favourable and juveniles disperse from the nests, local numbers can swell.

This species is listed as Vulnerable on the IUCN Red List. Kowari numbers have dropped because of careless placement of livestock around waterholes and bores. Stock tramples the prime habitat and burrows of Kowaris. Global warming also poses a threat: under some scenarios, rainfall in desert areas is expected to be even less predictable than at present. Under sustained limited rainfall, local populations of Kowaris can crash.

Genus *Dasykaluta* (Kaluta or Little Red Kaluta, 1 species)

Distribution: Pilbara, Carnarvon Basin, Little Sandy Desert, Western Australia

Habitat: Sandy plains, clay plains, scree slopes, and hummock grasslands, especially with dense spinifex thickets

Size: 14–18 cm (20–40 g)

Named *Dasykaluta rosamondae*, the Little Red Kaluta is a striking reddish, small carnivorous marsupial, renowned for its habit of hiding within spinifex hummocks. The scientific name alludes to the beautiful Rosamund Clifford, red-haired mistress to King Henry II of England, who was hidden by the King behind an elaborate maze. The Little Red Kaluta was first classified as an antechinus in 1964. However, in 1982, studies of penis shape by dasyurid biologist Patricia Woolley suggested this animal to be distinct. Genetic work conducted at the same time concurred, and Australian taxonomist Mike Archer raised a separate genus, *Dasykaluta*, just for this little animal. Recent DNA work suggests that *Dasykaluta* is sister to the Dibbler (*Parantechinus apicalis*).

Kalutas are russet brown to coppery above and below, with rather rough fur. The head and ears are short. The tail, which may be thickened at the base, is slightly shorter than the head–body length. This species feeds voraciously on a range of invertebrates and vertebrates, including lizards. Kalutas are inquisitive, flicking their thick tail in the air while investigating the surroundings. Like antechinuses and phascogales, Kalutas undergo an annual male die-off at the conclusion of a synchronised annual mating season. Mating occurs during September and gestation lasts ~7 weeks. Females give birth to up to eight young in November and wean them after ~3.5–4 months. By February–March, juveniles disperse from the nest and are sexually mature shortly before the mating season begins in September. Males die annually, but some females survive to breed in a second season. Unusually for small carnivorous marsupials, Kalutas are mostly active during the day. Both males and females typically enter torpor each night. Females usually rest throughout the night with their body temperature similar to the surrounding soil (21°C). Males transition in and out of torpor and maintain a higher resting temperature (~25°C). This difference between the sexes may be related to the fact that males are more active than females and possess a larger home range. Some studies have captured hundreds of individual Kalutas across the duration of the research. This suggests densities of more than 1.9 individuals per hectare: a high density relative to other arid-dwelling dasyurids. Yet other studies have

Fig. 2.6. The Little Red Kaluta (*Dasykaluta rosamondae*) eats mostly invertebrates, such as this unfortunate centipede, which it tackles with little fear. Source: Jiri Lochman.

returned much lower captures, so abundances may be patchy across the species' range in Western Australia.

This species is classified as Least Concern on the IUCN Red List, and is found in several protected areas.

Genus *Dasyurus* (quolls, 4 species Australia, 2 species New Guinea)

Distribution: Disjointed and patchy ranges in coastal and near coastal areas of mainland Australia; Tasmania. New Guinea

Habitat: A wide range of forested (dry to wet) habitats, coastal scrub, beaches, heathland, savanna, grasslands; sometimes around human settlements

Size: 45–131 cm (0.2–5.0 kg)

The distinctive quolls were first named by Kerr in 1792. This was more than 20 years after footprints, and, shortly after, an animal itself, were noted by Joseph Banks in his journal dated 1770, as the pioneers entered Sydney Cove. Quolls were known as tiger cats by early colonists, who considered them one of their greatest pests. As naturalist John Gould once penned, quolls were the 'dreaded enemy of poultry.' One animal was captured after killing nine ducks and three roosters. The local newspaper report of the incident proclaimed that quolls kill 'from a naturally unsuitable desire of shedding blood'. Initially, quolls were placed in the civet genus, *Viverra*, but just a few years later, French taxonomist Geoffroy Saint-Hilaire recognised them as marsupials rather than placentals and dutifully gave them their own genus. Other species were soon discovered: in 1800 the Eastern Quoll (*Dasyurus viverrinus*) in Tasmania; in 1841 the Western Quoll (*D. geoffroii*) in Western Australia; and in 1842 the Northern Quoll (*D. hallucatus*) in northern Australia. Two New Guinean quolls have also been recognised: continuing the imaginative naming tradition of their Australian counterparts, the New Guinea Quoll (*D. albopunctatus*) was described in 1880, then much later the Bronze Quoll (*D. spartacus*) in 1988. Recent DNA-based trees position quolls as sister to the Tasmanian Devil (*Sarcophilus harrisii*).

The quolls vary in size from medium to large carnivorous marsupials. They are superficially cat-like in appearance, albeit with longer, broadly pointed snouts and more elongated bodies. Their fur colour typically ranges from fawn to brownish-orange through to dark greyish. There are distinctive

Fig. 2.7. Quolls (*Dasyurus* spp.) are expert hunters and capable climbers, especially the Spotted-tailed Quoll (*D. maculatus*) shown here. They will take possums, gliders and birds from tree nest holes or on branches, sometimes high up in the canopy. Source: David Nelson.

whitish spots on the bodies of all quolls, which, in one Australian species, the Spotted-tailed Quoll (*D. maculatus*), extend along the tail. The tails are tapering and about as long as the head–body length. Tails can sometimes be bushy. Quolls are active mostly at night, but several species will move about during the day, especially in the depths of winter. They are mostly active on the ground or on fallen logs, but are ready and capable climbers and can hunt birds and possums in tree hollows. Quolls are fierce predators of a range of mainly vertebrate prey, tackling animals larger than themselves without fear. Their vertebrate prey includes a range of birds, mammals (bandicoots, gliders, possums, rabbits, wallabies) and reptiles (skinks, snakes). Invertebrate prey is also taken, particularly in summer. The largest species, the Spotted-tailed Quoll, is considered a hyper-carnivore with very high proportions of vertebrate meat in the diet.

Breeding and movements of some Australian species are fairly well known. This is because they are large enough to fit with a radio-collar that can provide months of data before the battery needs replacement. The gestation is ~3 weeks. Females have a rudimentary pouch that typically houses six to eight nipples. In some species, females can reproduce several times in the event of a lost litter. Females repeatedly become receptive during a breeding period that may last a few months in total. Females may mate with multiple males and mating bouts can last up to 17 hours. After a couple of months suckling, young are deposited in the nest by the mother as she goes out to forage alone. Young are fully weaned by ~4–6 months of age. In most cases, males and females live several years and up to 5 years, respectively. Some populations of one species, the Northern Quoll, undergo a male die-off at the end of breeding. At least part of the male decline in this species is due to exhaustion. Males continually visit several females, covering many tens of kilometres a day, to check if females are ready to breed. The home ranges in this species are large: up to 5000 ha for males. Female home ranges may be one-fifth of this size. Quolls rest by day in numerous dens throughout their territory. These sites are conspicuously marked by latrines (i.e. piles of poo), which are often on prominent high points of boulders. These are used to communicate with other quolls to mark territory boundaries.

Populations of quolls tend to be sparse, and many are in decline. There are several reasons. There is probably competition and predation from domestic cats, foxes and dogs (*Canis familiaris*). Cane Toad (*Rhinella marina*) poisoning and consumption of poison baits targeting foxes and dogs also threaten some quoll populations. Add these factors to extensive habitat fragmentation, deliberate shooting and road mortality, and it is unsurprising that quolls are declining. Two species are categorised on the IUCN Red List as Endangered and the remainder as Near Threatened.

Genus *Ningaui* (ningauis, 3 species)

Distribution: Central, western and southern Australia

Habitat: Desert habitat, spinifex hummock grasslands, low shrublands, woodlands, scrub, heathlands

Size: 10–17 cm (3.5–14.0 g)

Mike Archer described the genus *Ningaui* only recently, in 1975. At that time, he named two species. A third species was named in 1983 based on both skull morphology and genetic protein differences. Unfortunately, this last animal is indistinguishable externally from one of its relatives: the two look-alikes occur together in some areas, making it hard for biologists to separate them in the field. Recent DNA phylogenies have resolved the ningauis as sister to the Kultarr and dunnarts.

Ningauis are small, with reddish-brown to greyish fur above and paler fur underneath. The ears are moderate in size and the fur can appear coarse. The tails are thin, tapering and about as long as the head–body length. These small animals are ferocious predators and will tackle a variety of prey including centipedes, spiders, millipedes, cockroaches, butterflies, moths and bugs. They also consume vertebrates such as skinks if opportunity arises, and bite readily when handled. Most water is taken from prey and supplemented by licking dew in arid environments. Females have four to eight nipples in a pouch, which is a simple, bare depression in the belly. Breeding is typically in spring, with young suckled in spring/summer and weaned ~3 months later. These activities are timed to coincide with the greatest likelihood of rainfall and availability of prey to support the mother's production of milk. Juveniles leave the nest after weaning and reach maturity by ~10 months. These animals can reproduce more than once in a season, but, in some populations, females and males may not survive to breed a second time. Ningauis are active by night. In one species, there is evidence of frequent torpor day and night. This must be important to relieve heat loss stress in such a small mammal. By day, ningauis hide in abandoned burrows left by lizards or spiders. By night, they hunt for food, climbing with dexterity among spinifex leaves, using their tail to grip and balance. Recapture rates and densities of these animals are low, suggesting they have mobile home ranges. During or before breeding, individuals may move hundreds of metres each night. Males typically move further than females. Ningauis, and other desert dasyurids such as dunnarts, are notoriously difficult to catch in metal box (Elliott) traps. Pitfall traps are usually much more successful.

Fig. 2.8. A small, but ferocious, carnivorous marsupial of the Australian desert country, the Wongai Ningaui (*Ningaui ridei*), like many of the desert dasyurids, is a mobile predator in a patchily resourced landscape. The animals mostly shelter during the day in burrows that they have usurped from lizards or spiders. This one has captured a centipede, which it has quickly dispatched with a bite to the head. The species rarely drinks free-standing water, obtaining most of its water from a selection of juicy prey items. Source: Jiri Lochman.

All three ningaui species are classified as Least Concern on the IUCN Red List. Ningauis have broad distributions, with some populations occurring in protected areas. Nevertheless, too-frequent and hot burning of landscapes, which destroys hummock grass refuge areas, sheep grazing and sand mining, are threats to local populations.

Genus *Parantechinus* (Dibbler, 1 species)

Distribution: South-west Western Australia, including Fitzgerald River National Park, Boullanger, Whitlock and Escape Islands

Habitat: Scrub- and heathland plant communities

Size: 23–26 cm (40–125 g)

The Dibbler was originally classified as a *Phascogale* when it was named in 1842. More than 100 years later, it was given its own genus, *Parantechinus*. However, in 1964 it was considered to be no more than a colourful antechinus, and was therefore moved into the genus *Antechinus*. In 1982, Mike Archer moved it back into *Parantechinus* on the strength of both genetic and morphological information, and it remains there today. Recent DNA studies confirm the Dibbler's distinctiveness, but its relationships with other dasyurids remain uncertain. It has variously been resolved as sister to the Kaluta, the false antechinuses or the three-striped dasyures (the latter of New Guinea).

Dibblers are small- to medium-sized carnivorous marsupials. The coat is coarse, brownish-grey and flecked with white, giving it a distinctive grizzled or frosted appearance. The belly fur is greyish-white, and often tinged strongly with yellow. This animal also has distinctive white eye-rings and a tapering, hairy tail. Dibblers eat a variety of invertebrates, but can tackle vertebrates such as House Mice on occasion. Invertebrate prey items include spiders, cockroaches, beetles, earwigs, bugs, flies, wasps, scorpions, grasshoppers and crickets. Dibblers breed annually in March. Females have an unenclosed pouch that will carry up to eight young. Young are

Fig. 2.9. An Endangered species of the coastal heath and scrub plant communities of southern Western Australia, Dibblers (*Parantechinus apicalis*) are 'facultatively semelparous' – male die-off after breeding occurs in only some populations in certain years. In the Dibbler's case, this may be related to poor environmental conditions rather than the inevitable hormone-related mass-deaths seen annually in all male antechinuses and phascogales. Source: Jiri Lochman.

dependent on their mother for 3–4 months, after which they leave the nest. They are sexually mature at 10–11 months. In one population on Boullanger Island, and only in some years, males exhibit a complete die-off. Yet in mainland populations, male Dibblers survive beyond their first breeding season in the wild. Studies suggest that die-off of Boullanger Island Dibblers is related to occasionally poor environmental conditions. This is quite different to the stress hormone-related annual die-off in antechinuses, phascogales and the Little Red Kaluta. Dibblers are active both by day and night.

Dibblers are classified as Endangered on the IUCN Red List. Indeed, the species was presumed extinct until a pair was collected by chance on *Banksia* trees in 1967 at Cheyne Beach near Albany in south-west Western Australia. Before European settlement of Australia, this species was far more widespread. Land clearing for agriculture, which has been marked in the south-western region, has likely contributed to the steep decline, as has predation by feral cats. On the mainland, the species is under further threat as its habitat is impacted; this area covers less than 5000 km^2. On Boullanger and Whitlock Islands, populations are healthy (with hundreds of individuals in total), but even these have declined in recent decades. Dibblers have been introduced to nearby Escape Island and are persisting there also. Current threats include feral cats and foxes, fire and possibly House Mice on the islands. The root-rotting plant disease *Phytophthora* also reduces available habitat. Indeed, at Cheyne Beach, *Banksia* heathland has succumbed recently to this scourge, and Dibblers have also disappeared there as a result.

Genus *Phascogale* (phascogales, 3 species)

Distribution: Collectively found in coastal/near coastal areas patchily around Australia

Habitat: Dry sclerophyll forests and woodlands in tropical, subtropical and temperate regions

Size: 21–50 cm (38–310 g)

The first phascogale was described in 1793 by Friedrich Meyer. It was nominated as a new species of civet: a placental carnivore. The first civet had been described in 1758 by the redoubtable Carl Linnaeus. The genus *Phascogale* was recognised as unique and raised in 1824. Twenty years later, another species was discovered, this time by John Gould. This much smaller, redder phascogale was clearly different to the Brush-tailed Phascogale (*Phascogale tapoatafa*) named earlier. The type specimen was given to Gould by his trusty collector John Gilbert who at the time was stationed at a military station in Western Australia. Gilbert commented at the time of collection that he was indebted to a domestic cat, which had captured it during the night. Other soldiers in attendance commented that, before capture, the phascogale was regularly seen raiding the storeroom. Recent DNA, morphology and field studies have confirmed that a northern form, once considered a subspecies of the Brush-tailed Phascogale, should be recognised as a new species. There is also variation within Brush-tailed Phascogales, suggesting that further cryptic (i.e. hidden) species are waiting to be described.

The squirrel-like phascogales are medium-sized carnivorous marsupials. They are striking in appearance: fawn-brown to greyish in fur colour with large ears and eyes, large feet and a long tapering tail. The last half to two-thirds of the tail ends in a distinctive brush of dark hairs. These animals spend most of their time in trees. They are extremely agile, capable of acrobatic leaps of up to 2 m or more between tree limbs. They will also feed on the ground, hunting opportunistically. Prey can include small birds, mammals and reptiles, but the diet mostly comprises invertebrates such as spiders, grasshoppers, cockroaches, beetles, earwigs and flies.

Fig. 2.10. Female phascogales (*Phascogale* spp.), such as this Brush-tailed Phascogale (*P. tapoatafa*), typically leave their six to eight young in the nest once they become too large to carry in the pouch. But sometimes the youngsters have other ideas and will hitch a ride on mum, clinging on tenaciously as she moves about in the treetops. Source: Jiri Lochman.

In semi-urban areas, phascogales are known raiders of chicken coops and aviaries – they attack with daring and viciousness. They also eat nectar from flowering trees when it is available. Phascogales are mostly active during the night, but will investigate food sources during the day on occasion. Like antechinuses, phascogales breed annually over the course of a few weeks during winter. At the end of breeding, all males die due to stress-induced immune system malfunction. Gestation is about a month long. The six to eight young suckle for up to 4 months before being weaned and, at 6–7 weeks, are deposited in the nest by the mother while she forages. She will return periodically during the night for the young to suckle. Young disperse from the nest in mid-summer. Females have home ranges of up to 40 ha, but ranges can extend as little as 5 ha in prime habitat – plenty of old standing trees supporting tree hollows for nesting. Male home ranges are extensive, up to 100 ha.

Two species are classified as Near Threatened and one as Vulnerable on the IUCN Red List. Cats and foxes prey on phascogales. The feral menace plus the phascogales' large home range, low density and annual die-off make all members of the group vulnerable to local extinctions. Habitat clearing, land degradation and fragmentation are ongoing threatening processes.

Genus *Planigale* (planigales, at least 4 species Australia, 1 species New Guinea)

Distribution: Northern and eastern Australia. Southern and south-eastern New Guinea

Habitat: Cumulatively, nearly every terrestrial habitat, from rainforests, savannas and grasslands to weed-infested urban areas. Commonly found on cracking clays on floodplains, grassy areas and interdunes and blacksoil plains

Size: 10–15 cm (2.5–17.2 g)

The Common Planigale (*Planigale maculata*) was named by naturalist John Gould in 1851, based on an animal caught in New South Wales. In the first half of the 20th century, another three species were named. By 1976, when Australian taxonomist and palaeontologist Mike Archer revised the group, five species had been named. Archer suggested a close relationship between the Common and Papuan Planigale (*P. novaeguineae*). This relationship has now been confirmed by DNA studies, but the taxonomic status of these two species is unclear, and so is that of the rest of the group. DNA and morphology studies indicate that planigales from northern Australia consist of several cryptic species. Trees of life recover planigales as a distinct group, most closely related to dunnarts, ningauis and the Kultarr.

Planigales are small, reddish-brown to grey creatures. They have thin, tapering tails about as long as, or slightly longer than, their head–body lengths. Their heads are flat, with a conical snout. One species, the Long-tailed Planigale (*P. ingrami*), is the world's smallest carnivorous marsupial and among the world's smallest mammals. Adults weigh as little as 2.5 g. Planigales are mostly active at night. They are ferocious predators and savagely attack a range of mostly invertebrate prey that may be larger than they are. Upon catching an animal, planigales hang on and will not let go, even if they are dragged off by the fleeing prey. They will also eat small lizards and mammals (such as House Mice), remorselessly attacking the head until the animal is killed or disabled. Invertebrate prey includes grasshoppers, spiders, moths, ants, slaters, beetles and centipedes. They also eat carrion (i.e. dead animals). Through the day, planigales shelter deep in soil cracks, under rocks, fallen bark, and even tin and human debris in some areas. On early winter mornings, they can be found basking in the sun. They regularly enter torpor – essential for conserving energy in such a small mammal. Indeed, in some cases, torpor may total 21 hours in a 24-hour cycle. At night, planigales may huddle in a group for shared warmth before venturing out to hunt. Females have a well-developed rear-opening pouch that contains eight to 12 nipples. The breeding season is long, up to and sometimes exceeding 6 months, and females may breed multiple times. Gestation is 2–3 weeks; young suckle for ~3 months, but after 6 weeks they are too large for the mother to carry and are left in a grassy nest. Nests are often littered with the discarded body parts of unfortunate prey. Movement patterns are known for only a couple of species, and indicate that home ranges are unstable. Individuals can move up to a kilometre or more in a few days, with males typically moving further than females, and both sexes using different rest sites each day. Overnight, animals may move hundreds of metres. Densities as low as 1 individual per hectare are commonly reported.

Fig. 2.11. Planigales (*Planigale* spp.) have flat heads and contortionist bodies that allow them to squeeze in soil cracks and among fissures in rocky habitat. Source: Anders Zimny.

All planigales are classified as Least Concern on the IUCN Red List in view of their typically large distributions, yet feral cats, foxes and domestic dogs kill many planigales, and their habitats are threatened in many places by conversion of grassland and woodland to agriculture. Cane Toads were thought at one time to represent a novel, additional, threat to planigales. However, Common Planigales learn quickly to avoid poisonous Cane Toads. One study showed they initially avoided not only toads but also native frogs. However, they quickly learned to discriminate these taxa, and subsequently avoided toads but included frogs as prey items when opportunity arose.

Genus *Pseudantechinus* (false antechinuses, 6 species)

Distribution: Northern, central and western Australia

Habitat: Drier areas: including low open woodlands, tree and shrub steppes, rocky habitat, breakaways, scree slopes, red sand plains and sandstone outcrops

Size: 12–24 cm (14–50 g)

The first false antechinus was captured during the 1894 Horn Scientific Expedition to central Australia. It was described 2 years later by expedition zoologist and photographer Baldwin Spencer as a phascogale. Another species, also classified as a phascogale, was named by Oldfield Thomas in 1906. Both species were

Fig. 2.12. False antechinuses (*Pseudantechinus* spp.) are denizens of semi-arid and arid Australia. Many species nest and refuge in rocky outcrops, but in some regions, Fat-tailed Pseudantechinus (*P. macdonnellensis*) also use termite mounds for nesting. It is uncertain if the termite mound-loving populations of the red sand plains represent a different species to the animals of the rocky hills and breakaways. Do these simply represent different habitat preferences, or have populations within the species' broad range in central Australia become separated and genetically distinct over time? Source: Jiri Lochman.

later placed under *Antechinus* until Tate, in 1947, showed them to differ from known antechinuses, and provided them a unique genus – *Pseudantechinus* ('false antechinus'). There followed another species in 1954, two more in 1988 and a final, sixth species, in 2000. Genetic work on the group has paralleled morphological study. Trees of life recover false antechinuses as only distantly related to antechinuses. They are in fact sister to a group containing the three-striped dasyures from New Guinea and the Australian Dibbler, Kaluta, Kowari and mulgaras. However, the false antechinuses themselves are very diverse and warrant further taxonomic attention. The Sandstone Pseudantechinus (*Pseudantechinus bilarni*) and, to a lesser extent Woolley's Pseudantechinus (*P. woolleyae*), are distantly related to a more tight-knit group containing the remaining species. Cryptic species likely await formal recognition within this group.

The false antechinuses are small with largish ears and colourful fur that varies from orange-red to greyish. There is typically a chestnut patch of fur behind the ears. The tails in all but one species are usually thick at the base and may be carrot-shaped, serving as fat stores for use in lean times. Dietary information exists for only one species, the Fat-tailed Pseudantechinus (*P. macdonnellensis*). This species eats mostly insects such as beetles and grasshoppers, but it also hunts spiders, slaters and snails. Small rodents feature occasionally on the menu, and are either hunted down and killed or taken as carrion. Breeding behaviour is better understood for the false antechinuses as a whole. Animals breed once a year in winter, with young born in spring after a gestation of 38–52 days. Up to six young may be contained in the female's rudimentary pouch, and are suckled for up to 3.5 months. Juveniles then disperse from the maternal nest, reaching maturity at ~10 months. Both males and females often survive to breed more than once and for up to 4 years. False antechinuses are mostly active at night, but often sunbathe during winter, as do a range of other dryland carnivorous marsupials. This is an inexpensive way to stay warm and ease the burden of constantly regulating the body temperature. Movement patterns are little understood in this group, but some studies suggest that males have larger home ranges than females. Compared with other arid-dwelling carnivorous marsupials, false antechinuses appear to be quite sedentary. Their limited mobility may be related to animals' access to relatively stable and dependable food resources in their preferred, patchy rocky habitats.

Most species are classified as Least Concern on the IUCN Red List, in view of their large, arid-centred distributions. But one species is classified as Near Threatened (*P. mimulus*). This species occupies a small area, and there have been recent declines in habitat quality and probably population size at some sites. Frequent burning, habitat degradation by cattle and sheep, predation by feral cats, as well as consumption of poisonous Cane Toads, are known threats.

Genus *Sarcophilus* (Tasmanian Devil, 1 species)

Distribution: Tasmania

Habitat: A range of habitats but prefers open dry forest, scrub, grassy woodlands and pasture; also occurs in urban fringes

Size: 81–91 cm (7–14 kg)

Lieutenant G. Harris, general surveyor for the Van Dieman's Land Colony, was also an amateur naturalist. In 1806, he presented sketches and descriptions of two new 'opossums' to famed naturalist and botanist, Joseph Banks. These were the Thylacine or Tasmanian Tiger (*Thylacinus cynocephalus*) and Tasmanian Devil (*Sarcophilus harrisii*). Banks was duly impressed and presented the information to the Linnean Society in April, 1807. In 1808, Harris' formal description noted that locals called the animal 'native devil.' Devils were eaten by some colonists, who claimed they tasted like veal. Perhaps these folk were seeking retribution against an animal that, according to naturalist John Gould, 'assailed the sheepfolds and hen houses whenever opportunity arose.' As a result, colonists quickly drove the Devils from Hobart town into deeper parts of surrounding forest. The name 'devil' was well deserved. Harris kept a captive pair, noting 'they appear to be untameably savage'. John Wood, naturalist and writer, opined 'few animals deserved their popular titles better … the innate and apparently ineradicable ferocity of the creature can hardly be conceived.' Recognising the unique nature of the beast, in 1837 Cuvier erected the genus *Sarcophilus*, meaning 'flesh lover'. Recent DNA-based trees suggest that the Tasmanian Devil is sister to the quolls (*Dasyurus*). There are two genetic subgroups of Devils, one in the east and the other in the north-west of Tasmania, but they are not recognised as taxonomically different.

This is a very distinctive animal and the largest living carnivorous marsupial. Devils are about the size of a small-medium dog. They have black fur, sometimes with a reddish tinge. Typically, there are white fur markings on the chest and sometimes shoulders and rump. Devils are mostly ground-dwelling, but they are capable climbers, especially as juveniles. Devils eat live vertebrates or carrion. They ambush their prey and are pack hunters. A group of Devils will wear their prey down, often pursuing it for kilometres, before tearing it limb from limb and devouring every scrap that remains. Pound for pound, they have the most powerful jaws of any living mammal, and readily pulverise the bones of whatever they are eating. Their prey includes kangaroos and wallabies, possums, wombats and birds. They often eat communally – up to 22 individuals have been seen eating a single cow carcass. And they are noisy feeders. Snorts, barks and growls may quickly escalate to screams of outrage at the prospect of missing out on their share. This, combined with noisy bone crunching, is reportedly enough to 'raise the dead'. Devils usually den in underground burrows or concealed woodland, but they sometimes rest inside logs in thick patches of scrub, or even underneath buildings. They maintain several dens within their home range.

Mating occurs during just 5 days across each 35-day oestrus period, typically from February to March. When the female is receptive, her pouch becomes red and her neck thickens. The latter will protect her from the savage bites that follow from potential suitors. Mating is a mixed bag. Screaming and biting is rampant, along with genuine affection. Females are hardly passive during the affair, aggressively asserting their preference for older, dominant males. Males in turn attempt to ward off other male suitors by guarding mated females. In the end, typically,

Fig. 2.13. Tasmanian Devils (*Sarcophilus harrisii*) have the most powerful jaws of any mammal their size. They will crush even the bones of their prey to powder. After they feed, not a scrap of the victim remains. Source: Jiri Lochman.

different males will be fathers of any given litter of up to four offspring. Females can enter oestrus a second or third time each year, in the event of failed litters. Gestation lasts 21 days. Fifteen to twenty young are born, but only four are destined to be nurtured in the pouch – it is a race to the death to get there. Many carnivorous marsupials exhibit the same strategy – it is evolution's way of ensuring the pouch is full. Young spend 16 weeks in the pouch after which they roam independently further and further from the den, until they leave for good around January. Most females start breeding at ~18 months of age, but males may not mate until they are 3 or 4 years old.

Tasmanian Devils live for ~6 years in the wild, under normal conditions. However, in the last two decades, conditions have been far from normal for the species. An infectious cancer, devil facial tumour disease (DFTD), has decimated many populations. The disease is restricted to Devils, is always fatal, and is spread by intimate injurious contact. This is a common occurrence for Devils; animals are routinely injured during both feeding and mating. The disease kills animals within a year of becoming adult, because the Devil's immune system cannot recognise the cancer. Importantly, this is before most Devils would normally mate. The Tasmanian Devil is one of the few social dasyurid marsupials, and this sociality has ensured rapid spread of the disease. The result is vastly reduced local populations, with losses of up to 95% of the original numbers. Surviving populations have a very young age structure. Since its discovery in 1996, the disease has caused an 85% overall decline in Devil numbers. There is a naturally resistant population in north-west Tasmania and there are moves afoot to isolate these animals as an 'insurance' population. The good news is that researchers are getting closer in their development of a vaccine. They have grown facial tumour cells in the laboratory and killed them. They then mixed the substance that killed the tumour with an inflammatory agent. The resulting concoction was injected into six Devils. The idea was that immune response would be stimulated by the inflammation. Along with the tumour killer, the Devil's body would be stimulated to mount its own defence to the cancer. It worked. Five Devils had an immune response, indicating that immunity had been stimulated to attack the tumours their bodies had previously ignored. The next step is a field trial involving hundreds of animals, and this should happen within years. In the meantime, there are encouraging signs that wild populations of Tasmanian Devils are evolving resistance to DFTD (see also Chapter 11, vignette).

Devils are large animals, so radio-collars can be placed on them for months at a time. This has returned detailed information about their movement patterns. Devils range freely throughout the winter and do not enter torpor, regardless of the ambient temperature. The species is mostly nocturnal and active ~8 hours a night. Devils have a large home range: 8–27 km^2. Some individuals move long distances. One animal moved 50 km in 48 hours on a return scavenging trip to a garbage tip. Animals are occasionally sighted during the day, especially in snow during deep winter.

The Tasmanian Devil is classified as Endangered on the IUCN Red List. Apart from DFTD, a range of threatening processes exists. Road mortality, persecution, habitat loss from land clearing and possibly even ingestion of poison baits (laid to kill foxes) reduce the number of Devils each year.

Genus *Sminthopsis* (dunnarts, 16 species Australia, plus 2 shared with New Guinea)

Distribution: Collectively, all of Australia; southern New Guinea

Habitat: Predominantly arid/semi-arid habitats such as rocky desert (gibber) plains, tussock, spinifex and shrub-dominated plains and sand dunes, but also tropical savanna, grassland, temperate woodland and open forest; two species are found in or near rainforest, closed forest communities

Size: 10–32 cm (8–70 g)

The dunnarts form a large genus, containing 18 species. Five species were discovered from 1838 to 1847 and placed under *Phascogale*. One of these was the Fat-tailed Dunnart (*Sminthopsis crassicaudata*), presented to John Gould by his collector Gilbert in 1844. Gilbert could only procure a single specimen. It was brought to him by a domestic cat when staying at the military station on Williams River. According to Gilbert, the feline had a habit of going into the bush and returning several times a night with a small mammal or bird in its mouth – a habit unfortunately shared by the several million-strong feral cat population in Australia today. Gilbert likened the fat in the tail of the beautiful little animal to that of a beaver, which made it almost impossible to skin. Oldfield Thomas, in London, recognised that the distinctiveness of dunnarts warranted a separate genus. He raised *Sminthopsis* to accommodate them in 1887. Numerous other species were to follow. Almost 100 years after Thomas raised the genus, Mike Archer conducted a major revision of the group, recognising 12 species, including a new one of his own, but he also drew attention to geographical variation in some species that suggested the presence of further, hidden species. Some of these cryptic species were described in the 1980s by Darryl Kitchener and colleagues, using evidence based on both DNA and morphology, but mystery still clouds dunnart taxonomy. Recent DNA-based trees of life do not support a grouping of dunnarts that excludes all other genera. The Long-tailed Dunnart is in fact sister to the Kultarr (genus *Antechinomys*). Another group of dunnarts is sister to the ningauis (genus *Ningaui*). Some of these groupings are supported by morphology and others are not. The taxonomic implications of this genetic muddle are not yet clearly understood, but research is under way in an attempt to provide resolution.

The dunnarts are small carnivorous marsupials with large ears and eyes. They typically sport a darkened eye-ring. Their sometimes fattened tails range in length from notably shorter than, to much

Fig. 2.14. Many carnivorous marsupials live mostly solitary lives once they have left the nest, until it is time to mate, but in many species, young still have plenty of time for play, such as these Little Long-tailed Dunnarts (*Sminthopsis dolichura*). Such bouts are crucial for the development of strength and hunting capabilities that will be needed if they are to survive as independent adults. Source: Jiri Lochman.

longer than, their head–body length. The feet tend to be long and thin. The dorsal body colour ranges from greyish through fawn to brownish-red, while the belly fur is typically whitish. Dunnarts are mostly active at night and ground-dwelling, and occur typically in drier habitats. During the day, they rest in nests, hollow logs, at the base of tussocks or in underground areas within burrows or soil cracks. The diet, like that of so many other carnivorous marsupials, is generalist and opportunistic, and includes beetles, moths, grasshoppers, spiders, termites, cockroaches, centipedes, and the like. Many species will also tackle vertebrates, eating small lizards and mammals. Pound-for-pound they are ferocious predators and will tear the heads off mice and other rodents if caught in pitfall traps with them. Like many other species of arid-dwelling carnivorous marsupials, they often require no water beyond what is in their prey. Many species use torpor daily to conserve energy. The dunnart mating season is long, running from July to March in many species. Females may breed more than once within a year and have six to 10 nipples. Gestation is short, as little as 11–12 days in some species. Females support lactating young for 2–3 months, followed by the typical pattern of weaning and juvenile dispersal. Females may mature in as little as 4 months, but in males maturity takes longer. Dunnarts are renowned for being difficult to catch in metal box (Elliott) traps, but several species can be captured more readily, even exclusively, in pitfall traps. In other species, even pitfall traps are ineffective. Many species have unstable home ranges, perhaps reflecting the need for dunnarts to be able to track and exploit unpredictable and patchy food resources. Despite this, or even because of it, in some areas dunnarts feature as the most commonly caught carnivorous marsupial species.

Several species are classified as threatened on the IUCN Red List, with two species listed as Vulnerable (Butler's Dunnart, *S. butleri* and the Sandhill Dunnart, *S. psammophila*) and another as Critically Endangered (Kangaroo Island Dunnart, *S. aitkeni**). Threats to dunnarts include too frequent burning of their habitat, predation by cats and foxes, and ingestion of poisonous Cane Toads. Removal or modification of habitat to permit stock grazing and agriculture are a problem for some species.

* Although listed as a full species by the IUCN, we consider this dunnart to be a subspecies of the Grey-bellied Dunnart (*S. fuliginosus aitkeni*).

Order Dasyuromorphia; Family Myrmecobiidae (1 genus)

Genus *Myrmecobius* (Numbat, 1 species)

Distribution: Patchy occurrence at a few sites in south-western Australia

Habitat: Eucalypt forest and woodland in moderately moist to semi-arid areas

Size: 34–50 cm (300–750 g)

The Numbat (*Myrmecobius fasciatus*) was named by Waterhouse in 1836, after an animal was captured within the Swan River Settlement in Western Australia. One subspecies from central and eastern Western Australia is now extinct. This was John Gould's 'anteater'. His collector, John Gilbert, described Numbats as: 'much like a squirrel, running in successive leaps. Every now and again they pause, raising the body and resting on the hind feet to assess danger.' The Numbat is a survivor of an ancient group. Unfortunately, no fossils have been found of the species, but DNA and morphological analyses suggest that Numbat ancestors last shared common ancestry with other marsupials, possibly the thylacines, some 30–40 million years ago (mya).

Numbats are striking, with red-brown fur above and paler fur below. On the back half of the animal, the upper fur is darkened (sometimes black) and on the rump there is a series of transverse white stripes/bars. The stripes extend onto the mid-back before breaking into white splotches and spots towards the shoulders and neck. The head is flat; the snout thin and pointed. The ears are long and conical. The eyes are large and there is a bar of black fur that runs from the base of the ear forward through the eye and onto the snout, on each side of the head. The body fur is coarse and shaggy (enhancing heat absorption) and the tail hairs are long and often erect. This gives the tail a brush-like appearance. The animal has five fingers on each hand, but only four toes on each foot. The tongue is long and thin, and can be extended at least 5 cm beyond the tip of the snout. Numbats feed exclusively on termites, particularly wood-eating species. They are medium–large carnivorous marsupials and require 15 000–20 000 termites a day to survive. Much time and effort is spent digging them out. Prey is detected by scent as the Numbat moves around a likely area, nose to the ground. When a termite gallery is detected, the Numbat digs rapidly

Fig. 2.15. The Numbat (*Myrmecobius fasciatus*) has a long tongue that it uses to expertly lick up its prey – termites. Each adult needs to consume up to 20 000 termites a day to survive. Source: Jiri Lochman.

with the paws, exposes the termites and then licks them up with its long tongue. Mating usually occurs during the first half of January. Immediately before and during this time, males wander far and wide in search of females. Gestation is 14 days, so young are typically born at the end of January. Females have four nipples and most carry the full complement of young. There is no pouch, but the mother's long belly and fur afford the young some protection when they are small. Young remain attached to the mother's nipples for ~6 months, after which she leaves them in a burrow and suckles them at night. By September, the young are making short excursions in the area immediately surrounding the burrow. By October they are weaned, and in December they leave the nest. After leaving the nest they are essentially solitary. The female range is ~25–50 ha. The male range is smaller than this in winter, but increases by an order of magnitude before mating. Numbats are only active during the day, when their termite prey is most active in the upper soil layers. They spend most time on the ground, but are capable climbers and will readily climb trees to access tree hollows as a night refuge. They descend to the ground head-first with back legs splayed, squirrel-like. They are often seen on their bellies, splay-legged, basking on cool winter mornings.

This distinctive and beautiful animal is classified as Endangered on the IUCN Red List. There are probably less than 1000 mature individuals left in the wild, occurring only in a couple of patchy locations in Western Australia (Dryandra and Perup). However, they have been successfully reintroduced at several sites in Western Australia, South Australia and western New South Wales. Feral foxes and cats are their main threats. There have been recent declines at some sites, but some methods of control, such as cat-proof fences and fox baiting, may allow Numbat recovery. Numbats once enjoyed a much more extensive range and they were well known to Aborigines, who used them for food. Unfortunately, changes to fire regimes (less frequent but hotter burns) and the introduction of foxes and cats following European settlement sent many local populations to extinction.

Order Notoryctemorphia; Family Notoryctidae (1 genus)

Genus *Notoryctes* (marsupial moles, 2 species)

Distribution: Central and north-western Australia

Habitat: Sandy habitats, including linear dunefields, dune complexes and sandy washouts of temporary watercourses within scattered spinifex hummock grasslands

Size: 8–15 cm (30–70 g)

The peculiar marsupial mole was discovered in central Australia in 1888 by Edward Stirling. He named it in the following year. A new genus, *Notoryctes*, was erected just 2 years later in 1891. Recent DNA analyses suggest that larger animals found near the type locality may be genetically divergent to those from other locations, but the significance of this is presently unknown. More-detailed sampling of these elusive animals is required. A second species of marsupial mole was named by Oldfield Thomas in 1920 to account for a smaller animal captured in the deserts of north-

Fig. 2.16. The bizarre marsupial moles (*Notoryctes* spp.) are among the least known of all carnivorous marsupials. They spend most of their life underground and use their powerful, large claws to efficiently 'swim' through the sand. Source: Jiri Lochman.

western Australia. Marsupial moles have an ancient heritage, apparently sharing common ancestry with other marsupial groups that date back some 60–70 mya. Fossils in this group suggest that ancestral moles were once tolerant of wetter habitats. Yet, today, the marsupial moles are specialised for a subterranean lifestyle in arid environments.

Marsupial moles are superficially similar to placental moles: both species live mostly underground, having reduced eyes and modified limbs. Marsupial moles have two large modified claws on the hands for digging. The tail is short and the fur sleek and creamy-yellow. The nose pad is large and oval-shaped. There is a range of skull and skeletal morphology differences between the two species. Also, there are differences in size, narrowness of claws and the tail. Little is known of marsupial mole diet, other than from examination of gut contents of some museum specimens and attempts to maintain them in captivity. Gut contents contained a range of invertebrates and some plant material, including grass seeds. The invertebrates included sawfly larvae, ants, beetle larvae, termites, butterflies/moths, grasshoppers, scorpions and spiders. Moles kept in captivity accepted a range of invertebrates and also geckoes. Larger food items are grasped in the claws of the hands before being chewed. Little is known about breeding. Females have two nipples, and instances of single and twin pouch young have been recorded. Little is known of activity patterns. Marsupial moles are presumably mostly active at night, when soil and surface air temperatures drop in their desert habitats. They seem to alternate between periods of frenetic activity and inactivity, when they appear to be asleep. Marsupial moles have been found on the surface during the day. Yet they were inactive, perhaps forced to the surface by flooding. The animals move in a sinuous, shuffling motion when on the surface, with their tails held off the sand. They require some soil moisture to dig tunnels efficiently but are capable sand-swimmers. Although sand-swimming comes at higher energetic cost than moving on the surface, they are afforded protection from predators by being underground. So, provided they can encounter enough prey to make underground movement feasible, this may be the standard in transportation. Nothing is known of range size, territoriality or social organisation in these mysterious creatures.

Both species are classified as Least Concern on the IUCN Red List. Most likely, marsupial moles are sparsely distributed, as judged by low numbers of captures that continue to be made across space and time. Some parts of their distribution are affected by human development, and animals have been uncovered during road construction.

New Guinean fauna (1 order, 1 family, 6 genera, 15 endemic species)

Order Dasyuromorphia; Family Dasyuridae (6 genera, 15 species)

Genus *Murexia* (murexias, 5 species)

Distribution: Central and northern New Guinea

Habitat: A range of forest types including primary, secondary and regrowth forest from sea level to 3600 m. Found in beech forest, mossy forest, pandanus forest and subalpine grassland

Size: 22–55 cm (17–134 g)

Before 1984, the murexias were thought to be New Guinean *Antechinus*, but some timely research on penis shape and size showed them to be quite distinctive! At that time, antechinuses in Australia also included what we now recognise as a range of genera. This motley crew consisted of the Kaluta, false antechinuses and the Dibbler. In the decades that followed, DNA studies suggested that the murexias were quite distinct. In DNA trees, they were positioned as sister to the Australian antechinuses

Fig. 2.17. Like their Australian relatives the antechinuses, the New Guinean murexias (*Murexia* spp.) are often at home in the trees. They are elusive and this rare photo sequence shows a Long-nosed Dasyure (*Murexia naso*) as it scampers high up in the tree canopy. Source: John Hornbuckle.

and phascogales. In 2002, Australian taxonomist Steve Van Dyck conducted a major morphological revision of antechinuses and murexias. Based on considerable body size and shape differences among murexias, he consigned them to five separate genera. The most recent DNA work suggests that they are perhaps best left under the umbrella of a single genus, *Murexia*. However, there are some relatively deep genetic differences both within and between species, suggesting that the murexias began diverging early, perhaps 12 mya. Therefore, a range of cryptic taxa is yet to be formally recognised.

The murexias are fairly uniform in head–body colouring, from brownish to grey, with slightly warmer tones on the rump. Chestnut patches feature behind the ears in a couple of species. One species has a broad black dorsal stripe that runs from the nose to the tail base. The tails are typically brownish, about the same length as the head–body, and thin and tapering with ventral crests in several species. However, one species is black-tailed and another has a whitish tip on the tail. Murexias eat primarily insects, such as beetles, bugs, moths, butterflies, grasshoppers, crickets, cockroaches and flies, and other invertebrates such as spiders and worms. They will occasionally take vertebrates, including other mammals and birds. Trapping studies of some species have found females in different stages of lactation, and juveniles at different ages, suggesting they breed at various times throughout the year. Female murexias have four nipples in an unenclosed pouch. Gestation is from ~19–22 days and suckling continues for ~3.5 months. Juveniles mature at 8–14 months. Murexias are probably mostly nocturnal. Some species are ground-dwelling but others are partly or mostly arboreal. Little is known about the movements of these animals, other than from the results of spool and line tracking after release of captured animals. This method employs a role of thread attached to the captive that unravels, leaving a trail as the animal rapidly disappears. These studies suggest that murexias are agile animals that may travel hundreds of metres in search of food, sometimes nesting in hollow trees, fallen logs or ground holes (see also Chapter 7).

Four species are classified as Least Concern on the IUCN Red List, with a note that some species contain complexes of cryptic species and the distributions are uncertain. One species, the Broad-striped Dasyure (*M. rothschildi*), is classified as Near Threatened on the IUCN Red List. This species is known from less than 10 locations, some of which are threatened with habitat loss through conversion of forest to agricultural land, oil palm plantations and rural gardens. This species is not yet known from any protected areas.

Fig. 2.18. The back stripes of three-striped dasyures (*Myoictis* spp.) are more prominent in some species than others. This *Myoictis* is devouring a rainforest skink. Source: Auscape.

Genus *Myoictis* (three-striped dasyures, 4 species)

Distribution: Northern and southern New Guinea; patchily in central New Guinea

Habitat: A range of forest types including primary, secondary and regrowth forest from sea level to 1800 m. One species, *M. melas*, can be found in disturbed habitats, such as old gardens

Size: 14–24 cm (head–body only) (61–255 g)

A three-striped dasyure was the first carnivorous marsupial collected for science in New Guinea. In August 1828, local Papuans carried a small male marsupial aboard a visiting Dutch vessel anchored at Triton Bay. It was proudly presented to Heinrich Macklot, Salomon Müller and Heinrich Bole: members of the Natural Science Commission for the Netherlands Indies. The Papuans called the animal 'Insinsie', telling the Europeans it lived in forest on the ground and hunted mostly at night. Twelve years later, after travelling to Timor, Java and Sumatra, Müller returned to Leyden, where he raised the strange New Guinean animal as a new species of *Phascogale*. *Phascogale* had been previously nominated by his boss and Museum Director, Coenraad Temminck, in 1824 to accommodate the Australian Common Brush-tailed Phascogale. However, in 1858 astute British taxonomist John Gray nominated the genus *Myoictis* to account for these distinctive New Guinean three-striped dasyures, naming a new species in the process. Only two species were recognised in the genus for almost 150 years. Then in 2005, a morphological and DNA revision indicated that one existing subspecies, recognised by George Tate in 1947, should be raised to species level and also identified a further cryptic species. Thus, there are four recognised species under *Myoictis* today, and one of these is likely composed of further cryptic species. DNA phylogenies suggest that the most closely related carnivorous marsupials to the New Guinean three-striped dasyures are Australian: the Dibbler, Kowari, mulgaras and Kaluta. They shared common ancestry perhaps until 10–12 mya.

The three-striped dasyures are so-named for the three longitudinal brown-blackish stripes that typically extend down the back from head to rump. These stripes are more noticeable in some species than others. Occasional specimens are uniformly dark, but, generally speaking, all four species are colourful (light brownish-grey through orange to dark brown) and distinctive. The tails are striking. They are long and thickish at the base with a tapered tip, which is white in one species. Tail form varies among species from brownish and moderately hairy to orange-brown and decidedly bushy. There is little ecological information on these medium-sized carnivorous marsupials. Diet is unknown for all species, although likely to include both small vertebrates and invertebrates. In two species, females have four nipples in the pouch and in the other two species, females have six nipples. Three-striped dasyures likely breed at various times throughout the year. In captive colonies, animals have been observed mating, but no offspring have arisen from the unions. This is typical of endeavours to breed captive carnivorous marsupials! Three-striped dasyures are believed to spend most time on the ground. They move about mostly during the day, but home ranges are unknown for all four species.

Two species are classified as Least Concern, the other two as Data Deficient, with very limited known distributions, on the IUCN Red List. Possible threats include predation by cats and dogs and poisoning after consuming Cane Toads.

Genus *Neophascogale* (Speckled Dasyure, 1 species)

Distribution: Weyland Range to Hagen Range, Central Range of New Guinea

Habitat: Montane and tropical moss forests and subalpine grasslands

Size: 33–43 cm (212 g)

In 1909, the second Dutch South New Guinea Expedition visited the remote Snow Mountains. Conditions were treacherous: expedition leader Hendrikus Lorentz was almost killed after slipping off

Fig. 2.19. Speckled Dasyures (*Neophascogale lorentzii*) are mysterious denizens of high-altitude moss forests in New Guinea. This rare photo has captured one on the move. Note the diagnostic speckled fur and conical-shaped head. Source: Auscape.

a steep rock wall, breaking a rib and lying incapacitated. A Papuan carrier, who remained with him in the snow, froze to death while waiting for help. But, scientifically at least, the expedition was deemed a success. One of the specimens obtained was an adult female Speckled Dasyure (*Neophascogale lorentzii*) replete with four pouch young, captured at an altitude of 2600 m. The new species was coined *lorentzii* to honour the intrepid collector, and was classified originally within the genus *Phascogale* by Fredericus Jentink, Director of the Leyden Museum, in 1911. It was later transferred to a unique genus, *Neophascogale*, by Georg Stein in 1933. Stein was a German collector and self-educated naturalist who in 1931 had been sent on an expedition to New Guinea by the Harvard Museum of Comparative Zoology. His prime focus was bird collection, but he also named several mammals, nominating *Neophascogale* on his return from the wilds to accommodate the distinctive Speckled Dasyure. Recent DNA trees recover the Speckled Dasyure as sister to another genus of New Guinean carnivorous marsupials, *Phascolosorex*. These two genera together form a clade that is sister to the Australian quolls and Tasmanian Devil. These groups last shared a common ancestor perhaps 10–12 mya. The ancestor evidently moved from northern Australia to New Guinea, presumably dispersing during one of the intermittent low sea level phases of that time.

This animal is very distinctive: medium-sized with silvery, speckled fur and very long claws. It has a conical shaped head and long snout. The eyes are smallish; the ears moderate. There is no information on what these animals eat in the wild, but in captivity they survive well enough on strips of beef, cicadas and beetles. This is an animal of the high tropical moss forests and is typically only found above 2000 m. It is active during the day, spending part of the time on the ground and the rest up trees.

Speckled Dasyures are classified as Least Concern on the IUCN Red List. They can be locally common, especially at higher altitudes. However, they may have been removed from some areas. For example, an exhibition conducted on the eve of the Second World War caught 96 specimens. This was between 2400 and 3300 m in the region of Lake Habbema and Mount Wilhelmina. Yet a return to this area in 1998 yielded no captures whatsoever. The people of New Guinea hunt this medium-sized carnivorous marsupial with bows and arrows for food.

Genus *Phascolosorex* (Narrow-striped Dasyure and Red-bellied Phascogale, 2 species)

Distribution: Birds Head and across the Western and Central Range of New Guinea

Habitat: Mature and disturbed tropical moist forests at elevations of 900–3600 m

Size: 23–42 cm (132–145 g)

Originally an orphan from Florence, Odoardo Beccari became a conscientious student, excelling at the universities in Pisa and Bologna. After graduating, he spent a few months at the Royal Botanical Gardens in Kew where he met renowned naturalist Charles Darwin and botanist William Hooker. Thus inspired to travel, he became a naturalist and collector in remote Malaysia, Indonesia and New Guinea between 1865 and 1878. In May–June 1875, Beccari visited Hatam and Mori in the Arfak Mountains of New Guinea, spending more than a month collecting, where he was hindered by incessant rains and swollen torrents. He nevertheless reached a heady altitude of 2040 m, exploring the surrounding heights and returning to the museum in Genoa, Italy with much valuable zoological material. His booty included two new species of carnivorous marsupial. The first was designated by German Wilhelm Peters and Italian Marchese Doria in 1876. The second species, named in honour of Doria, was raised by an impressionable Oldfield Thomas in 1886, after visiting the Genoa Museum while researching for his 1888 catalogue of the marsupials in the British Museum. Both species were originally

Fig. 2.20. One of the two species of carnivorous marsupials shared between Australia and New Guinea, the Red-cheeked Dunnart (*Sminthopsis virginiae*). Source: Anders Zimny.

placed under the genus *Phascogale*. The genus *Phascolosorex* was raised in 1916 by German mammalogist and Curator at the Berlin Zoological Museum, Paul Matschie. However, it wasn't until a review by English-born American zoologist George Tate in 1947 that the two species were formally shifted into the new genus. The common names remain confusing because one retains a reference to phascogale, a genus actually limited to Australia. The other refers to a dasyure, which is also used as a common name for the sister taxon, *Neophascogale*. The name dasyure also appears within the more distantly related *Murexia*, which is in fact a sister group to the Australian *Antechinus*. This is a vivid example of how confusing common names can be! DNA studies suggest that *Phascolosorex* and *Neophascogale* have probably been separated from common ancestry for 5–7 million years.

Phascolosorex species are distinctive, bearing a single dark mid-dorsal stripe that extends from head to rump. The body fur is otherwise brownish-grey sometimes tipped with silver, and there is a yellowish–reddish tinge on the belly. The Narrow-striped Dasyure (*P. dorsalis*) varies in colour and size across its range and almost certainly contains cryptic species. Insects, including beetles, bugs, grasshoppers, crickets, moths and cockroaches, are eaten by this species, and spiders are also a favourite. It also hunts vertebrates, including frogs, lizards and mammals. Narrow-striped Dasyures apparently breed throughout the year because pregnant females have been found in every season. Females have four nipples. There is no discrete pouch: just a fold of skin partially covering a depression in the belly where the young attach for 4–4.5 months before weaning. The ecology of the Red-bellied Phascogale (*P. doriae*) is largely unknown but, like the Narrow-striped Dasyure, it is probably ground-dwelling and active during the day. Home range and social organisation are unknown in both species.

Both species of *Phascolosorex* are classified as Least Concern on the IUCN Red List. They probably have stable populations, but they are affected by predation from domestic dogs in some areas. Other threats come from forest disturbance to accommodate shifting agriculture and timber extraction. New Guinean nationals hunt them for food in some places, which may limit local numbers.

Three other genera of carnivorous marsupials are shared by Australia and New Guinea – the quolls (genus *Dasyurus* – two species are endemic to New Guinea), the planigales (genus *Planigale* – one species is endemic to New Guinea) and the dunnarts (genus *Sminthopsis* – two species are shared with Australia). In each case, most species are Australian, so their generic accounts are presented in the section on Australian genera.

American fauna (3 orders, 3 families, 13 genera, 58 endemic species)

Order Microbiotheria; Family Microbiotheriidae (1 genus)

Genus *Dromiciops* (monito del montes, 3 species)

Distribution: Chile and Argentina

Habitat: Temperate rainforest

Size: 15–26 cm (16–32 g)

Its Spanish name means 'little mountain monkey', but these marsupials are not closely related to any primate. Monitos are in fact the only surviving members of a once diverse marsupial order. They are often referred to as living fossils. The family was named in 1887 to categorise a range of 20-million-year-old Patagonian fossils, but it wasn't until 1955 that scientists realised the family wasn't extinct. The first Monito del Monte (*Dromiciops gliroides*) to be captured was initially considered an opossum (Family Didelphidae). However, it was later recognised as a living member of the old family after similarities were found between monito skulls and the Patagonian fossils. Yet relationships between monitos and other living marsupials remain puzzling. It is widely accepted – based on DNA, morphology, biogeography and fossil evidence – that monitos are more closely related to Australian (or australidelphian) rather than to other South American (or ameridelphian) marsupials. However, where they sit in the tree of life compared with other Australian marsupials remains controversial. The most recent DNA research suggests that monitos are sister (i.e. most closely related) to a group containing all other Australian marsupials, but shared ankle bone morphology and chromosomal DNA arrangements imply that monitos are actually nested within the Australian marsupials. Genetic research in 2008 showed there to be three strongly

Fig. 2.21. The monito del montes (*Dromiciops* spp.) are at home among bamboo clumps where they hunt for mainly invertebrate prey in the spring, which provides crucial protein needed to fuel mating. In the summer, however, they switch their diet to include fruit, which provides extra sugar to be stored as fat to supplement bouts of winter torpor. Some populations are well studied; note the metal ear tag on the individual pictured that scientists are using to monitor its growth across the year. Source: Marcela Franco.

divergent groups of monitos, and in 2016 these were recognised formally as three different species.

Monitos are small, with rounded ears and a short snout. The silky, dense body fur is brownish-grey on the back, and off-white on the belly. Monitos have long, semi-prehensile tails and short, strong limbs. Their hands and feet are large, with disk-like pads at the tips of fingers and toes. Monitos are excellent climbers, with opposable thumbs and toes. They spend nights moving among the trees within dense, humid forest, often in bamboo clumps, in search of prey. Prey includes various types of insects (beetles, grasshoppers, crickets, butterflies, moths, bugs, wasps and ants), fleshy fruits and occasionally nesting birds and small rodents. The type of prey taken depends on the season. Insects provide additional protein during the spring breeding period. Fruit provides a sugar hit to promote fat accumulation at the base of the tail. This in turn supports bouts of winter torpor, which are initiated at ~4.5°C and can last for 1–5 days at a time. Monitos breed in spring, and two to four tiny young are born in November. They remain in the mother's well-developed pouch for 2 months. Thereafter, young remain in a nest of bamboo leaves and moss for another month. They make nightly forays on their mother's back, until they reach independence at the end of April. Monitos are the only known disperser of Mistletoe (*Tristerix corymbosus*) seeds. This behaviour is a simple mutualism: an ecological relationship in which both partners benefit from their association.

Because of the recent taxonomic revision, monito del montes require reassessment before IUCN Red Listing. *Dromiciops gliroides* is listed as Near Threatened, but one of the two newly described species, Mondaca's Monito del Monte (*D. mondaca*), may be endangered due to its limited distribution in a fragmented area. The genus suffers from habitat clearing, which can drastically affect local population numbers.

Order Paucituberculata; Family Caenolestidae (3 genera, 7 species)

Genus *Caenolestes* (shrew-opossums, 5 species)

Distribution: Slopes of the Andes in Venezuela, Colombia, Ecuador and Peru

Habitat: Primarily subtropical and temperate montane forest to dense scrubland and pasture

Size: 19–28 cm (25–53 g)

When the Dusky Shrew-opossum (*Caenolestes fuliginosus*) was discovered by Tomes in 1863, it was believed to represent the link between Australasian and American marsupials. Its upper teeth were polyprotodont; that is, it had several pairs of similar-sized incisors. The lower pair of incisor teeth formed large, forward-slanting tusks, similar to those of Australian diprotodonts such as kangaroos and koalas. However, as knowledge of marsupials grew worldwide, the shrew-opossums were recognised as 'pseudodiprotodonts' – a diverse group of now otherwise extinct marsupials. Today, the caenolestids are positioned more narrowly as a group of mostly extinct marsupials that had already appeared in the

Fig. 2.22. The shrew-opossums (*Caenolestes* spp.) are opportunists that eat a variety of insects and their larvae. They locate their prey by smell, poking their noses into moss and leaf litter. Once the prey is secured it is held between the forepaws while the animal sits up on its hind legs and tail to eat. Source: Reed Ojala-Barbour.

fossil record some 50–60 mya. Caenolestids clearly are an ancient group. Current knowledge has them diverging from other marsupials very early, some 80–100 mya. Indeed, recent DNA trees position this group as sister to the rest of the marsupials, including both the didelphimorphians (American marsupials) and the australidelphians (Australian and New Guinean marsupials and monito del montes).

Members of the genus *Caenolestes* are small marsupials that are shrew-like in appearance. They have small eyes and long, tapering noses. Their fur is dense, typically medium to dark greyish-brown on the back and lighter off-white to drab brown on the belly. The tail length is about equal to, or slightly longer than, the head–body length. The tail is only slightly hairy and tapers gradually to the tip. Habits of the group are little known, but all species appear to feed primarily on a range of invertebrates including caterpillars, centipedes, spiders, beetles, grasshoppers and scorpions. Occasionally, vertebrates and fruits are eaten by some species. Interestingly, the shrew-opossums sometimes co-occur with true shrews (genus *Cryptotis*) and may use the same habitats. Indeed, shrew-opossums likely prey opportunistically on the smaller *Cryptotis*. Breeding is generally poorly understood in *Caenolestes*, but occurs in August–September for some species. These animals spend much time scurrying about on the forest floor in search of food. They hunt along trails, tunnels and cavities under tree roots bordering streams. They also frequent steep mountain slopes. They are active at night or in the early morning and evening. Shrew-opossums probably spend most of their time on the ground, being poorly adapted to climbing.

Three species of *Caenolestes* are classified as Vulnerable on the IUCN Red List (the Condor Shrew-opossum, *C. condorensis*, Blackish Shrew-opossum, *C. convelatus* and Sangay Shrew-opossum, *C. sangay*). Deforestation to accommodate new roads and agriculture are known threatening processes.

Genus *Lestoros* (Incan Shrew-opossum, 1 species)

Distribution: Eastern slopes of the Andes in Peru and Bolivia

Habitat: Various, from cloud forest and wet elfin forest to disturbed scrub

Size: 19–26 cm (20–35 g)

The Incan Shrew-opossum (*Lestoros inca*) was first discovered in 1915 at an altitude of over 4 km above sea level in the deserted ancient Incan city of Machu Picchu, near Cuzco, Peru. The specimens were sourced during an American National Geographic Society expedition, captured by renowned zoological collector Edmund Heller. Such was the fear of interception by wartime German submarines that Heller divided his 30 captured specimens, with some sent to England for identification and a reserve series held in Washington for safe keeping. However, the shipped specimens arrived safely during 1916, where the new species was named by British taxonomist Oldfield Thomas. Thomas originally described the beast under another genus name, *Orolestes*, but it was later discovered that this name was already in use for a dragonfly. Thus, another genus name was proposed, *Cryptolestes*. As it turned out, this was being used for a beetle. Finally, *Lestoros* was chosen, and it remains to this day.

Incan Shrew-opossums look superficially similar to shrew-opossums of the genus *Caenolestes* (not to mention true shrews), but there are several features of the skull and teeth that set them apart. These include a gap between the upper incisor and canine teeth, a small or absent first premolar tooth and a large, lower third incisor tooth. Phylogenetic trees produced from combined genetics and morphology indicate that the Incan Shrew-opossum and Long-nosed Shrew-opossum (*Rhyncholestes*) form a sister pair. These two, in turn, are sister to a group containing the other five shrew-opossums (genus *Caenolestes*). Incan Shrew-opossums are small marsupials, with small eyes and a long, narrow nose. Their fur is greyish-brown to slate grey with a tapering, slightly haired tail that is about as long as the head–body. The habits of this little animal are poorly known, but the diet includes a range of mostly invertebrate prey. Females have been observed lactating in November and March. They have four nipples in the pouch and occasionally a fifth on the abdomen. Incan Shrew-opossums scurry on the forest floor among tangled tree roots during the evening, night or early morning, rarely ascending above ground.

This species is classified as Least Concern on the IUCN Red List. It occurs in a wide range of forest habitats, river drainages and protected areas.

Fig. 2.23. Long-nosed Shrew-opossums (*Rhyncholestes raphanurus*) have small eyes and a long, narrow snout. Their eyesight is poor, but sense of smell and hearing acute. They have long, forward-protruding lower incisor teeth that they sometimes use to impale their predominantly invertebrate prey. Source: Agustin Iriarte.

Genus *Rhyncholestes* (Long-nosed Shrew-opossum, 1 species)

Distribution: Chile and Argentina

Habitat: Temperate rainforest

Size: 16–22 cm (23–40 g)

The Long-nosed Shrew-opossum (*Rhyncholestes raphanurus*) was named in 1924 based on a specimen captured in January 1923 by American taxonomist Wilfred Osgood. Osgood caught the female *Rhyncholestes* in dense temperate forest at the mouth of the Rio Inio at the south end of Chiloe Island, Chile. After a week of intense trapping, Osgood managed to secure just two specimens: an adult male and female. The traps could only be checked every third day and the male specimen by that time had begun to decay and its skin could not be saved. After a further 5 weeks of trapping in and around the type locality, no further specimens could be found. Indeed, only three specimens of Long-nosed Shrew-opossum existed in museums until the 1970s. The two currently recognised subspecies were originally designated in 1987 as separate species based on differences in canine teeth from a single specimen of each taxon.

These are small, shrew-like marsupials, with brownish grey fur, a long, narrow snout and small eyes. Unlike all other caenolestids, the Long-nosed Shrew-opossum stores fat seasonally in its tail, which can become quite swollen. Indeed, the species name *raphanurus* means 'radish-tail'. Fat storage in the tail is an adaptation to unpredictable food resources in highly seasonal (cold or dry) climates; in lean times, fat can be reabsorbed from the tail and used as an energy source. A range of carnivorous marsupials have this ability. The list includes fat-tailed mouse opossums (*Thylamys*), the Patagonian Opossum (*Lestodelphys halli*) and the monito del montes (*Dromiciops*) from South America. Some Australian carnivorous marsupials use the same trick, including many dunnarts (*Sminthopsis*) and false antechinuses (*Pseudantechinus*). Interestingly, Long-nosed Shrew-opossums have been captured in the depths of winter, on snow pack. They appear to be active year round. So, despite the ample fat storage in their tails, they likely do not enter hibernation or even torpor for prolonged periods. Long-nosed Shrew-opossums are active at night and likely restricted to the ground. They are good diggers and may even spend some time under the ground in search of their primarily invertebrate prey. They supplement their diet with plants and fungi. Lactating or pregnant females have been captured from December to March, and also in May, October and November. Yet no young have ever been observed. Female Long-nosed Shrew-opossums have no defined pouch and probably use a nest to rear their young.

Rhyncholestes raphanurus is classified as Near Threatened on the IUCN Red List. It is a little known species, subject to a range of threatening processes such as logging and agriculture. The species has been found in remnant forest but appears most closely associated with primary Antarctic Beech (*Nothofagus*) forests: a habitat that is unfortunately rapidly disappearing from much of the species' distribution.

Order Didelphimorphia; Family Didelphidae (9 genera, 48 species)

Genus *Chacodelphys* (Pygmy Opossum, 1 species)

Distribution: North and north-east Argentina in Chaco and Formosa provinces, and Misiones

Habitat: Humid ecoregions featuring grasslands with tall grasses, but plausibly also forests associated with these areas

Size: 12 cm (10 g)

In 1920, a passionate bird biologist, Alexander Wetmore, travelled to South America to study and collect Northern Hemisphere migrants and complete work for his PhD on the birds of Puerto Rico. He was a meticulous field worker, taking detailed notes of daily activities, collection methods and fauna observations. Field conditions were often rough and one night he recorded that bed bugs gave him little rest: 'I got up twice, turned on a light and the first time killed 41 and the next time 18. The pillows here are filled with a sort of seed; mice or rats tore a hole in mine and would not let it alone. So finally I threw it in a corner on the floor.' The hardships of a field biologist! Wetmore's remit for his employer, the Bureau of Biological Survey, included collecting a broad range of material. During his time in Argentina, Chile, Peru and Uruguay, he made extensive mammal collections, featuring 150 skins and skulls of rodents and other mammals. One of these skins was to prove very important. It was collected from marshy grassland savanna abutting forest with jungle undergrowth so thick that Wetmore's team needed machetes to penetrate it. A decade later, in 1930, Harold Shamel of the Smithsonian Institution named a new species on the strength of that single skin and skull, placing it within the genus of mouse opossums (*Marmosa*). He emphasised its small size, mouse-like colour and short tail, but made no detailed examination of the skull. After a sequence of other taxonomic changes that saw the species placed within fat-tailed mouse opossums (*Thylamys*) and then gracile opossums (*Gracilinanus*), it was finally positioned in its own genus, *Chacodelphys*, by Robert Voss and colleagues in 2004. For many years the specimen was considered a juvenile because of its small size, but it was in fact a fully grown adult male. The Pygmy Opossum (*C. formosa*) is the smallest known opossum, weighing just 10 g.

The fur on the back and head is brown, with buff-coloured flanks and pale brown belly fur. There are marked dark brown eye-rings, from nose to eyes, and the cheeks are reddish brown. The ears are very small. The tail is shorter than the head–body length and is naked, dark brown above and paler below. To this day, little is known of the species' biology. Indeed, no other specimen of the animal was obtained until the 2000s – it had apparently disappeared for 80 years after discovery. Surveys of the area conducted during the intervening period failed to find it, possibly because of its low density and difficulty of trapping the animal in standard metal box traps. The animal's diet presumably comprises almost exclusively insects and other invertebrates, at least as far as can be judged from its molar teeth which have distinct, self-sharpening edges. Specimens have been recovered from owl pellets, suggesting the beast may be active at night when owls are on the prowl. One specimen was captured in a grassland pitfall trap, suggesting it spends at least some time on the ground.

The species is listed as Near Threatened on the IUCN Red List. It occurs in a range of probably fragmented populations that are threatened with clearing for agriculture.

Genus *Chironectes* (Water Opossum, 1 species)

Distribution: Mexico, Central America, Venezuela, Colombia, Ecuador, Peru, Bolivia, Brazil, Paraguay, Uruguay, Argentina

Habitat: A range of tropical and subtropical habitats with access to permanent water

Size: 52–83 cm (510–790 g)

The elegant and highly distinctive Water Opossum (*Chironectes minimus*) was named based on a single specimen collected in 1780 by German geographer and zoologist Eberhardt von Zimmermann. Further specimens were examined by Zimmermann's renowned contemporaries: Boddaert, Pennant, Link, Shaw and Kerr. The latter named another subspecies

Fig. 2.24. The Water Opossum (*Chironectes minimus*) has distinctive webbed feet that it uses to paddle through the water and long, slender fingers to probe into crevices to catch unsuspecting crustaceans, molluscs, fish and frogs. Source: Thiago Semedo.

in 1792. Two further subspecies were named in the early 1900s.

The Water Opossum is the only marsupial adapted for semi-aquatic life. Both morphological and DNA trees suggest that Water Opossums are most closely related to other large opossums: *Lutreolina* (lutrine opossums), *Didelphis* (black- and white-eared opossums) and *Philander* (four-eyed opossums). *Chironectes* is large and striking. The fur on the back is a pale, silvery grey. There are four, transverse dark grey–black bars at the shoulders, mid-back, hips and rump. The mottled coat breaks up its outline, thus protecting it while swimming from aerial predators. The underside of the animal is white. The tail is long and naked, except for some fur at its base. In keeping with its semi-aquatic habits, the fur of the Water Opossum is short, dense and water-repellent. The feet are fully webbed, with long, tactile toes and fingers, tipped with pads and reduced claws. Both females and males have a backward-facing pouch, although in males it is less developed. When in the water, the pouch can be drawn tightly closed to prevent young from getting wet while their mother swims. The pouch protects retracted testes in swimming males.

This is a truly carnivorous marsupial, consuming a wide variety of prey. It mostly eats aquatic or semi-aquatic vertebrates (especially fish and frogs) and invertebrates (crustaceans, molluscs and crickets), but it will supplement this diet with fruits and aquatic plants. *Chironectes* catches prey by swimming with the webbed hind feet in alternate strokes and holding the arms extended. In this fashion, its long, grasping fingers can reach into crevices to catch unsuspecting victims. The species is opportunistic and will take other prey if available. One animal was discovered by surprised biologists swimming up to and eating several bats caught in a mist net suspended over a river. Female Water Opossums dig 60-cm-long tunnels in riverbanks just above the water line. At tunnel's end is a comfortable nest lined with leaves and grass. Litter sizes vary from one to five. Young and mother are tolerant of low oxygen levels. This allows for long dives by the nursing mother in search of prey. Across the species' range, young can be found at all times of the year, suggesting that breeding is not seasonal. After ~48 days, the young detach from the mother's nipple but are still nursed on milk. Juveniles climb on the mother's back in transit, as in many other carnivorous marsupials. Juveniles have also been seen swimming alongside their mother. Water Opossums are nocturnal and usually most active for the first half of the night. They are excellent swimmers, but also jump and climb with agility. Their linear home ranges along river courses may reach up to 4 km in males, but perhaps only a quarter of that in females.

This species is broadly distributed and considered Least Concern on the IUCN Red List. However, it is threatened in some areas by deforestation and water pollution caused by human activities, such as gold mining.

Genus *Lestodelphys* (Patagonian Opossum, 1 species)

Distribution: Central and southern Argentina

Habitat: Semi-desert shrublands and steppes

Size: 19–24 cm (60–100 g)

The Patagonian Opossum (*Lestodelphys halli*) was first discovered in Cape Tres Puntas, Argentina, at the far southerly latitude of 47°. It was captured by collector T. H. Hall during mid-winter in a fox trap baited with meat, and sent to the Western Australian Museum. From there, it was transferred by exchange to the British Museum where it was named *Notodelphys* by renowned British taxonomist Oldfield Thomas in 1921. In naming the new genus, with the species name *halli* denoted in the collector's honour, Thomas wrote that its occurrence at such high latitude was: 'most

Fig. 2.25. The only member of its genus, the Patagonian Opossum (*Lestodelphys halli*) occurs further south (to 47°S) than any other living marsupial. The animal shown here is grooming after a bout of feeding. This opossum has a voracious appetite, storing energy as fat in the swollen tail to cope with the low temperatures and irregular food supplies that occur seasonally in its stronghold in Patagonia. Source: Dario Podesta.

unexpected, this locality being far to the south of any hitherto recorded for the family.' He also noted: 'the animal has a shortened muzzle which gives it great biting power … as insects are rare and fruit non-existent in its far southern habitat, this opossum has had to acquire peculiar habits and no doubt lives largely on mice and small birds.' The genus *Lestodelphys* was later nominated by Tate in 1934 to replace Thomas' *Notodelphys*, which had already been used for both a marine crustacean in 1847 and a pouched South American frog in 1854. The Patagonian Opossum is the only species in the genus. In DNA and morphology trees, the Patagonian Opossum is sister to the fat-tailed mouse opossums (*Thylamys*). Indeed, the Patagonian Opossum is superficially similar to the Elegant Fat-tailed Opossum (*Thylamys elegans*), which occurs in nearby Chile.

The Patagonian Opossum has dark grey fur on the back with lighter greyish-fawn fur on the sides and rump. The belly is white and a throat gland is present. Hands and feet are pure white. A darker stripe extends from the top of the head to between the eyes. The eyes have subtle darkened fur around them. The head is robust. Ears are rounded and pinkish. The tail is short, only two-thirds the length of the head–body. The tail stores fat and can become very swollen, almost club-shaped. Females lack a pouch. An extraordinary 17–19 nipples are found on the abdomen, with eight or nine on each side and one in the middle. Knowledge of diet in the wild is limited, but captive animals prefer vertebrate prey including lizards, birds and rodents. These are summarily dispatched with a precise bite to the neck and consumed head-first. In captivity, the Patagonian Opossum will also readily eat invertebrates such as beetles, grasshoppers, moths/butterflies, scorpions and spiders. *Lestodelphys* enters torpor irregularly, if trapped in cold conditions. The species can reabsorb fat from the thickened tail in lean times. Nothing is known of breeding. Captive animals are active night and day, with peak activity bouts after sunset and before sunrise. Patagonian Opossums are agile and fast moving, capable of jumping at least 30 cm into the air. However, they spend most time on the ground. They are expert diggers and excavate underground chambers that they use for respite during the day.

This species is classified as Least Concern on the IUCN Red List. Habitat modification threatens some populations, but the species occurs over a large area both in and outside of protected areas.

Genus *Lutreolina* (lutrine opossums, 2 species)

Distribution: Colombia, Venezuela, Guyana, Bolivia, Brazil, Paraguay, Uruguay and Argentina

Habitat: A range of habitats near water: grasslands, savanna woodlands, pampas, marshes and swamps

Size: 40–76 cm (200–910 g)

The Big Lutrine Opossum was originally described by Felix de Azara in 1801. Azara was a Spanish military engineer with no formal biological education, who had been sent to Buenos Aires in 1781 to map the boundary between Portuguese and Spanish possessions in South America. However, because of a perpetual lack of interest from Portuguese officials, the border survey was continually postponed. So Azara kept himself busy by describing the mammals, birds, geography and indigenous culture of the region. He was industrious, describing some 75 species of mammals, most of which were previously unknown. However, Azara used only common names, so they were not formally recognised until others applied scientific nomenclature based on all his hard work. Such was the case for the Big Lutrine Opossum, the third of Azara's six opossum species to be named. The animal was formally described by French taxonomist Desmarest in 1804 and named under the genus *Didelphis*, denoting it with the species name *crassicaudata* to account for its thick tail. It was not

Fig. 2.26. The otter-like Massoia's Lutrine Opossum (*Lutreolina massoiana*) is among the most carnivorous of America's marsupials. It eats a range of vertebrates, including other opossums, rodents, fish, frogs, lizards and snakes. These animals appear to be immune to the venom of some snake species. Source: F.R. Gonzalez.

placed in a distinct genus until the indefatigable Oldfield Thomas, more than a century later, nominated *Lutreolina* (meaning 'otter-like') in 1910.

Until recently, the Big Lutrine Opossum was the only *Lutreolina* species recognised, its disjunct distribution encompassing Colombia, Venezuela and Guyana in northern South America and several other countries in central-eastern South America, but in 2014, a new species was recognised based on DNA and morphology. This is Massoia's Lutrine Opossum (*L. massoiana*), which is apparently restricted to Yungas mountain forests in Bolivia and north-western Argentina. In recent DNA-based trees, *Lutreolina* has been positioned as sister to a clade containing opossums (*Didelphis*) and four-eyed opossums (*Philander*).

The otter-like lutrine opossums are large with uniformly brown-yellowish orange fur. The tail is thick and slightly shorter than the head–body length. The tail is densely furred on the basal half. The remainder is naked-looking and a broad brownish-black band is followed by a yellowish-white band towards the tip. The snout is short and strong, as are the legs. The fur is soft and dense. Although lutrine opossums live near water and are active in it, their pelt is not water-repellent. These animals are among the most carnivorous of America's marsupials. They have very strong jaws that enable them to tackle a range of vertebrate prey, including other opossums, rodents, fish, frogs, lizards and birds. They also regularly eat snakes, and appear immune to the venom of species such as pit vipers, but they also take a range of invertebrates, such as crustaceans, beetles, termites, wasps, grasshoppers and spiders, and occasionally consume fruits and carrion. The Big Lutrine Opossum builds a globe-shaped nest made of grass in tree hollows, abandoned burrows or grasslands. Litter size varies from seven to 11 young. The timing of breeding varies geographically, and occurs typically from September to April, more occasionally in winter. Lutrine opossums are nocturnal, fierce and aggressive. They are equally adept whether swimming, running or climbing. In captivity, Big Lutrine Opossums have been observed to ferociously attack human intruders, using their tail as a spring to launch themselves up to 60 cm into the air. This species will travel hundreds of metres a night when active, but its home range is small, typically less than 1 ha. In some areas, populations may be quite dense, with up to two animals per hectare.

The Big Lutrine Opossum has a large distribution and is listed as Least Concern on the IUCN Red List. In southern areas, wetlands are drained for agriculture and this poses a threat. Massoia's Lutrine Opossum was only recently discovered, but is also listed as Least Concern on the IUCN Red List. This species is endemic to Yungas forests that are becoming increasingly fragmented due to intense human disturbance, but it fortunately also occurs in several protected areas.

Genus *Metachirus* (Brown Four-eyed Opossum, 1 species)

Distribution: Mexico, Central America, Venezuela, Colombia, Ecuador, Peru, Bolivia, Brazil, Paraguay, Argentina

Habitat: A range of forest habitats from sea level to 2100 m; also in orchards and yards near human settlements

Size: 52–65 cm (284–480 g)

The Brown Four-eyed Opossum (*Metachirus nudicaudatus*) was discovered in Cayenne, French Guiana and named by French naturalist Geoffroy Saint-Hilaire in 1803 as a species of *Didelphis*. The species name, *nudicaudatus*, means 'naked tailed'. The new genus *Metachirus* was not applied until German naturalist Hermann Burmeister raised it in 1854, after spending 2 years studying natural history in Brazil. There are currently five subspecies of Brown Four-eyed Opossum listed, all based on morphology. However, DNA analyses conducted since 2003 suggest there may be five or more species within *Metachirus*. Not all of them match the morphologically based subspecies. A revision is therefore under way, based on combined DNA and morphology to resolve the confusion. The Brown Four-eyed Opossum is the largest pouch-less species within the Family Didelphidae. In recent DNA-based trees, *Metachirus* has been positioned as sister to a clade containing the Water Opossum (*Chironectes*), lutrine opossums (*Lutreolina*), four-eyed opossums (*Philander*) and other opossums (*Didelphis*).

Metachirus is a distinctive beast, with red-yellow-brown fur on the back and darker fur along the midline and rump. The belly is yellowish-white to brown. The body fur is dense, short and soft. The ears are brown and naked. Broad, reddish-dark brown eye-rings extend down towards the nose and back to the ears, merging on top of the head. A broad, dark stripe also runs from the middle of the head down towards the tip of the snout. This merges with the dark eye-rings to enclose yellowish-white spots above each eye, giving this species its common name. The tail is long, slender, naked and brownish. This large species has a generalist diet. It eats mostly invertebrates, including cockroaches, beetles, ants, termites, spiders, bugs and wasps, but also takes a range of vertebrates, such as lizards, small mammals and birds. The Brown Four-eyed Opossum is an important seed disperser of several forest fruits. It builds dens at the base of trees, sometimes with multiple tunnel entrances in and around the roots, and may also construct spherical nests composed of leaves and twigs on the ground, concealed with litter. These opossums occasionally nest in rocky crevices. Using a curled tail, they can transport nest material back and forth. Females have no pouch. There are nine nipples, four on each side and one in the middle. Average litter size ranges from five to eight. Gestation is 20–28 days and young remain attached to teats for 75–80 days. There are a further 30–45 days spent in the nest before weaning. Juveniles leave the nest at ~130 days. Timing of breeding and lactation vary with locality, but across the collective range, Brown Four-eyed Opossums breed throughout the year. They are ground-based and active at night, and are excellent runners, periodically leaping large distances. Radio-tracking studies indicate that this species moves on average more than 500 m a night, with home-range size estimated at 1–8 ha. The Brown Four-eyed Opossum can occur in high densities, up to 0.5–6 individuals per hectare.

The Brown Four-eyed Opossum is listed as Least Concern on the IUCN Red List. However, the species urgently needs taxonomic revision. DNA studies suggest that deep divergences occur between populations over small geographic distances, and some of the cryptic species within *Metachirus* may need protection.

Fig. 2.27. The Brown Four-eyed Opossum (*Metachirus nudicaudatus*) is exclusively ground-dwelling and an adept runner, moving in high speed leaps and bounds. These animals are found in a range of forest habitats and even in and around human settlements. Source: Thiago Semedo.

Genus *Monodelphis* (short-tailed opossums, 23 species)

Distribution: Collectively, much of South America except southern Chile and Argentina

Habitat: A range of forest habitats – lowland to upland and dry to wet; some species in swamps/marshes and others in highly fragmented habitats

Size: 10–29 cm (7.5–150 g)

This speciose genus has a particularly tortured taxonomic history, moving through more than a dozen different generic placements over nearly two and half centuries of study. The first short-tailed opossum was discovered in 1776 by German zoologist Philipp Müller and placed in Linnaeus' placental shrew genus *Sorex*. Several more short-tailed opossums were discovered in the 1700s and 1800s and assigned to numerous genera, including the civet (*Viverra*) and the American marsupial genus *Didelphis*. By the early 1900s, 10 species were recognised, some placed by Thomas and Osgood under French surgeon and naturalist Rene Lesson's 1842 *Peramys* and then eventually under *Monodelphis*, which had been coined by British zoologist Gilbert Burnett in 1830. Numerous other species were recognised by 1950. Between 2004 and 2013, further revisions recognised another handful of species, but none of these works included both morphological and DNA data on all species. A recent (2014) revision noted that, although 21 species are recognised in the genus, some may be questionable. After this revision, two new species were named, in 2015 and 2017 (see also Chapter 3, vignette).

The genus ranges from eastern Panama to northern Argentina in various habitats. Collectively, the short-tailed opossums are variable in fur colour, reproduction (one species is a suicidal breeder, the others are not) and behaviour (some are active at night, others during the day). The short-tailed opossums vary from small- to medium-sized. Their fur colour ranges

Fig. 2.28. Many carnivorous marsupials are obsessive groomers. This Amazonian Red-sided Opossum (*Monodelphis glirina*) is cleaning its snout and vibrissae (whiskers). The whiskers are an important sensory organ used by such species to hunt and detect prey. Source: Thiago Semedo.

Fig. 2.29. In a striking example of convergence, although distantly related and living on different continents, *Monodelphis* spp. (like *M. domestica*, pictured) of the Americas look remarkably like Australian antechinuses. Some *Monodelphis* spp., such as the Orange-sided Opossum (*M. dimidiata*), also exhibit a male die-off after breeding like that shown throughout the antechinus group. Source: Thiago Semedo.

from silvery-grey through light brown and reddish-brown to uniformly dark grey. Some species have three narrow longitudinal blackish stripes running down the back. In others, there is a broad strip of greyish fur along the back. The tails are tapering, some thick at the base, others not. Tail length ranges from about the same as the head–body length to only about half the head–body length. The snouts are of moderate length and conical. The eyes and ears are moderate to small in size. For two-thirds of named species, there is no information on dietary composition, but what is known suggests that members of this genus mostly feed on invertebrates such as bugs, ants, moth/butterfly larvae, spiders, beetles, grasshoppers, scorpions and millipedes, with vertebrates being hunted opportunistically. Vertebrate prey include frogs, lizards and small snakes. Fruit supplements the diet in several species. Breeding is only known for about half a dozen species. One of these, the Orange-sided Opossum (*M. dimidiata*), breeds once a year and undergoes a male die-off at the close of the breeding season, as in the Australian *Antechinus*. For the remainder of species for which breeding information is available, mating occurs at various times of the year and females have six to 11 young. Nests are made from a range of materials, including leaves, grasses and human refuse. Nests can be positioned in tree forks, tree hollows, rock crevices or fallen logs. Most species for which activity data are available (about eight species), are active only during the day. For a couple of others, nocturnal activity seems to be preferred. Most species are believed to be ground-dwelling, although a couple of species have been found nesting up to 1.5 m above ground in trees, so they must have some climbing ability. Data on movement is minimal. Some species have been found to move tens to hundreds of metres per day and densities of some populations are estimated to be high, from 0.5 to four individuals per hectare.

Most species are classified as Least Concern on the IUCN Red List, but some have not been formally assessed. One species is listed as Vulnerable (Reig's Short-tailed Opossum, *M. reigi*) and another as Critically Endangered (possibly extinct) (Single-striped Opossum, *M. unistriata*). Threats exist for a range of species. These include: logging, burning, habitat fragmentation, conversion of land to agriculture, and urbanisation. Although two new species have recently been named (2015 and 2017), this speciose genus may benefit from further taxonomic assessment, and therefore the status and threats posed to some species groups are uncertain.

Genus *Philander* (four-eyed opossums, 7 species)

Distribution: Mexico, Central America, Brazil, Paraguay, Argentina, Colombia, Venezuela, Peru, Bolivia, Ecuador, the Guianas

Habitat: A range of undisturbed and disturbed forests, ranging in altitude in both dry and wet areas; some species occur in swampy/marshy inundated areas

Size: 50–68 cm (148–910 g)

The first four-eyed opossum was described in 1758 by Carolus Linnaeus, and was recognised then as *Didelphis opossum*, a different form of his *Didelphis marsupialis*, with which it shares a similar distribution. Just 4 years later (1762), French zoologist and natural philosopher Mathurin Brisson raised a unique genus, *Philander*, to accommodate *D. opossum*, but this generic name was not applied by Oldfield Thomas more than a century later in 1888 when compiling his famed *Catalogue of the Marsupialia and Monotremata in the collection of the British Museum (Natural History)*. Instead, species were placed in the subgenus *Metachirus* within the genus *Didelphis*. Wilfred Osgood followed the example of Thomas, when raising a new Peruvian species in 1913. Debate about the taxonomic position of *Philander* continued into the 1970s and 1980s, until finally it was accepted as a genus in its own right. The group has been scrutinised further in recent times, with three species being nominated in the last decade, but a full revision is required for the Gray Four-eyed Opossum (*P. opossum*), which will probably see some of its subspecies raised to full species, expanding the ranks of *Philander*.

Fig. 2.30. Four-eyed opossums (*Philander* spp.) are formidable predators, sometimes even taking primates as prey. They regularly nest in tree hollows. This animal has been surprised hiding in a fallen hollow log. Source: Thiago Semedo.

Four-eyed opossums are large, grey-black opossums with conspicuous white or cream-coloured spots above each eye. The ears are bi-coloured or black and naked-looking. The tail is either about the same length as the head–body or slightly longer. The basal two-thirds of the naked-looking tail are typically blackish and the last one-third, towards the tip, is white. Detailed ecological information is available for only two species: the South-eastern Four-eyed Opossum (*P. frenatus*) and Gray Four-eyed Opossum. These species mostly eat invertebrates and vertebrates, supplementing their diet opportunistically with fruit. Invertebrates taken include beetles, ants, spiders, millipedes, flies, dragonflies, scorpions, cockroaches, bugs, grasshoppers, termites, snails and crustaceans, while vertebrates include rodents (squirrels, rats), short-tailed opossums, primates, bats, lizards, snakes, frogs and birds. Up to 30 different types of fruit have been identified as food. Fruit is taken more often in drier months, most likely as a water source. The nest consists of a bowl of dry leaves with entrance tunnels. These are positioned in tree hollows up to 10 m off the ground or in fallen logs, and even in the rooves of human houses. There is a well-developed pouch in adult females that can carry up to eight young. Litter size and timing of breeding seem to vary with location, but collectively these animals may breed throughout the year. Gestation is short (just a couple of weeks) and weaning occurs at 65–80 days. Most species are believed to be nocturnal, although some may be active by day. These animals are found mostly on the ground, but they are capable climbers, jumpers and swimmers. Home-range sizes vary widely, from 0.2 to 12 ha, partly in relation to habitat type, and individuals may move several hundred metres per night. Densities have been predicted to average 0.2–1.25 individuals per hectare, but in some places may reach two per hectare.

Six species are classified as Least Concern and one as Data Deficient on the IUCN Red List. Threatening processes include habitat alteration such as logging and oil drilling. Some species are tolerant of human disturbance.

Genus *Thylamys* (fat-tailed mouse opossums, 11 species)

Distribution: Brazil, Chile, Peru, Bolivia, Argentina and Paraguay

Habitat: Varied, including open and closed forests at low to high altitude that provide dry to humid conditions; grasslands, savannas, scrubs, thickets, semi-arid and arid areas, deserts

Size: 14–29 cm (11–66 g)

The first of the fat-tailed mouse opossums was named by French zoologist Anselme Desmarest in 1804, under the genus *Didelphis*. Three more species were named in the 1800s under the same genus. Oldfield Thomas named several species and subspecies in the early 1900s, under British zoologist John Gray's (1821) genus *Marmosa* (mouse opossums). By the late 1960s, there were 11 species and the genus *Marmosa* was still being applied. The genus *Thylamys* has since been used to denote the unique genus of fat-tailed mouse opossums; it had been first used by John Gray in his 1843 catalogue of British Museum specimens, when

Fig. 2.31. Fat-tailed mouse opossums (*Thylamys* spp.) such as the one pictured are capable of reabsorbing fat stored in their tail for use as an energy source in lean times. This is useful for coping with life in unpredictable, dry climates. The same trick is used by some of the Australian dunnarts (*Sminthopsis* spp.), which are also found in dry areas. Source: Pablo Jayat.

referring to the Elegant Painted Mouse, *Thylamys elegans*, which was George Waterhouse's (1839) *Didelphis elegans*.

Members of this genus range in coat colour from silvery-grey to reddish-brown above, and have whitish/off-white fur below. All species have large ears and eyes with dark eye-rings. Fat-tailed mouse opossums have pale cheeks and some indication of a darker head stripe that runs down between the eyes and onto the snout. The hands and feet are pale. The tails are thickened and longish, reaching as much as twice the head–body length in one species. These are small- to moderate-sized carnivorous marsupials. Ecological information is available for only about half the named species. The diet appears to be mostly invertebrates, supplemented opportunistically with vertebrates and fruits. A range of invertebrates has been found in scat contents, including beetles, spiders, millipedes, ants, moths/butterflies and flies. Some species hunt small vertebrates, including lizards, rodents, small birds and bird eggs. The diet is rounded out in some species with fruits, leaves and seeds. Fat-tailed mouse opossums construct nests of leaves, twigs, hair, feathers, grass and other items, and position them in tree holes, rock crevices or under shrubs. Some species will raid a bird's nest, consuming the eggs/young and then use the nest for their own purposes. Guinea-pig galleries are sometimes used to hide nests. Up to 15 nipples have been reported in females of some species. Young can range from just three or four to a full complement of 15. These animals often breed more than once, at various times, during the year. However, most breeding activity seems to be aligned with the wet summer, when most food is available to support growing young. The genus appears to be nocturnal, and several species can enter daily torpor. In such cases, torpor frequency varies with ambient temperature and food availability. In captivity, where food is provided daily, they may not enter torpor at all. Fat-tailed mouse opossums are mostly ground-dwelling, but several species climb and jump well, even using their tail for assistance, and can be found in small bushes. Home range is little known, except in a couple of species, and varies in the few individuals studied from 0.08 to 2.3 ha. Densities of some species have been estimated as large: from 0.4 to over 20 individuals per hectare, depending on the habitat and time of year.

Most species are classified as Least Concern or Data Deficient on the IUCN Red List. One species has been classified as Vulnerable (Karimi's Fat-tailed Opossum, *T. karimii*), with habitat clearing and conversion of habitat to monoculture (soy beans) cited as threats.

Genus *Tlacuatzin* (Tlacuatzin, 1 species)

Distribution: Apparently endemic to Mexico, from the west and extending east, through Oaxaca to the Yucatán Peninsula

Habitat: Evergreen and deciduous forests, scrublands, grasslands and even cultivated regions in dry forest regions

Size: 21–33 cm (30–70 g)

Joel Asaph Allen, Curator of Birds and Mammals at the American Museum of Natural History, was one of America's leading naturalists in the late 1800s and early 1900s. He was a prolific writer, producing over 1400 scientific works. Like every self-respecting museum taxonomist, Allen made industrious use of numerous collectors, sending them into the dangerous wilds in search of new species on his behalf. One such collector was Dr Audley Cecil Buller, MD. At the behest of Allen and his museum, in 1889 Buller began an epic collecting trip through Mexico, ranging after several years over some 1600 km, collecting in total 300 mammal specimens, representing 55 species, 16 of which were new to science. One species of small opossum was collected by Buller on the Isthmus of Tehuantepec in southern Mexico during 1890. Buller was then aged 37, but would die aged just 41 years, at the conclusion of his Mexican fieldwork. Allen named the new opossum in 1893, the year before Buller's untimely death, classifying the little animal initially under the opossum genus *Didelphis*, and subsequently the mouse opossum genus *Marmosa*, where it remained for more than a century. In 2003, on the strength of both genetics and morphology data, a new genus was prescribed to accommodate the single species, *Tlacuatzin* – an Aztec word meaning 'opossum'.

The Tlacuatzin (*T. canescens*) is pale grey to greyish-brown on the back and head, with paler fur on the snout, chest and belly. The eyes have a dark ring and the ears are broad, round, naked and blackish. The tail is about the same length as the head–body. The tail has fur at the base but is otherwise naked, and greyish, with a paler tip. The diet of the Tlacuatzin consists mainly of invertebrates, particularly insects such as bugs, grasshoppers, butterflies, moths and beetles. Tlacuatzins also eat vertebrates, such as geckoes, and are renowned thieves of bird eggs. They also feed on cactus flowers when in bloom and eat wild figs. Tlacuatzins make spherical nests of leaves and stems, lined with grasses in a variety of locations, always above the ground. Breeding occurs year round, with litters of eight to 14 young. Mating occurs with an engaged male and female hanging suspended from a tree by the tail. The species is probably mostly nocturnal, spending the majority of time in trees. Tlacuatzins can be numerous in certain areas, reaching estimated densities of 0.8–8 individuals per hectare.

The species is classified as Least Concern on the IUCN Red List. The Tlacuatzin has a wide distribution, a large overall population, and is found in a range of protected areas. However, the species' prime habitat of deciduous forest continues to be rapidly cleared in many areas.

'Nearly' carnivorous marsupials

A feature of the creatures just described is that, while their diets are broadly carnivorous, many eat other types of food too. This dietary strategy is valuable if you are a small- to medium-sized mammal that needs to eat continuously when active to maintain an internal body temperature, and especially so if your food resources are unpredictable. Such generalism is

Fig. 2.32. A Virginia Opossum (*Didelphis virginiana*) sketched by famed German artist Gustav Mutzel in 1927. *Didelphis* species are large (up to ~5 kg), with a broad distribution and cosmopolitan diet. Some seem to have taken advantage of urban sprawl, and the food that humans inadvertently provide them, to increase their range. The Virginia Opossum is the only marsupial in the USA, occurring as far north as Canada on the North American continent. With prehensile tails and opposable 'grasping' thumbs, they appear as comfortable in trees as on the ground. Source: Gustav Mutzel.

Fig. 2.33. A 'nearly' carnivorous marsupial victoriously celebrating its taste for invertebrates. This is the Red Mouse Opossum (*Marmosa rubra*). Source: Jorge Brito.

the cost of being warm-blooded and ever-hungry. As a consequence of their often broad diets, a black and white cut-off for carnivory does not exist. Because diet is often determined by the seasonable availability of food, the total range of what is taken as prey can be extensive. It is fair to say that if we included every marsupial that sometimes ate animals as 'carnivorous', the list might include nearly all species. Exceptions would be many of the living 'tusk-toothed' (diprotodont) marsupials, such as kangaroos, wombats and the Koala, which are strict herbivores. To be so inclusive would be impractical, so we have drawn an arbitrary line in the sand. Yet several marsupials that we haven't mentioned so far sit quite near to that line. These are often described as omnivorous – species that eat plants, fruits and animals. It is worth briefly mentioning them in turn now.

In the Americas, we might include members of the genus *Didelphis*, such as the Virginia Opossum (*D. virginiana*). This is the only marsupial found today in the USA. Members of this genus are opportunistic and large (up to 3 kg). This opossum eats many fruits and seeds, but also invertebrates and vertebrates, and even takes carrion (i.e. dead animals). Other American marsupials also straddle the border of carnivory, eating plants, fruits, seeds and invertebrates. These include: mouse opossums (*Marmosa*), slender opossums (*Marmosops*) and other opossums (*Gracilinanus*).

From Australia and New Guinea, we might include on our list of borderline carnivores some or all of the bandicoots. For example, the Striped, Mouse, Pygmy and Papuan Bandicoots (genus *Microperoryctes*) are denizens of New Guinean tropical forests: all eat fruits, seeds, fungi, a range of soil invertebrates (including worms and insects) and sometimes even small vertebrates. Some members of the mostly herbivorous diprotodonts are also well worth mentioning. For example, consider the New Guinean Feather-tailed Possum (*Distoechurus pennatus*). This animal lives in lowland swamp and peat forests and eats fruit, nectar, seeds and a range of invertebrates. The striped possums (*Dactylopsila*) that occur in wet forests in New Guinea and Australia feed on fruit, nectar and various invertebrates, including wood borer larvae. The latter are favoured prey, impaled on the long, probing fourth finger of these possums before being eaten with relish. Finally, we have the Mountain Pygmy Possum (*Burramys parvus*) of alpine south-eastern Australia. This animal is not dissimilar in appearance to the strange monito del montes of South America. These pygmy possums eat fruits, nectar, pollen and a suite of invertebrates. They especially target Bogong moths (*Agrotis infusa*) that migrate from the western plains in spring, feasting upon these fat-laden prey in large numbers.

Now that we have introduced our carnivorous marsupials and where they live, let's investigate how they came to be there.

3

BIOGEOGRAPHY

Where are carnivorous marsupials found and how did they get there?

Alfred Wegener, a German polar explorer, geophysicist and meteorologist, formalised the theory of drifting continental landmasses in 1912. Yet the idea hearkens back to at least the 1500s. Wegener's idea, like many scientific theories, was initially controversial, but since the 1950s, most scientists hold with his notion that Earth's continental landmasses move or drift, sliding inexorably about on gigantic plates. There are eight major plates and many minor ones. Continental movement of just a few centimetres a year turns into thousands of kilometres when considered over millions of years. This extraordinary tectonic phenomenon has had a profound effect on the movement and evolution of animals and plants living on drifting continents over the last hundreds of millions of years, particularly terrestrial species with limited dispersal capacity, such as carnivorous marsupials.

Collectively, carnivorous marsupials are found today over a large geographic area – Australia, the Americas and New Guinea. However, most species (98.5%) occur on one landmass and nowhere else, which renders them both unique and vulnerable. Carnivorous marsupials have a deep history. In particular, continental land animals have experienced long periods of isolation and during this time they have travelled separate evolutionary paths. Within each landmass, there is a wide variety of habitat types, which over time have promoted species diversity. Australia, for example, is today predominantly (70%) arid/semi-arid rangeland, but there are also considerable tracts of open forest, closed forest and tropical, subtropical and temperate rainforest in near-coastal areas. Over the last several million years, the Australian continent has become drier and windier. This has reduced forest cover and promoted development of semi-arid regions and extensive deserts. Animal groups such as carnivorous marsupials have expanded into these new areas and diversified. Similarly, South America is an extremely diverse landscape. It contains all of the world's major ecosystems, including vast tracts of Amazonian rainforest, grassland, shrubland, desert, plateau and high mountain ranges (some glaciated). This range of ecosystems supports a diverse array of animals and plants, including carnivorous marsupials. New Guinea contains large tracts of mostly unexplored upland and lowland rainforest, which support a vast diversity of species: some known and many others awaiting discovery.

Common ancestors of the modern mammals (marsupials, monotremes and placentals) apparently existed ~180–220 million years ago (mya). The oldest known marsupial fossil, from China, has been dated at ~125 mya. So, to understand the origins and development of the carnivorous marsupials, we need first to consider what was happening in Earth's history between ~125 mya and 200 mya.

Around 175–200 mya, the supercontinent Pangaea that incorporated almost all landmasses on Earth, began to break apart. This ultimately sundered to give rise to two major continents – Gondwana and Laurasia. Gondwana contained Africa, South America, Arabia (in the west) and India, Madagascar, Iran, Tibet, fragments of south-east Asia, Antarctica and Australia (in the east). Laurasia included Europe,

Fig. 3.1. The great southern land, Australia – a composite satellite image. Most of the continent's interior is dry. Human populations are centred on the wetter eastern and south-western coasts, but carnivorous marsupials are found in great diversity across the entire landmass. Those species occurring in very dry areas have a range of adaptations that help them cope with the inevitable uncertainties in rainfall, and therefore food. Source: Reto Stöckl/NASA Goddard Space Flight Center.

the rest of Asia and North America. By ~130 mya, western Gondwana began to drift apart. Not long after this, ~125 mya, India, and a little later, New Zealand, began to drift away from other eastern Gondwanan continents. Australia–New Guinea began disconnecting from Gondwana ~80 mya, but this likely didn't proceed in earnest until ~55 mya. It was only by ~38–45 mya that Tasmania, Australia's southern connection to Antarctica, fully detached. This allowed ocean currents to flow between the two continents for the first time.

A near relative of China's 125 mya marsupial, aged 115 mya, is known from Texas in the USA. If marsupials originated in Europe or Asia and moved to North America, it likely happened via the Bering Land Bridge. From there, ancestral marsupials most likely moved from North to South America during the very late stages of break-up of Pangaea. This possibly occurred via a chain of volcanic islands that erupted from the shallow sea. If so, the ancestral animals may have island-hopped from North to South America on flotsam. From there, the marsupials must have travelled south through South America to other parts of Gondwana. Sometime after Gondwana separated completely from Laurasia, marsupials became extinct in North America. At this time, Australia probably did not have many mammals, other than some monotremes.

The Australian marsupial radiation

The age and distribution of fossils tell us that by 70 mya there were still likely to have been narrow bands of land between southern South America and Antarctica. Antarctica and Australia were also connected. These connections were probably crucial in allowing some marsupials to move gradually from South America, across Antarctica and into Australia. One such disperser must have made it to Australia, giving rise to Australian marsupials. Millions of years later, before Australia separated from Antarctica, some descendants moved across Antarctica once more, back into South America. These marsupials remain in South America today and are true 'living fossils' – the carnivorous monito del montes. The three extant species represent the only survivors of an otherwise lost order of marsupials. Although monitos may have been ancestral to the Australian marsupials, they became extinct in Australia. After both South America (~60 mya) and Australia (~40 mya) fully separated from Antarctica, the land animals on both continents underwent a long period of isolated evolution.

In the case of Australia, between 25 mya and 40 mya, land mammal diversifications mostly involved marsupials and smaller numbers of monotremes (although bats flew in occasionally from elsewhere). Indeed, it was only ~15–25 mya that the Australasian and south-east Asian tectonic plates collided, resulting in successive rafting migrations of land mammals via Indonesia and New Guinea. Some examples of these migrants are placental rodents (over the last 5 mya), as well as other taxa such as frogs and reptiles. During their northward drift, New Guinea and Australia were periodically separated by sea level fluctuations across the wide, but shallow, Arafura Sea and Torres Strait. These countries have only become fully separated by ocean channels in the last 6000 years. However, the mountains of New Guinea are young. They were mostly formed within the last 5 mya, so there was little mixing of Australian and New Guinean animals until the last 10–15 mya. In recent times, 40–60 kya (thousand years ago), Aboriginal people have also had an impact on the distribution and abundance of Australian marsupial faunas. They are at least partly responsible for a wave of extinctions of giant marsupials – the megafauna. Such was also true of native people's impact on mammals of other continents, but in the last 250–400 years, humans have had accelerating impacts. In Australia and elsewhere, European settlers devastated habitats and introduced a range of mid-sized carnivores (cats, foxes and dogs) and other mammals (cattle, sheep, pigs, rabbits, camels, goats and horses). These species have collectively had a major negative impact on Australia's unique mammals.

Australia's climate preceding and immediately following separation from Antarctica was warm and wet. This 'greenhouse' climate was ongoing, reaching a peak towards 40 mya, around the time that Tasmania completely separated from Antarctica. Over these millions of years, the warm, wet conditions favoured forest growth (perhaps predominantly rainforest). So, by the time Australia started drifting northwards, it was well forested, but after the northward drift began, global temperatures driven by changing ocean circulations began to gradually drop. Crucially, Australia and South America had both cut loose from Antarctica, allowing the Antarctic circumpolar current to establish. By ~35 mya, ice age regimes in the southern poles had begun. Temperature extremes became more pronounced and rainfall more erratic. So, paradoxically, as Australia drifted north towards the equator, it became cooler, windier and drier. The warm, wet forests gradually retracted. Drier 'sclerophyll' forest communities, better adapted to the erratic weather patterns, took hold. By 15 mya, the Antarctic ice cap was permanent and growing. Ice age conditions intensified. Rainfall became even more erratic. This marked the onset of aridification in Australia, with development of open grassland and saltbush plains.

Also ~15–20 mya, the collision of the Australasian and Pacific plates initiated a sequence of New Guinean mountain upthrusts that became especially prominent over the last 5 million years. Ultimately, this orogenic (mountain-building) activity created pockets of wet forest habitat on the rugged slopes and valleys of the new ranges. These environments must have provided critical refuges for wetter adapted animals and plants that ongoing drying in Australia was squeezing out of their once suitable habitat. By ~6 mya, the Antarctic ice sheet had increased dramatically in size. Drying conditions in Australia intensified further. Arid and semi-arid conditions began to take hold throughout much of central Australia. Between 3 and 5 mya, there was evidently a warming trend. Northern Australia experienced periodic heavy rainfall, but the interior of Australia remained arid. By ~2.5 mya, the north polar ice cap froze over completely. This was a tipping point – Pleistocene ice ages now held sway. Australia had minimal ice sheets during this period, but it was periodically colder and windier. This caused the continent to become even more arid, with more extensive formations of dune and desert environments. These areas were surrounded by semi-arid shrublands and grasslands, with woodlands confined mostly to the east and south-west. Rainforests hung on only in small pockets, mostly along the east coast. Within the last 500 000 years, the cycles of cold, dry weather in Australia more than doubled. Rainfall became more intermittent and aridity more profound. Wind, dust and dunefield formation took hold.

Australia's ongoing aridity, low relief and poor soil fertility have had profound consequences for its animals and plants. Eroded mountains mean limited, sluggish river systems. Australian soils are salty. They have suffered from repeated incursions of inland seas, and continual erosion by dryness and wind. In the face of this environmental onslaught, some marsupials retracted to wetter areas. Other groups became isolated and speciated, causing several explosions in diversity as they moved back into drier habitat areas. Yet other groups (indeed in some cases entire families) could not adapt and became extinct. Most of the patchy, but informative, fossil record suggests that it was not until ~20 mya that more arid-adapted faunas began to appear. The greatest period of diversity for arid-adapted marsupials did not begin until 5–10 mya. In fact, many genera of arid-adapted carnivorous marsupials appear only a few mya ago in the fossil record. DNA studies suggest that their origins fall mostly within the last 10 million years. During this period, habitat shifts probably occurred multiple times. This arid-adapted group constitutes the majority of Australia's carnivorous marsupials today. It includes the mulgaras, Kowari, Kaluta, false antechinuses, ningauis, planigales and most dunnarts. Such species needed to employ behaviours that conserve water and energy in a very unpredictable environment. Today, we see many such strategies in these groups, such as fat storage, high

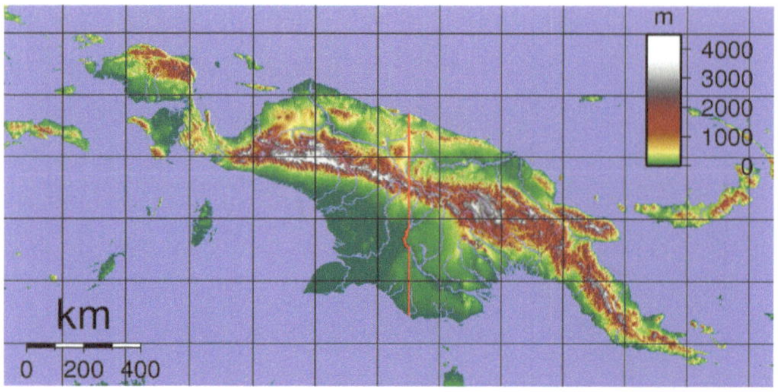

Fig. 3.2. A topographic map of New Guinea. Note the high, snow-capped mountains of the central Highlands, reaching more than 4 km high. With a range of rainforest types and also savanna, New Guinea remains mostly unexplored by scientists. Its dark, mysterious forests surely contain many new species of plants and animals, including carnivorous marsupials. The small knob of green at the bottom of the map is the northern tip of Queensland, the closest part of the Australian landmass, just a couple of hundred kilometres away. In the last few million years during low sea level intervals, rafts of carnivorous marsupials periodically moved north from Australia into New Guinea, perhaps seeking refuge in its warmer, wetter forests. Source: Sadalmelik/Wikimedia Commons.

mobility, nocturnal activity, having a generalist diet and using daily torpor. About 10–12 mya, the ancestors of carnivorous marsupials such as *Murexia, Phascolosorex, Neophascogale* and *Myoictis* retracted to shrinking wetter areas in the montane rainforests of New Guinea. Some dunnarts and antechinuses became established in the north-east Queensland Wet Tropics, signifying a shift back to wetter environments for some members of these otherwise drier- and open-forest adapted genera. A few groups of Australian carnivorous marsupials are much older and survived the successive environmental changes. These recalcitrants include the Numbat (*Myrmecobius*) and highly specialised marsupial moles (*Notoryctes*). The moles today are desert specialists, but fossil forms are known from much wetter periods (20 mya) during Australia's history.

The New Guinean marsupial radiation

New Guinea is more mountainous and considerably younger than Australia. Much of the mountain upthrust we see today began only after Australasia had drifted northwards to collide with the Pacific plate. The majority of the central and northern highlands of New Guinea became established in the last 2–4 million years. Today, these mountains extend largely unbroken east to west for 1800 km and are up to almost 5000 m high. This is considerably higher than Australia's tallest peak, Kosciuszko, at 2228 m. Most likely, little of New Guinea was above sea level and land-connected to Australia before 20 mya, so there was scant opportunity for overland movement of Australian carnivorous marsupials to New Guinea much before 10–15 mya. From this time, 2–10 mya, global sea levels probably dropped periodically – enough to allow some of the Australian marsupials to move into New Guinea. In particular, connections existed from time to time over much of the last 2 million years, with sea levels dropping by as much as 140 m. Vast areas between northern Australia and New Guinea were then exposed as open plains. This opened the way for successive migrations of fauna. The plains probably supported stands of savanna-woodland habitat, interspersed with patches of closed, wetter forest that acted as ecological filters. At times this habitat promoted movement of closed forest-loving species and at other times favoured drier, woodland- and savanna-adapted species. Migrating species doubtless were periodically isolated by subsequent inundations as sea levels rose. They may have speciated during isolation, then moved about again as sea levels dropped. Most likely, this allowed some wetter-adapted carnivorous marsupial groups to move into new niches and away from the ongoing drying habitats of Australia.

The mountains of New Guinea were at times some of the wettest places on Earth, supporting true rainforests of huge extent: over 700 000 km^2. Some of the carnivorous marsupials from New Guinea today appear to be limited to these extremely wet environments. They include *Phascolosorex* and *Neophascogale*. Ancestors of these two genera, and also the antechinus-like *Murexia*, may have moved across from Australia during one of the dispersal phases, ~10–12 mya. Further south in New Guinea are large tropical savanna woodlands, more similar to the vegetation of northern Australia. It is here that several species of otherwise Australian carnivorous marsupial genera are shared with New Guinea: dunnarts, planigales and quolls.

There are probably numerous cryptic carnivorous marsupials hiding in New Guinea that have evaded scrutiny from scientific eyes. This is both because the country still remains poorly explored and because its rugged terrain lends itself to scenarios of evolution in isolation. This is true both altitudinally and also across its vast array of lowland and upland habitat

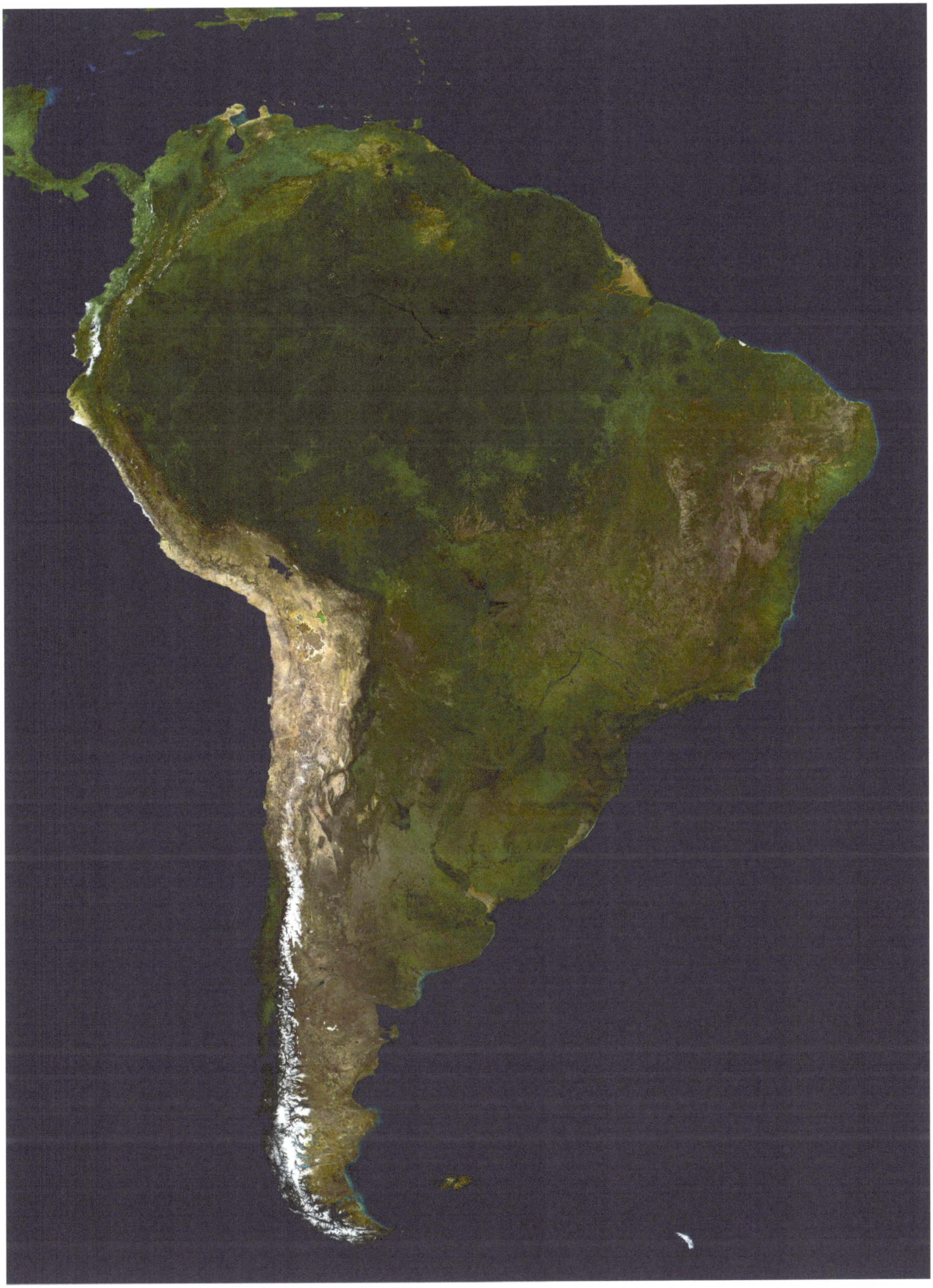

Fig. 3.3. South America and its diverse array of environments are shown clearly in this orthographic projection. The great Andes mountain range, with peaks thrusting almost 7 km high, runs virtually the length of the continent on the west coast. In the north is the extensive Amazonian rainforest basin. In the south are the drier plains of Patagonia and the snow-capped peaks of the southern Andes. Source: NASA.

types. New Guinea is admittedly less diverse in known carnivorous marsupials (and other fauna) than other areas supporting large tracts of rainforest, such as the Amazon in South America. This is most likely due mainly to its smaller size, younger age, and smaller founding stock of carnivorous marsupials. Its source populations of migrants from the Australian south were likely intermittent. In short, the refuge that New Guinea offered came too late. It was too hard to reach for most of Australia's carnivorous marsupials that would have otherwise been favoured by its wet, mountainous niches.

The South American marsupial radiation

In South America, after separation from Antarctica, there was also an explosive radiation of marsupials. This diversification continued largely in isolation from 60 mya (at detachment) until ~3–7 mya. Around this time, North America (drifting south) collided with South America (drifting north). This collision formed a land bridge known as the Isthmus of Panama, which is, today, part of Central America. It permitted mixing of North American and South American terrestrial faunas for the first time in more than 100 million years. During this fabled 'Great American Biotic Interchange', most of the mixing occurred from north to south. Northern placentals probably out-competed many southern marsupials for access to food resources. Indeed, only three surviving South American forms succeeded in moving north, and only one of these was a marsupial – the Virginia Opossum (*Didephis virginiana*). Today, South America stretches across a range of latitudes from tropical to temperate and supports a diverse array of habitat types. Like Australia, South America has habitats ranging from very wet to very dry. And like New Guinea, South America has some very high mountain ranges, notably the Andes. This is the longest mountain range in the world, running along the western edge of South America and reaching an elevation of almost 7000 m.

About 66 mya, there was a mass extinction event – many scientists believe an asteroid (or perhaps a comet) slammed into the Earth just off the coast of what is now Mexico. This had a major effect globally, quickly wiping out ~80% of all animal life. There were doubtless initial major upheavals in weather – earthquakes, tsunamis, and the like – but there was also a huge dust cloud of rock debris following the impact that probably lasted for months. It blocked out much of the sun's radiation and much life on Earth that could not generate its own energy, perished. This would have included most green plants, with flow-on effects up the food chain. Among Earth's lost species were those incredibly diverse and successful reptiles – the dinosaurs. At the time of the asteroid impact, dinosaurs ranged in size from 40 cm to the largest terrestrial land animals ever: 30–40 m long and weighing 80–100 tons. At this time too, the modern mammal groups had been around for tens of millions of years, but they were mostly small- to medium-sized. With the dinosaurs gone, niches became available for the mammals to diversify. And they did just that – spectacularly so!

Not long after this cataclysmic mass extinction event, South America became separated from Antarctica, ~57–60 mya. The climate at this time was warm and wet. Much of South America was probably covered in forest. The shrew-opossums and monito del montes were already likely established in the wet forests of South America at this point. Temperature and rainfall gradually decreased in the period 30–55 mya, as South America drifted north on its huge tectonic plate. The plants during this period probably included many forest plants that we would see in the tropics or subtropics today such as Antarctic beech (*Nothofagus*), which are indicative of colder climates. By 30 mya, there were probably temperate forests in some areas. Between 25–35 mya, temperatures dropped globally as a result of glaciations. These occurred because the circumpolar current was established after Australasia separated from Antarctica. From 15–25 mya there was a warming trend. During this period, we might have seen the existence of common ancestors of some of the carnivorous marsupial groups, such as the short-tailed opossums (*Monodelphis*). Subtropical plants would have become well established, but then temperature and rainfall again began to decrease. Seasonal temperature differences would have become more extreme, resulting in more arid environments, especially in southern South America, Patagonia. The uplift of the Andes mountain range, which was happening from 5–20 mya, would have accelerated this change, with the rising Andes acting as a barrier (rain shadow) to humid westerly winds. As a consequence of these events, southern South America became increasingly arid, and temperatures dropped generally as a result of periodic Antarctic glaciations. For South

American marsupials, the transition from wet, closed habitat to drier, more open habitat seems to have occurred in one direction. This differs from the Australian situation. At least some species within otherwise arid-adapted Australian genera seem to have moved back into wetter areas. In South America, shifts of some groups towards drier habitat probably occurred ~15–25 mya. These shifts were responses to the retraction of rainforest and the expansion of wooded savanna and grassland habitats in southern regions ~25–35 mya. The Patagonian Opossum (*Lestodelphys*) and fat-tailed opossums (*Thylamys*) represent this biotic shift from wet to dry. Today, these taxa are mostly distributed across middle to southern South America, where these drier habitat types predominate.

VIGNETTE: SHORT-TAILED OPOSSUMS: A HISTORY OF CONTINENTAL CONQUEST

By Silvia Pavan

Center for Conservation Genomics, Smithsonian's National Zoo and Conservation Biology Institute, Smithsonian Institution, USA

As mentioned earlier, the distribution of Australian and American marsupials has been heavily influenced by historical events and biological processes, such as long and short distance dispersals, extinctions, and millions of years of isolation. In addition to those historical factors, within each of these continents, living species also have their ranges determined by ecological factors. Therefore, the persistence of each group in a given area depends on their ability to thrive in the face of local conditions.

In landmasses with an extremely high diversity of environments, as we see in South America, few groups are able to maintain their ranges at a continental scale. Short-tailed opossums (genus *Monodelphis*) are one such example. *Monodelphis* comprises the world's most species-rich marsupial genus, with 23 currently recognised species, and is collectively distributed from eastern Panama to northern Argentina. These are small (<200 g) predators that inhabit the litter layer and other ground-level microhabitats.

Unlike most other American marsupials, which are restricted to either humid forest or drier habitat types, *Monodelphis* is found in a wide range of environments, including lowland rainforest (e.g. Amazonia and Atlantic Forest), open vegetation areas (e.g. Chaco, Cerrado and Caatinga), temperate grasslands of southern South America, and high-altitude landscapes in the Andes, Atlantic Forest and Pantepui. In fact, the only parts of South America currently not inhabited by these fascinating animals are the coldest and most arid areas of Chile and southern Argentina.

Despite a collective continental distribution for the genus, most *Monodelphis* species are, individually, able to tolerate only a limited range of habitats or ecological conditions. Therefore, most species are actually narrowly restricted to particular vegetation formations. This peculiarity makes the group particularly useful to study the past history of biomes (large areas with similar ecological characteristics), because the species' history might closely track landscape units over time. Learning about the patterns of distribution and diversification history of short-tailed opossums should therefore inform us about the biogeographic processes that have shaped the continental fauna, and ultimately help unravel the complicated evolutionary history of South American habitats.

Genetic studies suggest that the early ancestors of short-tailed opossums were forest dwellers that lived in the late Miocene (somewhere between 6 and 8.5 mya) in both Amazonia and the Atlantic Forest. As we see today, these two humid-forest biomes – the first in the north and the second in the eastern part of South America – are completely separated by a composite of drier habitats that range across the middle of the continent, and are broadly referred to as the South American Arid Diagonal (sometimes also called the Dry Diagonal). If *Monodelphis* ancestors inhabited a broad forested area that included both Amazonia and the Atlantic Forest, this implies that the Arid Diagonal did not completely separate these forested regions during the entire Neogene (from 2.5 to 23 mya), as it does today. Therefore, the history of short-tailed opossums suggests that the Arid Diagonal was

Fig. 3.4. Emilie's Short-tailed Opossum (*Monodelphis emiliae*). Note the grizzled greyish mid-body colour which helps distinguish it from other species in the genus. Source: Thiago Semedo.

reduced in comparison with what we see today, or at least connected by humid-forest passes. Indeed, historical forest corridors across the Chaco, Cerrado and/or the Caatinga have long been suggested by geologists and botanists, and faunal connections between Amazonia and the Atlantic Forest have also been inferred from other terrestrial groups, including other mammals, lizards and frogs.

Further patterns of distribution observed among species of short-tailed opossums render more insights about the dynamic history of South American landscapes. The astonishing geographic gap of more than 2000 km that now separates Gardner's Short-tailed Opossum (*Monodelphis gardneri*) in the Peruvian Andes from its closest living relatives – *M. americana* and *M. iheringi* – in eastern Brazil, is presumably explained by a dispersal event from the Atlantic Forest to the Andes. It seems that *M. gardneri* now occurs in the Peruvian Andes because its ancestors have journeyed from eastern to western South America, possibly through forest corridors that connected these presently non-adjacent biomes. This disjointed distribution pattern is not uncommon for birds, but is rare among non-flying vertebrates, with the few other non-bird examples including some treefrogs and rodents. The disjunctions among terrestrial groups are traditionally explained by humid-forest connections between the Andes and the Atlantic Forest. The timing of those passes has not been estimated for other groups, but, based on the history of the short-tailed opossums, these forest corridors are thought to have existed ~3.5–5.5 mya.

In addition to *Monodelphis gardneri*, other species are thought to have descended from ancestors that dispersed from Amazonia or the Atlantic Forest – the suggested centres of diversification for the group – to conquer other areas all over South America. For example, the three species currently inhabiting the dry forests of the Arid Diagonal (*Monodelphis domestica*, *M. kunsi* and *M. sanctaerosae*) are each descended from different Amazonian or Atlantic Forest ancestors that may have entered these open landscapes between 1 and 4 mya. *Monodelphis reigi*, endemic to the intriguing table mountains of northern South America, also known as the 'Lost World' of Pantepui, apparently descended from an Amazonian ancestor somewhere between 1.2 and 2.4 mya. In contrast, the ancestors of *M. palliolata* moved from Amazonia to successfully settle in coastal Venezuela and the Llanos during the last million years.

Fig. 3.5. A sub-adult Amazonian Red-sided Opossum (*Monodelphis glirina*). Source: Thiago Semedo.

But what makes *Monodelphis* so special and allowed them to succeed in so many habitat types? The answer to this question probably lies, at least partially, in their extraordinary diversity of phenotypic (i.e. observable) body characters. No other group of New World marsupials has developed more extreme variation in appearance than short-tailed opossums. For example, there is remarkable difference in fur colouration. The striking variety of coat colours ranges from species with unpatterned fur on the back (without any conspicuous markings), as seen in the Gray and the Pygmy Short-tailed Opossums *M. domestica* and *M. kunsi*, to species with boldly patterned coats. The latter include reddish sides contrasting with grizzled-greyish mid-back colouration in the Touan Short-tailed Opossum (*M. touan*), reddish head-and-rump contrasting with a grizzled-greyish trunk in Emilie's Short-tailed Opossum (*M. emiliae*), and longitudinal stripes along the back that are distinctly darker than the background colouration in the Northern Three-striped Opossum (*M. americana*).

At first sight, all *Monodelphis* species look superficially alike, with short tails, short limbs, rounded ears and relatively small eyes. Their external body type is singular, and reflects a general adaptation to terrestrial and semi-fossorial (i.e. digging) habits. However, careful analysis reveals great morphological variety, along with their remarkable behavioural distinctions. For example, the teat number variation exhibited by female short-tailed opossums encompasses the entire range known for the Family Didelphidae as a whole, presumably as a result of the disparate reproductive strategies observed within the genus. In as much as natural history data are scarce or absent for most species, we can at least conclude that short-tailed opossums are also unusually diverse in terms of behaviour and reproductive strategies. At least one species, *M. domestica*, is mostly active during the night, but several others are known to be partially day-active, such as *M. americana*. Although most species are known to reproduce several times during their life cycle, as generally observed for other mammals, representatives of the Southern Short-tailed (or Orange-sided) Opossum *M. dimidiata* die shortly following their first and only mating period.

This exceptional variation in body characteristics and behaviour might also help explain how numerous distinct species of short-tailed opossums are able to co-occur in the same areas. The Atlantic Forest is the top hotspot for the group, with as many as six species having overlapping ranges in south-eastern Brazil (*M. americana*, *M. dimidiata*, *M. iheringi*, *M. pinocchio*, *M. scalops* and *M. unistriata*). Remarkably, four species are known to occur together at two localities (*M. dimidiata*, *M. iheringi*, *M. pinocchio* and *M. scalops* at Riacho Grande, São Paulo; *M. americana*, *M. dimidiata*, *M. pinocchio* and *M. scalops* at Parque Nacional do Itatiaia, Rio de Janeiro). As might be expected, if all of those species inhabit the same area, their niches (i.e. living spaces) must be ecologically marginal to each other so they can share resources and co-occur. However, how exactly this partition occurs remains a mystery, because natural history and behaviour data are completely missing for most species.

The limited knowledge we have today of short-tailed opossums points us to a plethora of intriguing questions in many fields, including taxonomy, evolution, ecology and physiology. One thing we know for sure about *Monodelphis* is that we still have a lot to discover about the secret lives of these amazing conquerors.

4

RELATIONSHIPS

General morphology: body plans of carnivorous mammals

Not many people would confuse a mammal with a bird, or a bird with reptile, but difficulties in identification arise when animals look increasingly alike. We might expect related animals to look alike owing to their shared inheritance, but it is also common for the overall form of animal body parts to look similar even though they do not share a close evolutionary history. This phenomenon is called **convergence**. It demonstrates how environments can shape evolutionary change in similar ways over time. There are many examples in nature. Consider the wings of birds, bats and insects. These structures are superficially similar and all enable flight via uplift.

Fig. 4.1. A new Australian marsupial species – the Tasman Peninsula Dusky Antechinus (*Antechinus vandycki*). The species was discovered in a tiny stand of fragmented temperate rainforest in south-eastern Tasmania and named in 2015. This carnivorous marsupial looks like, and behaves similarly to, the shrew-opossums of South America and the murexias of New Guinea, even though the three groups last shared common ancestry perhaps 90 million years ago. Source: Gary Cranitch.

However, invertebrates and vertebrates have not shared a common ancestor for 500 million years or more. Furthermore, birds are merely 'glorified reptiles': survivors of the mass extinction event that wiped out the dinosaurs ~66 mya. In contrast, bats are flying mammals. These two groups, sauropsids (reptiles) and synapsids (mammals and relatives), diverged some 320 million years ago (mya).

Within the mammals, there are also many examples of convergence, in particular between marsupials and placentals. These two major modern-day mammal groups last shared a common ancestor ~160 mya. Many placentals and marsupials share similarities in overall shape, how they move, hunting techniques and prey types. In short, their ecological roles in the food web are similar. The distinct difference between the two groups remains their mode of reproduction and arrangement of sexual organs. Consider these examples of convergence between placentals and marsupials: wolves (placental) and Tasmanian Tiger (marsupial); cats (placental) and quolls (marsupial); moles (placental) and marsupial moles (marsupial); shrews (placental) and shrew-opossums (marsupial) as well as some antechinuses (marsupial); flying squirrels (placental) and sugar gliders (marsupial); rabbits (placental) and bilbies (marsupial); otters (placental) and lutrine opossums (marsupial).

Within the carnivorous marsupials, too, there are many examples of convergence. South America and its precious cargo of marsupials drifted loose from Antarctica, which was still connected to Australasia, some 60 mya, and yet there are many similarities in body form and ecology between America's carnivorous marsupials and their Australian counterparts. The shrew-opossums (South America) share similar form, habitat, prey type and hunting strategies to some antechinuses such as those in the dusky antechinus species complex (Australia). The short-tailed opossums (South America) are similar to *Myoictis* and *Phascolosorex* (New Guinea). Some short-tailed opossums even share with *Myoictis* three longitudinal black body stripes. The fat-tailed opossums (South America) are remarkably similar in appearance to the dunnarts (Australasia). Similarities between the two genera run right down to the large ears and eyes, dark eye-rings and thickened tails. Both groups are adapted to drier environments but contain members that are found in wetter habitat types. Four-eyed opossums (the Americas) resemble mid-sized quolls (Australasia). The Patagonian Opossum (*Lestodelphys halli*) of South America is similar in appearance and some habits to the Australian mulgaras. The South American monito del montes are similar in form and physiology to the Mountain Pygmy Possum (*Burramys parvus*) of Australia.

Fig. 4.2. A fine example of convergence. Although they last shared common ancestry perhaps 80 million years ago, (a) *Thylamys* of the Americas look remarkably similar to (b) Australian *Sminthopsis*, with small bodies, large ears and large eyes surrounded by dark rings. Sources: (a) Pablo Jayat, (b) Steve Wilson.

Interestingly, diversity of body form is more limited in carnivorous mammals than in their herbivorous counterparts. This is true of both placentals and marsupials. There is only so much stress the arm and leg bones of four-limbed predators can withstand. There are also serious energy costs to obtaining living prey. These constraints ultimately impose fairly strict limits on size and shape. Consider that the largest carnivorous marsupial that ever lived was South American and bear-like (*Proborhyaena*). It tipped the scales at perhaps 150–200 kg or so. The Australian Marsupial Lion (*Thylacoleo carnifex*) was probably only slightly smaller. Herbivorous mammals of the time, needing to consume large quantities of lower-nutrient plants, may have reached 50 times that size: a whopping 10 tons.

Despite the constraints imposed by the general carnivore body plan, the mostly insectivorous and carnivorous marsupials are certainly speciose and come in an array of forms and sizes. The carnivorous marsupials of Australasia can be described as small- to medium-sized, with separate fingers and toes. They generally have small (rather than tusk-like) lower incisor teeth. The bushy, carrot-shaped, or gradually tapered tail of most species is slightly shorter or longer than their head–body length. Excluding the Numbat (*Myrmecobius fasciatus*), which has reduced dentition, members of the Australasian Order Dasyuromorphia typically sport four pairs of needle-like incisors in the upper jaw and three pairs in the lower jaw. The canine teeth are distinctly curved and well developed. The premolar and molar teeth are sharply serrated, bearing distinctive cusps. Larger forms, such as the Tasmanian Devil (*Sarcophilus harrisii*), have very strong molar teeth that can pulverise bone. The smaller, insectivorous species have molar teeth featuring efficient shearing cusps for cutting up invertebrate cuticle (the hard external body covering). The Australian marsupial moles (*Notoryctes*) are of course odd-bods: specialist sand-swimmers

Fig. 4.3. Variations on a theme: the Water Opossum (*Chironectes minimus*). Note the webbed feet and 'Gollum-like' long, slender fingers with flattened disc pads. These adaptations allow the species to be well and truly at home in the water – paddling along with their feet and exploring submerged rock crevices with tactile fingers for food. Source: Thiago Semedo.

with alterations both externally (e.g. both species have bald, stubby tails) and in skull morphology. The American carnivorous marsupials generally satisfy the above body plan descriptions. However, one aquatic species, the Water Opossum (*Chironectes minimus*), has webbed feet, and the shrew-opossums have prominent tusk-like lower incisors. The carnivorous marsupial snout is typically long and conical-shaped. The four legs are mostly similar in length; a notable exception is the Australian Kultarr (*Antechinomys laniger*), which has long hind legs. The tail is never fully prehensile. However, the monito del montes (*Dromiciops*) use their grasping tails routinely when climbing, and other species in both Australia and the Americas do the same. Four-eyed opossums use their strong tails to transport nest material.

Carnivorous marsupials run the gamut from fully terrestrial to fully arboreal, and foot shape varies accordingly. Tree-living species tend to have broader feet than terrestrial forms. Monitos have uniquely opposable thumbs and rough pads on the underside of hands and feet to help them grip tree trunks.

Fig. 4.4. A colourful exception to the general drab-coloured carnivorous marsupial rule. This is a striking dark colour morph of Australia's Eastern Quoll (*Dasyurus viverrinus*). Source: Jiri Lochman.

There are no very large carnivorous marsupials. Indeed, no living species exceeds 14 kg, but they cover a diverse range of sizes with the smallest (being a tiny planigale) weighing in at just 2.5 g. Although no carnivorous marsupial is large, many are unusually ferocious, punching well above their weight. Within the genus *Dasyurus* (quolls), the two New Guinean species are the largest indigenous predators found there. The four Australian quolls are even more formidable and only eclipsed in size by another dasyurine: the Tasmanian Devil. The decades-old Warner Brothers cartoon that depicts the latter as a slavering, whirling dervish, is not far off the mark.

Males and females of any given carnivorous marsupial species are typically similar with respect to shape and fur colour, but not so size. Males are often larger than females, sometimes three times as big. Marked size differences between sexes in animals exhibiting a male die-off such as antechinuses and Northern Quolls (*Dasyurus hallucatus*) may be the result of evolutionary selection for larger, wider-ranging males in species with a promiscuous mating system. Likewise, when females need to rear their young alone, it pays to be smaller and more energy efficient.

Sometimes species of carnivorous marsupials look quite different, even to the casual observer. Other times, even different genera can look superficially similar. As described earlier, species with different evolutionary paths can converge on the same body form because of similar responses to similar environmental pressures, but there are also other sources of confusion that can make accurate identification difficult. Mammals are typically brown or grey, occasionally with some spots or stripes. They are often nocturnal and, for small species, being able to hide is important. Their sense of smell is usually well attuned, but colour vision is often limited. There has been less evolutionary fine focus on how they appear to one another. So, generally, they appear drab. Comparatively, mammals are much less colourful than birds and harder to identify. Skull and tooth features can be used to tell them apart, but this can be subtle and often you don't want to get that close! And yet, knowing one species from another is crucial to our understanding and management of the group. So what tools can we use to determine and discern carnivorous marsupial species?

What is a 'species', and how do you tell one from another?

Ideas about what constitutes a 'species' have changed over time. Species were once defined simply as groups of individuals that would freely breed with each other, but this definition proved to be rubbery when scientists found that what appeared to be different species would sometimes interbreed in captivity. However, organisms often behave differently under artificial conditions, so perhaps interbreeding then is not surprising. We now know that different species living together may also interbreed occasionally in nature, either sneakily or only under certain environmental conditions. This blurs the conventional boundary between species. What's more, many different species don't live in the same place as other members of the same genus, so they never have the opportunity to breed with each other in nature. For these species, testing to see whether they interbreed may not be relevant. Although sometimes puzzling and not necessarily decisive, breeding with each other – or at least having the potential to do so – is still used by many as a criterion of species status, at least for co-occurring species. Even in captivity, different species of carnivorous marsupials almost never interbreed, so for this group of animals the criterion of reproductive compatibility provides a good working definition of species limits.

As noted briefly in Chapter 1, studies of DNA have proven to be increasingly useful in identifying different species and in helping to unravel the relationships between them. The use of DNA protein products since the 1960s and the direct sequencing of DNA since the 1980s have been invaluable in finding cryptic (hidden) species. Some 'cryptics' have been overlooked by taxonomists who previously could rely only on morphology to distinguish species. And so seductive was DNA's siren song that some scientists proposed that DNA should be the sole criterion used in determining different species. With it we can construct detailed and informative relationship 'trees of life', or phylogenies, between organisms (Fig. 4.7). Each tree of life we construct is admittedly a 'best guess', but it is at least constrained by probability. In this way, the most likely groupings of species, or 'clades', are shown with quantitative emphasis, which is very attractive to scientists.

Fig. 4.5. The fat-tailed false antechinuses (*Pseudantechinus* spp.), such as this Ningbing Pseudantechinus (*P. ningbing*), may well harbour cryptic species. It will require voucher specimens from various locations, a study of their DNA and skull morphology, along with careful comparison to the original museum holotype specimens, to disentangle their true relationships. Source: Jiri Lochman.

In the last 20 years, we have filled our databases with so much DNA that many major relationships among organisms have been clarified, but should we rely on DNA alone to determine species? DNA is certainly a powerful tool in understanding relationships, but to rely on it solely can make management difficult. This is because, ultimately, species will be best conserved if we can recognise them by eye in the field. What's more, if we name species as different just based on DNA differences, we will greatly increase the number of species. At worst, such an approach may devalue current conservation efforts on iconic species because a torrent of newly 'discovered' cryptic species may 'flood the market'. Today, as we saw, there is still vibrant debate among scientists about what constitutes a 'species'. For many taxa, this issue seems to be solvable only in shades of grey, rather than black and white. This makes diagnosing species, and thus conserving them, a challenge. There is inevitably a subjective arbitrariness to the process. Yet, despite the rules of science, which are so critical to its success and distinctiveness, we must acknowledge there are always shades of grey in the enterprise. It is something scientists simply must live with; species distinction is no different.

Most scientists now agree that genetics is only part of the species story. Where possible, multiple lines of evidence should be used to determine species' status. There may be differences between populations in DNA and these patterns may be consistent using different parts of the DNA. These genetically different populations may in turn form different strongly supported groupings ('clades') in the tree of life. Individuals of each population may also be 'monophyletic' – that is, all more related to each other than they are to any individuals in other populations. To be sure, such DNA data are persuasive. Yet DNA evidence still needs to be combined where possible with other data types. The onus is on scientists to find such information to test the genetic hypotheses that DNA can generate. For example, we can test how well these genetic hypotheses are explained by differences in how an organism looks – its morphology. Sometimes, other aspects of its ecology

Fig. 4.6. The time of breeding each year in antechinuses is so synchronised and predictable for a species at any given location that it can provide clues you are dealing with different species. Several species in the genus have been discovered by researchers walking along trap lines and noticing that some captured females were in breeding condition while others were not. Within a species, all females at a location should enter breeding together. After study with DNA and comparison of subtle fur colouration differences, it was discovered they were actually two separate species! Here is the newest species in the antechinus genus, the Tasman Peninsula Dusky Antechinus (*A. vandycki*) from remote south-eastern Tasmania – it was named only in 2015 and we know almost nothing about the animal other than that it urgently needs our protection to avoid extinction. Source: Gary Cranitch.

can also be used as evidence of its species status. Observations about mating behaviour can be important in determining whether putatively different species interbreed or not. There are often a suite of behaviours that species use in pre-mating rituals. Some species, such as antechinuses, are so competitive that if two species occur together they always breed at different times of the year. Such breeding 'asymmetry' in co-occurring antechinuses is an important piece of evidence to suggest that we are dealing with two species rather than one. Indeed, several species in the genus *Antechinus* have been discovered using precisely this evidence. Thus, DNA may be an often-used starting point in present-day studies of species recognition, but it is **only** a starting point and should be tested as other sources of data become available.

Systematics

The living marsupials include more than 330 species, and some of their systematic relationships in the tree of life remain controversial. The group as a whole has been closely scrutinised using DNA in recent decades, with numerous attempts made to resolve genetic and morphological conundrums. The result is a general consensus about many relationships across both types of evidence. Yet some notable problems remain. For most of European taxonomic history, morphology has been used to determine relationships among organisms. Genetic techniques did not become available until about the 1960s and they were not widely used until the 1970s–1980s. During the 1980s, a suite of studies looked at protein products produced by genes and assessed how they differed both within and between species. These studies were very useful for finding cryptic (hidden) species and sometimes even genera. By the mid-late 1980s, researchers were examining differences in the DNA code itself – in particular using mitochondrial DNA (mtDNA). Mitochondria are the 'power packets' that provide animal cells with energy, and all contain DNA. Use of this source of DNA has proved to be fruitful because mtDNA is typically passed only along the female line. Thus, it is not mixed up during mating between males and females each generation. This fact, and its fast rate of mutational change, mean that differences that accumulate between species are easier for scientists to track using mtDNA. The problem with mtDNA is that it typically evolves as a single unit, so when using it this means we are looking at only a fraction of the genetic evidence that is available within an organism to make our conclusions. This is dangerous – patterns in mtDNA may not be representative. In other words, mtDNA alone may not clearly represent what is going on when reading evolutionary relationships between species. So, over the last two decades, scientists have sought to examine differences in nuclear DNA (nDNA) as well as mtDNA. A common problem with this combined approach is that nuclear DNA evolves much slower than most parts of mtDNA. Differences between species accumulate more slowly. Pound for pound, nDNA is perhaps less useful than mtDNA for resolving recent relationships, but using both DNA components together is better than relying on either alone. nDNA can resolve deeper (older)

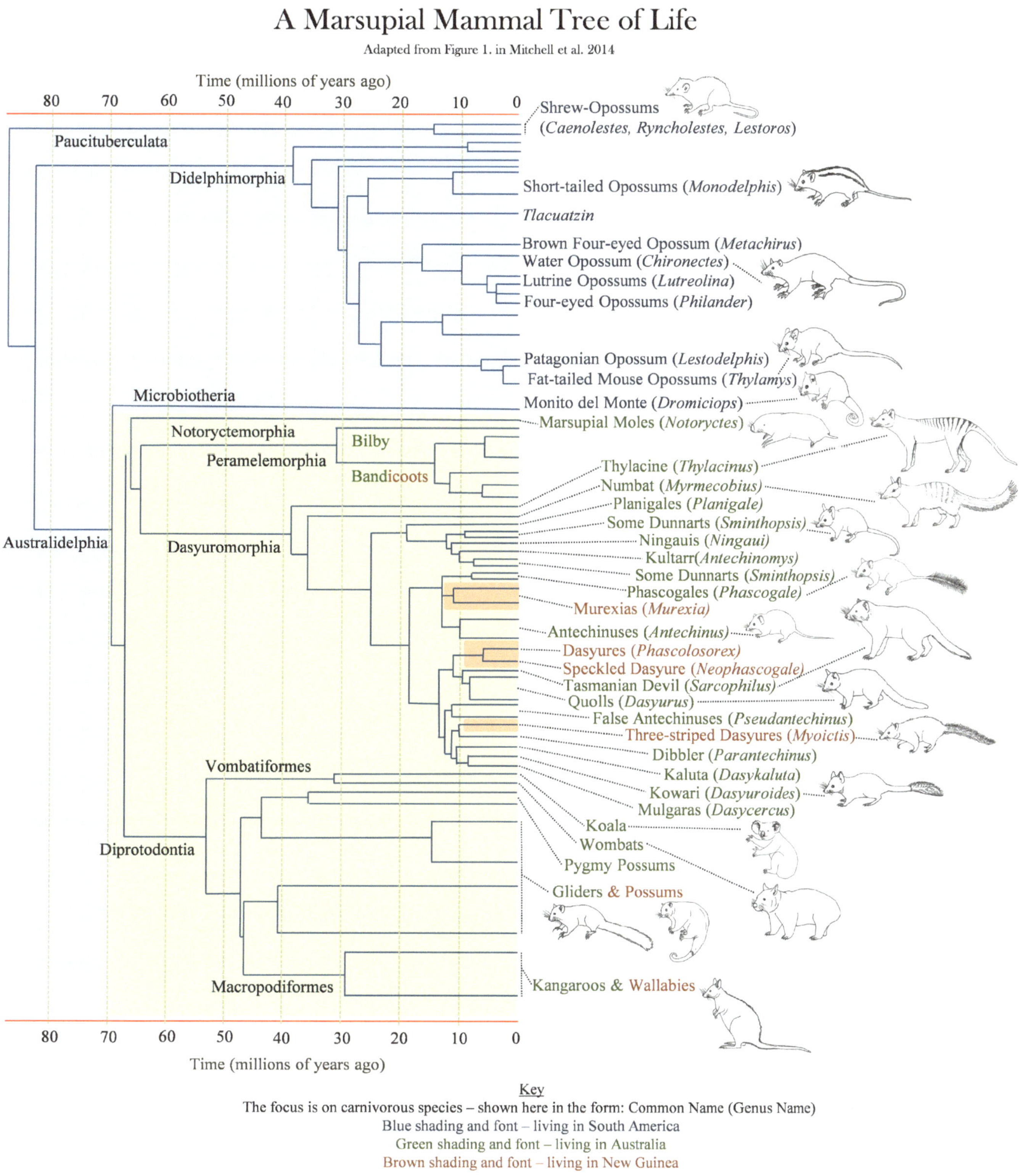

Fig. 4.7. A hypothesised marsupial tree of life, or phylogeny. The horizontal axis represents historical time, with the tips of tree branches, to the right of figure, equating to the present-day. As you move to the left along the timeline, tree branches collapse or 'coalesce', representing hypothetical ancestors that existed some time in the past. Our focus here is on relationships among carnivorous marsupials, so some of the other marsupial groups are simplified to aid readability. Source: Andrew Baker (adapted, with permission, from Figure 1 in Mitchell *et al.* 2014).

relationships between species that may better equate with morphology. MtDNA can clarify more recent evolutionary relationships.

The combined DNA data approach has been in vogue for the last couple of decades for taxonomic studies, particularly in the last 10 years. Especially when considered in the light of morphological differences, this provides a 'total evidence' approach. It is even better if these data can be combined with ecological evidence (behaviour, breeding) of the groups in question. The challenge occurs when data types don't agree – which do you then believe, and why? Another challenge is that each tree of life produced by any researcher is only a prediction. The computer programs we use to make such 'phylogenies' are only as good as the mathematical predictions we have made to model how the DNA evolves over time. In short, these trees are hypothetical. We take a snapshot of present-day differences between taxa and make our best prediction of what may have given rise to what, over time. The trees produced are fascinating to be sure, as well as informative, but what datasets we use and how we combine and model them can have a significant impact on the final result.

What major relationships have we discovered among carnivorous marsupials?

First, let's briefly summarise the groups of carnivorous marsupials shown in the hypothetical tree of life (Fig. 4.7). The marsupials are traditionally classified into two major cohorts: Ameridelphia and Australidelphia. The Ameridelphia are today confined to the Americas and consist of two major orders: Didelphimorphia (opossums) and Paucituberculata (shrew-opossums). The shew-opossums are all carnivorous, whereas the opossums contain a mix of both carnivorous and herbivorous forms. Only one member of the Ameridelphia is found in the USA, the Virginia Opossum (*Didelphis virginiana*). The rest are found in South America, Central America and Mexico. The australidelphians consist of five present-day orders. Two of these – Dasyuromorphia (carnivorous marsupials) and Notoryctemorphia (marsupial moles) – are carnivorous groups. Two other orders – the Diprotodontia (koalas, kangaroos, wombats and possums) and Peramelemorphia (bandicoots and bilbies) – are mostly herbivores, although bandicoots include variable amounts of animal protein in their diets. A fifth order within the Australidelphia is the Microbiotheria (monito del montes). Most scientists agree that marsupials leaving South America colonised Australia, via Antarctica – and this occurred at least once or twice. Monitos most likely evolved in Australia, after their ancestors invaded from South America, possibly ~70 mya. Thereafter, monitos apparently moved **back** to South America (via Antarctica), after which South America detached from Antarctica, taking the early monito del montes along with it. Monitos, or their ancestors, became extinct in Australia, and thus far no clear fossil of them has been found there.

Most scientists agree that the monito del montes form part of the australidelphian group, but their relationships with Australian marsupials are clouded. Some think the monitos are best placed as sister (i.e. most closely related) to the entire Australian marsupial group. Others believe that monitos are nested within the australidelphians, perhaps as sister to the Diprotodontia, with which they share similar sperm morphology. The ancestor of these two groups is hypothesised to have had both an opposable thumb and prehensile (i.e. grasping) tail.

Some other deep patterns in the marsupial tree are also contentious. For example, it is uncertain whether shrew-opossums or opossums are sister to the remaining marsupials. It is even contentious that the ameridelphians form a discrete clade to the exclusion of australidelphians. Further, some recent trees have recovered the shrew-opossums as sister to all other marsupials, nested within which are the opossums. These are in turn positioned as sister to all australidelphians. In yet other studies, the relative position of shrew-opossums and opossums with the Australian marsupials is reversed. Such deep, long branches of ancient groups are notoriously hard to decipher in trees of life. We don't have DNA from the many intermediate forms between the deep ancestors and today's living representatives. This may otherwise break up the long branches that defy our best mathematical sorting strategies. One thing is clear, however: the two orders that comprise the Ameridelphia are old. They hearken back perhaps 80 million years or more into the deep recesses of time. Relationships among the Paucituberculata suggest that, of the three living genera (all of which are carnivorous), the long-nosed and Incan shrew-opossums are sisters (*Rhyncholestes* and *Lestoros*). Together, these form a clade that is sister to the remainder of the shrew-opossums within the apparently diverse genus *Caenolestes*. Within the Didelphimorphia, there are several carnivorous genera. Of these, *Monodelphis* perhaps shared common ancestry with the Tlacuatzin (*Tlacuatzin canescens*) some 20–30 mya. Four other carnivorous

Fig. 4.8. A divergent and distinctive order of the cohort Ameridelphia – the shrew-opossums (Paucituberculata). The lower incisor teeth in shrew-opossums are tusk-like and point forwards, like those of the Australian Diprotodontia. This similarity in dentition between two so evolutionarily divergent groups caused much confusion for early taxonomists. Source: Reed Ojala-Barbour.

didelphid genera form a group – the Brown Four-eyed Opossum (*Metachirus*), Water Opossum (*Chironectes*), lutrine opossums (*Lutreolina*) and four-eyed opossums (*Philander*). These genera appear to have relationships going back ~5–15 mya. The last groups of carnivorous marsupials within the Didelphidae are the fat-tailed mouse opossums (*Thylamys*) and Patagonian Opossum (*Lestodelphys*), which appear to be sisters, sharing common ancestry ~5–10 mya.

Let's turn our attention now to carnivorous marsupials within the Australidelphia. The marsupial moles are typically recovered as an old group. They have been positioned as sister to a group containing another sister pair comprising the 'carnivorous marsupials' (Dasyuromorphia) and the bandicoots and bilbies (Peramelemorphia). Alternatively, the moles have been sorted as sister to either one of these orders. The marsupial moles are niche-specialist survivors of times when Australia was warmer and wetter. Their ancestry recedes to a time before Australasia was even separated from Antarctica.

The Dasyuromorphia of Australia and New Guinea is extremely diverse, containing ~78 living species of carnivorous marsupials. Our knowledge of evolutionary (systematic) relationships among dasyurids is due in no small part to the pioneering work of Mike Archer and colleagues. Recent analyses from Archer's research group based on fossils, morphology and DNA suggest that the Order Diprotodontia ('two forward teeth') diverged from polyprotodont marsupials ('many forward teeth') ~65 mya. Of the polyprotodonts, the bandicoots and bilbies, marsupial moles and dasyuromorphians apparently diverged from each other 59–64 mya. By ~56 mya, the four modern Australasian marsupial orders were distinct. Relationships among polyprotodonts remain clouded. However, the separate grouping (monophyly) of Dasyuromorphia is clear-cut, both morphologically and genetically. Relationships between the families within the dasyuromorphs, however, are unclear. Morphology and DNA cannot resolve whether the dasyurids are best positioned as sister to the thylacines or the

Fig. 4.9. The relationships of the 'living fossil' monito del montes (*Dromiciops* spp.) to American and Australian carnivorous marsupials remain obscure. Source: Agustin Iriarte.

Numbat. These groups may have shared common ancestry ~25–38 mya.

When worlds collide: DNA versus morphology

Morphological variation within most carnivorous marsupial groups has been well studied. However, over the last 30 years, genetic analyses have provided an avenue to resolve morphologically based groupings of organisms across the globe, including the carnivorous marsupials. Recent decades of work by dozens of researchers have dutifully re-sorted relationships among carnivorous marsupial genera. Even so, some generic level relationships within the modern representatives of carnivorous marsupials still remain uncertain, despite (and in some cases because of) large research effort. In many such cases, the morphology and DNA evidence clash. For example, the Subfamily Dasyurinae is morphologically diverse, ranging from the large genera (quolls and Tasmanian Devil) to the much smaller Kaluta and false antechinuses. This group includes endemic (i.e. geographically restricted) New Guinean genera such as Three-striped and Speckled Dasyures and *Phascolosorex*, but just where the three New Guinea genera fit within Dasyurinae is controversial stuff. DNA points towards them being sister to quolls and devils, to the exclusion of the Kowari, mulgaras, Dibbler, false antechinuses and Kaluta, but morphologically, this is surely unlikely. Possibly, the New Guinean genera all fall outside this mostly Australian group. If true, one option would be to recognise three tribes within Dasyurinae: Dasyurini (a predominantly Australian clade), Phascolosoricini (for Speckled Dasyure and *Phascolosorex*) and Myoictini (three-striped dasyures only). This represents the cutting edge in tree-of-life research in dasyurids. Meanwhile, within Phascogalinae, the separation of antechinuses, phascogales and murexias has been supported in various DNA studies, but their inter-relationships remain unclear. Murexias are well known to be highly variable in body form. Based on morphology, five genera were raised to encompass what is only recently considered (based on DNA) to be a single genus. Some of the different body forms within *Murexia* will surely be associated with new species hiding in the dense and forbidding forests of New Guinea. As for the Subfamily Sminthopsinae, relationships of the speciose dunnarts remain enigmatic. DNA studies over the last decade suggest that dunnarts are not correctly grouped together with respect to the Kultarr and ningauis.

Despite these contentions, DNA work has uncovered and helped to resolve a range of cryptic species in many of the carnivorous marsupial genera. The next decades will surely see an explosion of species diversity within the carnivorous marsupials. A swag of new antechinuses, whose stories of discovery are reported in the vignette at the end of this chapter, attest to this. If an explosion of new species is predicted in Australian dasyurids and American didelphids, then a veritable treasure trove most surely lies hidden in the many uncharted forests of New Guinea. Work to date suggests that most New Guinean dasyurid genera harbour cryptic taxa, notably within *Murexia*, the Speckled and Three-striped Dasyures. All five species of *Murexia* likely contain cryptic species; the morphologically diverse Short-furred Dasyure (*M. longicaudata*), probably contains several unnamed species. Discovery of dasyurids in New Guinea has lagged

Fig. 4.10. The Australian dunnart genus (*Sminthopsis*) may not be monophyletic – their evolutionary relationships with each other and the Kultarr (*Antechinomys*) are uncertain, even after decades of study. Source: Anders Zimny.

behind Australia because New Guinea remains relatively unexplored due to geographic impenetrability and political instability. What's worse, New Guinean dasyurids appear harder to catch than Australian forms. Animals such as *Murexia* take on average three to four times the trapping effort to catch compared with their Australian cousins, the antechinuses. Nevertheless, taxonomic reassessment is under way on some of these groups, and the future promises to swell the ranks of endemic New Guinean dasyurids to well beyond the 15 species currently recognised.

These days, resolution of new carnivorous marsupials requires taxonomic revision based on morphology, with prior groups informed by DNA. If the circle of knowledge between the DNA and the animal's appearance is not closed, the new species cannot be sustainably managed. This task typically requires collaboration between geneticists and (now vastly outnumbered) taxonomists. If this wasn't challenge enough, as indicated earlier, animal morphology often is at odds with genetics. This clouds the best path forward, but unless new species are resolved with total evidence, workers at the coalface cannot identify them. Crucial foundational work on their biology cannot progress. Challenges abound, but the excitement of venturing into the unknown trumps all. Hopefully, the next decades of carnivorous marsupial discovery will prove to be the most exciting and fruitful of all.

VIGNETTE: DISCOVERED TODAY, EXTINCT TOMORROW? CONSERVING CRYPTIC ANTECHINUSES

By Andrew Baker

School of Earth, Environmental and Biological Sciences, Queensland University of Technology, Brisbane, Australia

Antechinuses are well known to scientists, being commonly caught in field surveys. People living in bushy areas know them as furtive, night-time raiders of kitchen pantries and nesters inside lounge-room couches. So it was astonishing when our research group found a new antechinus species in 2010, right on the doorstep of Brisbane: a major Australian capital city!

Here is the story of discovery.

In the spring of 1975, Steve Van Dyck, then a dewy-eyed cadet working under Curator Mike Archer at the Queensland Museum, collected a female antechinus at Jolly's Lookout, west of Brisbane. She had eight pouch

Fig. 4.11. The recently discovered Buff-footed Antechinus (*Antechinus mysticus*). This male, the holotype, was captured raiding the kitchen cupboard of Queensland Museum Mammal Curator Steve Van Dyck. Source: Gary Cranitch.

Fig. 4.12. The threatened Silver-headed Antechinus (*Antechinus argentus*). This animal was captured at high altitude near the escarpment edge at Kroombit Tops National Park in eastern Australia. Source: Gary Cranitch.

young, and Steve took her home under observation as the babies matured. One of the male progeny died aged 9 months and was later registered in the museum collection as belonging to the ubiquitous Yellow-footed Antechinus (*A. flavipes*). However, Steve found the animals drably peculiar, and recorded as much in an urgent notebook scrawl, inserting a bold '?' in place of a species descriptor.

The game was afoot, but the entire genus needed revising. Steve spent much of the following 25 years doing so, naming a new antechinus species from far north Queensland and clarifying the taxonomy of several others. The mystery on Brisbane's doorstep was destined to remain unsolved for decades.

Fast-forward to early 2010. While working together on an Australian mammal field guide, Steve and I finally sat down to discuss the mysterious local mammal. In short order, we devised a research project that would see us share supervision of an honours student, Thomas Mutton. In the winter of that year, Thomas and I collected additional specimens of the strange beast, not only from Steve's original site at Jolly's Lookout but also the north-coast hinterland, west of Maleny. Other similar-looking specimens were sent to us from almost 1000 km north, at Eungella, in mid-east Queensland.

We now needed genetic corroboration that we were dealing with something different. And Thomas promptly provided it: specimens of a Yellow-footed Antechinus and the new mammal, both captured during a ravenous raid of Steve's rural kitchen cupboard, were a whopping 10% different in their DNA!

We then laid out the genetically validated new antechinus specimens from Brisbane, Maleny and Eungella alongside a range of Yellow-footed Antechinus from throughout the state. The external differences between the sister species, such as foot colouration and nipple number, were thrown into eye-popping, jaw-dropping, stark relief. Voila! A new Australian marsupial species: the 'Buff-footed' Antechinus (*A. mysticus*).

After this discovery, our research group felt a detailed study of the entire genus (then 11 species) was in order. Using a total evidence approach combining DNA, morphology (body size, shape and colour) and breeding studies, between 2012 and 2014 we discovered a further four new species of antechinus. This brings the total number of antechinus species to 15 and represents a 50% increase in known diversity within the genus over just 4 years.

The research required a crucial foundation that extended well beyond our research group – two centuries of morphology-based taxonomy of antechinus and several decades of pioneering genetic study. In total, there were dozens of scientists involved, spanning several generations, without whose combined toil these results would never have been unearthed.

Importantly, of the five new antechinus species, three are in the process of being listed as threatened and a fourth suffers from habitat fragmentation. This is striking, because none of the existing 10 species of antechinus were known to be threatened before 2010, when our research began.

One of our new species, the Silver-headed Antechinus (*A. argentus*), is apparently restricted to just a few scattered, high-altitude populations in south-eastern Queensland. The populations are genetically divergent, suggesting the species has been trapped on these isolated mountaintops for tens of thousands of years. This little animal may have one of the smallest distributions of any Australian mammal – and we literally stumbled across its existence.

In April of 1992, a fresh recruit of the Queensland Forestry Service, Luke Hogan, traversed the steep, winding road and numerous creek crossings that lead up to Kroombit Tops National Park, 400 km north-north-west of Brisbane, south-eastern Queensland. Hogan laid out a few lines of small mammal traps and fortunately caught two antechinuses from The Lookout on the eastern escarpment. One adult female eventually made its way to Queensland Museum Mammal Curator Steve Van Dyck. Catchily christened JM10945, this was a significant specimen because of its location, but otherwise appeared to belong to the ubiquitous Yellow-footed Antechinus. There it remained, in a museum specimen jar until 2012, when it was re-examined and tentatively identified as one of the newly discovered Buff-footed Antechinus, based on morphology.

Since 2011, we had been trying to catch antechinus at Kroombit Tops, to confirm their status using genetics and taxonomy. The first four trips (November 2011; July, August and September 2012) yielded no captures of antechinus from a total of 830 traps set across 10 sites, including 125 traps in the vicinity of The Lookout. Then, in December 2012, two juvenile antechinus were captured from near The Lookout (150 traps). These animals were predominantly silver-coloured from head to rump and looked remarkably different to the pale yellowish-brown preserved specimen collected by Hogan from the same location in 1992. At that time, previously unbeknown to us, Hogan had cleverly retained a genetic sample from his specimen; we subsequently acquired it. Our genetic analysis showed Hogan's original specimen and the two new juveniles were almost identical. But how could this be? Dramatic colour changes with age are not known in any other antechinus species. Clearly, we needed more adult animals to compare with the spectacularly silver juveniles.

So, in April 2013, we returned to The Lookout at Kroombit, and after considerable effort (550 traps) caught two adult antechinuses: a male and female. Genetically, these were very similar to the previously acquired DNA. Body colouration was also similar to the juveniles captured 4.5 months earlier: head and shoulders were silver, although the flanks and rump were subtly merging to olive-buff and the pale eye-rings more prominent. Given the low capture rates of antechinus at Kroombit Tops, we returned in June 2013 and undertook intensive trapping. A total of 2100 traps across four sites in rainforest, moist eucalypt forest and dry eucalypt forest yielded one female and six males from a single location, near The Lookout. These animals were genetically and morphologically almost identical to the animals captured in April 2013!

Thus, Luke Hogan's 1992 animal appears to be that rare, aberrant museum specimen that has responded poorly to preservation, becoming discoloured and washed out. The ravages of time and an uncertain

Fig. 4.13. (a) Remnant fragments of World Heritage Listed Gondwanan rainforest in eastern Australia are the favoured habitat of (b) the Endangered Black-tailed Dusky Antechinus (*Antechinus arktos*). Global warming is likely causing the cloud forest to retract, posing an extinction threat to this carnivorous marsupial, one of Australia's rarest mammals, which was only discovered in 2014. Sources: (a) Leanne White; (b) Gary Cranitch.

preservation/storage history apparently had their way with the specimen – all but obscuring its uniqueness to science. Yet, without Hogan's original specimen and the interest its strange appearance sparked, our research team may not have had the inspiration and desire to return to Kroombit so doggedly to solve this mystery.

Gene Mason, a PhD student in our team, has studied this population of Silver-headed Antechinus. His results from 2014 indicate this animal is extremely isolated to one small part of Kroombit Tops. He set more than 6000 traps over 7 months in 2014 while studying the breeding timing, growth and diet of this species and caught just five females and nine males. In such a small population, this cohort of adults may represent the grown young from only a few females in the previous year. What's more, the study incorporates two sites where animals have been caught in the past and, after a bushfire in 2013, one of the sites was burnt while the other was not. In 2014, only a single capture was made at the burnt site, with all other animals caught at the unburnt patch, which may now represent a critical refuge. Unfortunately, fire is common in this region; there was another bushfire in 2014 that burnt other parts of Kroombit Tops. Coupled with impacts of feral animals, such as cows, horses and pigs this new mammal species is under serious threat.

Another of our new antechinus species may be threatened with extinction due to climate change.

A large, shaggy antechinus captured in 1966 in a wire trap at Lamington National Park was originally thought to be part of an outlying northern population of Dusky Antechinus (*A. mimetes*, then called *A. swainsonii*), which is found patchily throughout south-eastern Australia. Two similar-looking animals – a juvenile and adult male – were found dead in nearby rainforest, in 1968 and 1971.

Then, in 1974, Steve Van Dyck of the Qld Museum decided to investigate further, camping for a single night in rainforest at Lamington's Binna Burra. Steve arrived at the campsite late and barely had time to deploy a couple of traps in the buttress roots of a large fig, right near the park entrance, before bedding down for the night. Incredibly, the next morning, one of the traps contained a large (120 g) antechinus that closely resembled the shaggy beasts caught a few years before. Steve took the animal back to the Museum where a barrage of diagnostic live photos were taken.

The following year, Steve decided to investigate another more extensive rainforest stand at O'Reilly's Guesthouse in the Green Mountains Section of Lamington National Park, ~13 km further west. After searching the dark forest on foot for several hours, he literally stumbled across another of the large antechinus at 10 a.m. The animal was scurrying in leaf litter in a stand of regenerating rainforest. On sighting Steve, the animal promptly ducked into a burrow constructed among the roots of a sapling. Not to be discouraged, Steve went after it and dug the animal out – the burrow was less than 10 cm below the ground with a 7 cm diameter tunnel that ended blindly after less than a metre.

The following day, another antechinus of similar appearance was found dead on a walking track ~3 km further south-east. Two years later, in 1977, another specimen was found dead and decaying in a cleared area not far from O'Reilly's Guesthouse, but it was to be another 13 years before another of the strange northern dusky antechinuses was seen. The year was 1988, this time more than 10 km away, in the eastern section of Border Ranges National Park. A female was caught at Brindle Creek and a male and female were caught from the headwaters of nearby Grady's Creek.

The creature then promptly disappeared again, this time for almost a quarter of a century.

Then, in 2012, our research team received a mysterious, grainy photo taken by an alert ranger depicting a large antechinus on the track at Best of All Lookout in World Heritage-listed rainforest at Springbrook National Park. We excitedly set out to catch one of these mysterious marsupials and compare them, once and for all, to Dusky Antechinus from Australia's south.

We deployed 300 traps for a few nights in a range of locations in and around the dense rainforest gullies at Best of All Lookout within Springbrook National Park. Our timing was good. We caught four of the beasts, all females, plus a solitary male 2.5 km further east on the caldera rim at Bilbrough Lookout. We conducted detailed studies over the year that followed and it turned out this northern population of antechinus was markedly different, both in appearance and DNA, to the southern Dusky Antechinus.

In fact, our DNA work suggests the newly named Black-tailed Dusky Antechinus (*A. arktos*) diverged from its southern Dusky cousins some 4 million years ago! At that time in prehistory, Australia was gradually drying out. Forests were becoming sparser and less connected. Ancestral Black-tailed Dusky Antechinus may have become

Fig. 4.14. The Tasmanian Dusky Antechinus (*Antechinus swainsonii*), Mt Field Tasmania. Source: Gary Cranitch.

isolated on their cool, wet northern mountaintops, initially trapped by a gradually changing climate and now threatened with extinction by a rapidly changing one.

From Emma Gray's PhD research on this species (2014–2017), we have found just a few disjunct, genetically fragmented populations, surviving only on the highest, wettest, World Heritage-listed cloud forests at the rim of an ancient, eroded volcano. The species is now rarely encountered and most likely is barely hanging on at very low density. Indeed, out of 10 000 traps set at various sites in 2014, Emma caught a total of just four females and six males. Alarmingly, the recorded history recounted earlier shows this species once occurred at several lower altitude sites further down the mountain where they were once more readily encountered. They haven't been seen there since the 1970s and recent decades have been the warmest in human recorded history. This suggests a combination of human disturbance and global warming has caused the animals to retreat to the highest, coolest and wettest parts of their range. But the animals have nowhere left to run. A continued warming trend may well seal their fate. We are desperately gathering information to get them formally protected with a federal endangered species listing. These animals may soon be lost from the wild. If so, the last resort would be captive breeding in a cooler, southern location to save them from extinction.

Our antechinus species discoveries have shown how much we still have to learn about diversity, distribution and management of carnivorous marsupials. Importantly, these species will form only a component of the cryptic diversity that remains to be discovered. The processes that have isolated animals in the past will potentially give rise to suites of new species. Places such as Kroombit Tops, Springbrook National Park, and other areas where geological knowledge suggests historical isolation, need to be explored more deeply to unearth the hidden diversity of life that they contain. Many animals inhabiting such areas will be rare and patchily distributed. It may take exhaustive searching to find them.

The antechinus discoveries also highlight the value of museums and the need to keep collecting. Our work was only possible because of the 'strange' specimens that were held in museums and the availability of many comparative specimens. These were essential in realising the uniqueness of each new species.

Citizen science also plays a crucial role. Dozens of researchers, hobbyists and members of the public have sent our team photos, tissue specimens and capture information. This has given us crucial insights into the distribution and breeding times of both new and existing antechinus species. The greater public interest, the more new species will be discovered.

In future, much new diversity will be discovered by comprehensive reassessment of each genus of carnivorous marsupial, with comparative collections made across their sometimes vast geographic ranges. Such work is taxing. It involves lugging heavy boxes of metal traps up and down dense, mountainous forests, but it is so rewarding when, after several thousand traps and years of effort, a new mammal species is discovered!

5

EXTINCT FAUNA

Fossil carnivorous marsupials and their ecology

There are 11 marsupial orders. Six of these are from the Americas and the other five are Australasian. Of the six American marsupial orders, three are now extinct. Of the five Australasian orders, one is extinct. Thus, seven of 11 known marsupial orders worldwide contain living members. The Australasian marsupials are now more diverse than those in the Americas. This may be because a range of placental mammals competed with marsupials in the Americas during their evolution. In contrast, in Australia, other than bats and a small number of monotremes (platypuses, echidnas), the marsupials had little mammalian competition. This is true at least until before the Australian and Asian plates collided. Then, a range of rodents entered and radiated through northern Australia, just a few million years ago. All 11 marsupial orders may have contained carnivorous forms, so we shall discuss each in turn.

The living American Order Didelphimorphia, the opossums, ranges from fully carnivorous to herbivorous forms that diversified mostly in South America. The South American fossil record of 50–55 mya shows that land mammal communities were dominated by the diverse didelphimorphians, along with a range of ancient ungulates, sloths, anteaters, and the like. The high diversity of didelphimorphians in the fossil record is impressive, given members of this group today mostly inhabit lowland tropical rainforest. What's more, the majority of good fossil deposits in South America are from higher latitudes, such as several sites in Argentina. The opossums then apparently underwent a reduction in diversity ~8–23 mya. Two factors are implicated. Rainforest began to

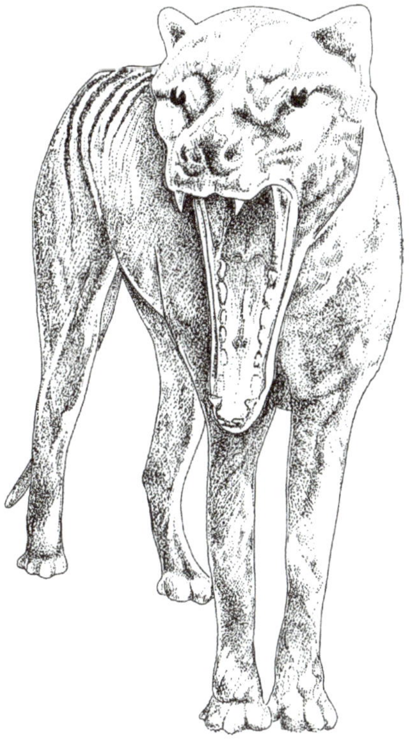

Fig. 5.1. A recently extinct carnivorous marsupial. This is an impression of 'Benjamin': the last Tasmanian Tiger or Thylacine (*Thylacinus cynocephalus*) in captivity. The work is adapted courtesy of a still from Australian naturalist David Fleay's video footage taken of the animal at Hobart Zoo in 1933. During the shoot, Benjamin bit Fleay on the backside. Evidently, the caged animal gave two threatening 'yawns', like the one displayed in this image, before sneaking around behind the photographer's curtain and biting him deeply on the buttocks. Fleay's resulting footage contained the last haunting images of this iconic mammal, at the time the largest of the carnivorous marsupials, on the verge of extinction. Benjamin died in captivity, left exposed to the elements on a cold winter's night in 1936. In the 1940s, Fleay attempted to capture a live wild Thylacine in the Tasmanian wilderness. He succeeded, sighting the captured animal from the next rise, but as he approached the animal escaped the trap. Shortly after, the species is believed to have become extinct in the wild. Source: Andrew Baker.

open up into wetland areas, which then became drained, so the opossums that preferred rainforest were disadvantaged. Another factor may have been the arrival of a range of predators from North America via stepping-stone movements across Central America. The latter region was in the process of being uplifted ~15 mya, so the timing of the arrival of new predators seems to fit. Today, this order is represented by the Family Didelphidae, which contains many carnivorous marsupials. Many genera within this family are also known as fossils dating over the last 20 million years. Indeed, the oldest didelphids with living representatives are recorded in the middle Miocene of Colombia, including two species of fat-tailed mouse opossums (*Thylamys*). Several fossil deposits have been found in the caves of Brazil and Argentina dating across the last 2 million years. These contain a range of carnivorous marsupials, including fat-tailed mouse opossums, lutrine opossums (*Lutreolina*), short-tailed opossums (*Monodelphis*), Patagonian Opossum (*Lestodelphys*) and four-eyed opossums (*Metachirus*).

The living South American Order Paucituberculata contains the carnivorous shrew-opossums. In Spanish, they have been dubbed *'ratones runchos'* (or 'mean mice'). Their forward-pointing teeth are formidable and similar to those of Australian diprotodontians. The fossil history of this group indicates that they were once diverse, with three extinct families comprising numerous genera and ~50 species. They are first recorded from ~50 mya in fossils from Brazil and Argentina, and early records are sparse. However, more-recent fossil deposits indicate a broad variety of shrew-opossum relatives, including very rich fossil deposits from Patagonia. Indeed, fossil skulls of shrew-opossum-like animals have recently been discovered in western Argentina in rocks aged ~25 mya. These beasts have the teeth of insect eaters and are very closely related to the paucituberculates. This suggests a likely high diversity of shrew-opossums and their relatives during this wetter period. The loss of diversity in paucituberculates is likely linked to the drying of habitats and loss of wetter forest caused by uplift of the Andes. The high, cold mountainous habitats created by mountain uplift may have provided wetter refuges for some of today's shrew-opossums, which include just three genera (*Caenolestes*, *Lestoros* and *Rhyncholestes*).

The third living American order is the Microbiotheria, represented today by three related species of carnivorous marsupial – the monito del montes (*Dromiciops*). Fossil forms from eastern Australia date back to ~55 mya and these may represent early immigrants into Australasia, originally from South America. There are also slightly earlier fossils from Antarctica, which was the connection point between South America and Australia. Presumably, this allowed early forms of microbiotherians to move into Australia. Subsequently, they travelled back to South America before it detached from Antarctica and drifted north. Fossil evidence shows that past distributions of the monitos occurred more widely in South America, including Chile, Argentina, Bolivia and Brazil.

Two extinct and little known American orders are less closely related to shrew-opossums. These are known from just partial skulls and jaws with teeth. First are the Groeberida, known from just two species, with parrot-like skulls and rodent-like teeth. The second group is the Argyrolagida. Beasts in this group probably hopped like small kangaroos.

The third extinct American marsupial order is the Sparassodonta. This diverse group included six

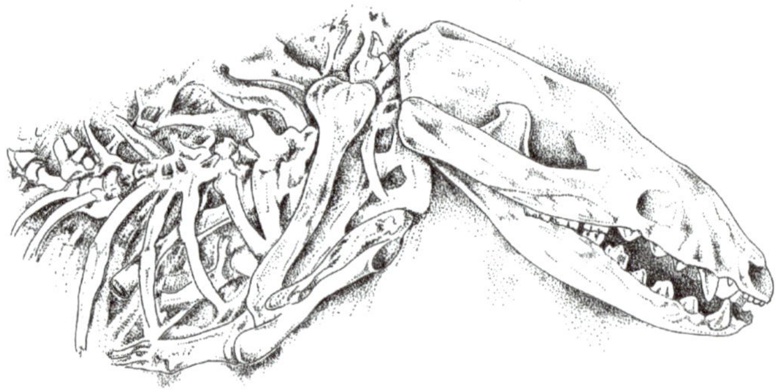

Fig. 5.2. An impression of the remarkably complete skeleton of a borhyaenid (*Lycopsis longirostris*). This 15 kg animal was likely an ambush hunter, stalking small-to-medium-sized prey. We know that it fed on mammals because fragmented rodent bones and teeth have been found in the stomach cavity of *Lycopsis* fossils. It was likely not a fast runner, but an ability to invert its hind foot and grasp branches might have allowed it to occasionally escape larger predators of the time. Note the very large head, a common feature of many borhyaenids. Early fossils, consisting only of skull fragments, suggested that borhyaenids were huge, larger than grizzly bears. However, more complete skeletons such as this one suggest they had unusually large heads and their total body weight may have only rarely exceeded 100 kg. The long, low-slung skull suggests these animals may have resembled members of the carnivorous placental mustelid family, such as weasels and martins. Source: Andrew Baker.

families that contained super-carnivores, including marsupial parallels to sabre-toothed tigers, bears, hyaenas and wolves. Some forms were very similar to the Australian thylacines. Sparassodonts may have included some of the largest carnivorous marsupials ever to walk the Earth, weighing in at a couple of hundred kilograms. The diversification of this group probably predated that of the Didelphimorphia. The eventual demise of the sparassodonts may have been due partly to competition with giant (100 kg!) carnivorous 'terror birds' and smaller but diverse didelphimorphians. The sparassodonts are well represented in the fossil record and include the first extinct marsupials ever described. In 1889, Florentino Ameghino described some fossils dating to 5–20 mya. They appeared to be carnivorous, with jaws like those of dogs or raccoons. In 1894, he placed them in the Family Borhyaenidae, positioned within the sparassodonts. Later fossil-based reconstructions suggested that the sparassodonts may actually fall outside what we know as marsupials. Instead, they may form a sister group, positioned more broadly within the 'Metatheria', Thomas Huxley's (1880) inclusive grouping of marsupials and their relatives. Metatherians are one of two base mammal groups, the other being the eutheria (placentals and their relatives). Regardless of their exact position, all experts agree that sparassodonts are an old group that is either marsupial or ancestral to them.

The fossil record gives ample evidence that sparassodonts were diverse and successful, and we now recognise five major groups within this order. The first, hathylacynids, are known mostly from jaw fragments. They were thylacine-, dog- or marten-like, and probably generalist carnivores. The second group comprised the prothylacynids. There is one excellent fossil with most of the body preserved. These animals were apparently climbers, thylacine- or hyaena-like in general appearance, and ~15 kg in size. We know they preyed on rodents, because a rodent bone and tooth were preserved inside one fossil! The third group contains the borhyaenids, including ~10 genera of dog- or thylacine-like animals, reaching up to ~30 kg. Some of these fossils are from ~50 mya, suggesting a successful group of capable, if short-legged, runners. The fourth group comprises the proborhyaenids. These were hyaena- or bear-like carnivorous marsupials. This group features powerful animals with ever-growing canine teeth that lived from 24–55 mya. There are four genera and some of the skulls of representative species are massive: up to 60 cm long. This group is mostly known from skulls and fragments. Early assumptions were that these animals were huge, but more recently scholars believe that they just may have had big heads for their size. It wasn't until 2002 that an almost complete skeleton was discovered: that of *Callistoe*. This generic name derives from Greek mythology, in reference to a nymph who was loved by Zeus. Zeus, fearing his wife Hera's wrath, changed the nymph into a bear to protect her identity. Yet, this animal was slender-snouted and probably most resembled a weasel in appearance. It was specialised for walking on the ground and likely a capable digger. One fossil shows evidence of a healed fracture in one finger, indicating frequent use of its hands. This animal probably weighed ~25 kg and may have preyed on armadillos, rodents, and the like.

The fifth sparassodont group is the most famous: the thylacosmilids. A fossil found in 1933 in north-western Argentina was named *Thylacosmilus atrox*. It had huge canines that, as for the proborhyaenids, continually lengthened. This marsupial super predator was likely similar in appearance to the Sabre-toothed Tiger (*Smilodon fatalis*), and represents an excellent example of convergence between marsupials and placentals, both in fossil form. In fact, super-long canines have evolved four or five times in mammalian carnivores. Early reconstructions suggested there were no incisor teeth in *Thylacosmilus*, but most experts wondered how they could then have torn meat off their prey. On closer inspection, the fossil skulls show that there was room for incisors, although none have been preserved. These animals had supremely powerful neck muscles, which is the case for many of the sparassodont groups, but, interestingly, like the sabre-toothed tigers, although their gape was large, their bite force was probably poor. This suggests that they used their very powerful necks in an unusual killing style. In *Thylacosmilus*, the upper canines were exceptionally long (30 cm or more), and compressed laterally, making them susceptible to breakage if the animal used them as primary attack weapons. However, if prey animals were first subdued, by stabbing precisely into the side of the neck, the curve in the canine teeth would bring their edge back under the carotid artery and jugular vein in the neck of prey. The tearing of these blood vessels would result in almost instantaneous death of immobilised prey. With short, powerful arms and semi-opposable thumbs, *Thylacosmilus* could easily pin down and restrain prey to perform this grisly dental work. Most likely, this carnivore was poorly suited to fast running, and may instead have been an ambush–short pursuit predator. Like the placental sabre-tooths, in acquiring supreme strength to overpower large prey, the highly specialised *Thylacosmilus* sacrificed speed and endurance. With poor stereoscopic vision and acute hearing, *Thylacosmilus* probably hunted mostly at night. It may have tackled very large prey, likely including other sparassodonts. There are several other genera in this group, and together they were the supreme predators of their age in South America.

The Australasian orders include the extinct and very odd Yalkaparidontia – an Aboriginal word meaning boomerang. These beasts were discovered in 1988 in the Riversleigh fossil deposits in north-western Queensland, Australia (see vignette, present chapter). They are known only from this region, and date back ~20 million years. Hundreds of isolated teeth, dozens of jaws, several skull fragments, and a single partial skull of *Yalkaparidon* exist. These are believed to represent two species. The teeth of these fossils are so strange that they must represent an ancient lineage. In recognition of their weirdness, the beasts are commonly called 'Thingodonts'. Their cheek teeth were evidently like those of moles, but their lower incisors protruded like those of diprotodontians. The skull is also primitive. They were possibly carnivorous, like the shrew-opossums. Their relationships with other mammals remain a mystery. Some researchers have suggested they are not marsupials at all, but an extinct sister group that is otherwise known only from scattered specimens of preserved teeth. Most others place these odd creatures within the marsupials, but cannot confirm whether or not they even lie within the Australidelphia. This is

Fig. 5.3. An impression of the sabre-toothed *Thylacosmilus* – a mother with cub cautiously moving along a tree trunk in an ancient forest. Note the super-long canine teeth and strong neck – important morphological features the creature likely used in concert to bring down large prey. Source: Jesse Hawley.

The Order Diprotodontia is extremely diverse and includes well-known living representatives such as kangaroos, wombats, possums and the koala. Today, with a few exceptions in the possum group, they are mostly herbivores (Chapter 2), but there are some fairly well studied fossil forms that certainly were not. One is the Marsupial Lion (*Thylacoleo carnifex*), a member of the extinct Family Thylacoleonidae. This family includes ~10 species in four genera, the earliest of which is known from ~25 mya. However, the best known members of the family are the youngest and largest of the group, and belong to the genus *Thylacoleo*. Up to the size of a large lioness (160 kg), these formidable predators were probably the dominant carnivorous mammals of their time. *Thylacoleo* ('pouch lion') is represented by several near-complete fossils from around Australia and lived 2 mya–46 kya. It was a true super predator that could probably take down prey much larger than itself, including the rhino-sized *Diprotodon* and giant kangaroos more than 2 m tall. *Thylacoleo* was highly specialised, with long, blade-like premolar teeth for cutting meat. Protruding, tusk-like incisor teeth were likely used as killing/stabbing instruments. Large, curved claws on the opposable thumbs doubtless helped them grab and manipulate their unfortunate prey.

The genus *Thylacoleo* was named by Gervais in 1852 and the type species, *Thylacoleo carnifex* ('pouched lion flesh-eater'), in 1858 by renowned palaeontologist Richard Owen from the British Museum. Owen described it as 'the fellest and most destructive of predatory beasts', but his assertion that *Thylacoleo* was lion-sized and carnivorous remained controversial for decades. Most early workers thought the animal was a melon-eating (!) herbivore. Yet recent extensive studies of its skull, jaws, teeth and skeleton suggest that, to the contrary, it was a ferocious predator. Its limbs were very powerful. With a highly reinforced skull, this animal would have been capable of dragging its prey up a tree in the manner of modern-day leopards. Studies on bite force in both living and fossil mammals indicate further that the Marsupial Lion had the strongest biting power of any mammal, living or dead. Interestingly, on a weight for weight basis, marsupial carnivores generally have higher bite forces than placental carnivores. This may be due to the relatively larger brain size in placentals, which gives less room for cheek/jaw muscle attachment.

Since 2009, several Aboriginal rock art paintings in the Kimberley, northern Western Australia, have been identified as *Thylacoleo*. However, the most recent studies suggest these may actually represent thylacines, because the paintings apparently were made well after the extinction of *Thylacoleo*.

There were other carnivorous kangaroo-like diprotodontians that include extinct members of the hypsiprymnodonts. Today, this family is represented only by the omnivorous Musky Rat-kangaroo (*Hypsiprymnodon moschatus*): a shy rainforest animal weighing less than 1 kg. However, the group was once more diverse, with representatives living from ~20 mya and including large forms up to 70 kg. The killer rat-kangaroo, *Ekaltadeta* – meaning 'powerful tooth' – was as large as a medium-sized wallaby. This animal had massive premolar teeth and was probably opportunistic in its diet, but tended towards carnivory. The even larger *Propleopus* may have been more truly carnivorous: consuming vertebrates and insects, as well as fruits and leaves. This animal sported molar teeth adapted for shearing and crushing, suggesting carnivory, rather than the grinding molars of modern-day herbivorous kangaroos. *Propleopus* likely moved on all fours, rather like dogs or wolverines. It was probably not a fast mover, but more likely an endurance, pursuit predator, like the Tasmanian Devil (*Sarcophilus harrisii*). Another extinct ancestor of the kangaroos, the so-styled *Fangaroo*, had large canine teeth, but, because these animals were more kangaroo-like in other ways, their impressive fangs may have been used to impress other fangaroos in sexual or defence displays, rather than for hunting.

The last order, the Australasian Dasyuromorphia, contains the most diverse assemblage of living carnivorous marsupials. Two living families are completely carnivorous. The most diverse of these, the dasyurids, contains ~75 species collectively between Australia and New Guinea. As we saw in Chapter 2, members of this predominantly Australian family range in size from the smallest living carnivorous marsupial, a planigale (*Planigale ingrami*) at just 2.5 g, to the largest, the Tasmanian Devil, at up to 14 kg the size of a small (but ferocious) dog. Another family contains just one living species: the termite-specialising Numbat (*Myrmecobius fasciatus*). The largest carnivorous forms were in a once-diverse, now unfortunately extinct, family: the Thylacinidae. One

Fig. 5.7. The emaciated captive Thylacine (*Thylacinus cynocephalus*) depicted in Joseph Gleeson's painting is an unfortunate reminder of the species' persecution at human hands. Gleeson was a US painter, and his is the only artistic work ever made of a Thylacine with cubs. The animals were held at the Smithsonian National Zoo. The pup (a female) portrayed here in the mother's pouch was sick on arrival and died shortly thereafter, on 12 September 1902. The cubs at foot were likely between 4 and 5 months old. Typically, the mother would not have tolerated a healthy pup of that age returning to the pouch so she must have been aware that the youngster was ill. Source: Wikimedia Commons.

species, the Tasmanian Tiger or Thylacine (*Thylacinus cynocephalus*), survived to modern times. Indeed, it was only recently (in 1936) sent extinct as a result of human persecution, but thylacinids were once a much larger family, with the oldest members dating back to 26–30 mya. As well as the Thylacine, 11 extinct species of thylacinid have been described. Up to five species may have coexisted in the Riversleigh World Heritage Area of north-western Queensland from 11 to 24 mya. There is a range of fossil genera scattered throughout eastern and (parts of) inland Australia, as well as the mountains of New Guinea. Australian Aboriginal cave paintings across the continent aged within the last 46 000–60 000 years also attest to the group's broad distribution, at least early in Australian human history. The Riversleigh thylacinids most likely inhabited forests. These areas were also occupied by a range of other carnivorous marsupials, including giant carnivorous rat-kangaroos (*Ekaltadeta*), marsupial lions (thylacoleonids), and dasyurids. The drying of Australia from 5 to 12 mya apparently led to the gradual replacement of forest with open woodlands, shrublands and grasslands, and these environmental changes broadly correlate with declines in the diversity of thylacinids. Some thylacinid genera of this period were medium-sized, such as *Nimbacinus*, which only reached ~5 kg but was nevertheless capable of bringing down prey larger than itself. Bite-force studies suggest that *Nimbacinus* was more similar in dietary habits and hunting capability to a living carnivorous marsupial, the Spotted-tailed Quoll (*Dasyurus maculatus*), than they were to the much larger, recently extinct Thylacine. Recent DNA trees suggest that thylacines are sister to all the other Australian carnivorous marsupials, sharing common ancestry with them as long as 30–40 mya.

The story of the last Thylacine evokes a grim fascination. Australia holds the unfortunate record of having the highest number of mammal extinctions of any landmass on Earth over the last two centuries. The loss of the spectacular Tasmanian Tiger is one of the country's most tragic losses. The Thylacine (or 'hyena') was first described in 1808 by Tasmania's Deputy Surveyor-General George Harris, just 5 years after European settlement in Tasmania. Harris placed the Thylacine within *Didelphis*: a genus originally created by Carolus Linnaeus to accommodate the American opossums. To Harris, the Thylacine appeared to be a 'dog-headed possum'. However, famed French taxonomist Geoffroy Saint-Hilaire had erected *Dasyurus* for the quolls in 1796, which he felt Thylacines more closely resembled. He placed the Thylacine accordingly under that genus in 1810. It wasn't until 1824 that Dutch zoologist and Leiden Museum Director Jacob Temminck granted the Thylacine its own genus, *Thylacinus*.

Thylacines were large and distinctive animals, weighing up to 45 kg and with a maximum skull size as large as a wolf. The snout of the animal was long and narrow, with a remarkably large gape (up to 70°). The eyes were large and dark; the ears rounded and moderate. The most striking feature of the Thylacine was the 13 to 22 dark brown horizontal bands that extended down the back from shoulder to rump. These overlaid otherwise fawn to greyish-brown fur. The tail was thick-based and tapering, extending stiffly from a curved rump. Unlike dogs, Thylacines could not 'wag' their tails. Indeed, Thylacines were more reminiscent of a kangaroo at the rear. Females had a backward-facing pouch that housed up to four young. The gestation was ~30 days, after which the growing young remained attached to the nipples for another 3 months. They left the pouch as independent juveniles at ~4 months. Activity probably took place mostly at night, although many sightings occurred during the day, and caves and rocky areas were probably important sheltering sites. They likely hunted both alone and in groups, using endurance to run down large prey such as kangaroos and wallabies. Smaller animals such as possums, bandicoots and rodents were also targeted. Thylacines fed almost exclusively on mammals, including on occasion, human livestock after these were introduced by European settlers.

The demise of the Thylacine was probably due to a combination of factors, all related to predation and hunting. Aborigines hunted Thylacines from 60 kya. Competition and predation from the Dingo (*Canis dingo*) on the mainland of Australia from 4 kya also would have limited their numbers. The Thylacine had a long generation time, with captive animals living up to 14 years, and large home ranges; as an apex predator, it probably also persisted at low density. Most likely, this combination of risk factors caused extinction of Thylacines on mainland Australia even before European settlement. The Tasmanian Thylacine population by around 1800 is estimated to have supported as few as 2000 to 4000 animals. Although this population was stable and close to carrying capacity for an island as large as Tasmania, the species' persistence at such small numbers was already tenuous, and its extinction inevitable as soon as it was perceived as a threat to the new settlers' flocks.

Consider the words of English naturalist John Gould concerning the Tasmanian Tiger in his famous book *The Mammals of Australia* in 1863: 'When the comparatively small island of Tasmania becomes more densely populated, and its primitive forests are intersected with roads from the eastern to the western coast, the numbers of this singular animal will speedily diminish. Extermination will have its full sway. It will then, like the Wolf in England and Scotland, be recorded as an animal of the past: though this will be a source of much regret, neither the shepherd nor the farmer can be blamed for wishing to rid the island of so troublesome a creature.'

The Thylacine's demise was due largely to the fear and suspicion of the animal harboured by early settlers. Thylacines were night-time raiders of livestock such as sheep, goats, pigs and chickens. Doubtless Thylacines fed on at least some livestock, but their damaging impact was greatly exaggerated. Aboriginal stories of babies being taken from campsites by Thylacines and reports of attacks on adult settlers only fuelled these fears, which led to the creature's persecution. Thylacines were considered vermin simply to be eradicated. So it was that, from 1888 to 1912, the Government Bounty scheme paid 1 pound per Thylacine carcass. Over the 24 years of the bounty, 2184 Thylacine carcases were deposited, but after 1900, bounty claims plummeted and the last successful one was submitted in 1909. Thylacines were now only rarely encountered. The last known captive animal died in 1936 at Hobart Zoo, freezing to death when left exposed to the elements on a particularly cold winter's night. Wild Thylacines probably hung on until ~1950. Today, despite occasional claimed sightings in remote Tasmanian wilderness areas, the Thylacine is considered by all authorities to be extinct.

Evolutionary patterns: I. From ancestral forms to the Pleistocene

On balance, the marsupial fossil record indicates that none of the seven living orders of marsupials can be much older than the mass extinction event that wiped out the dinosaurs ~66 mya. Possibly, few marsupials survived this event. Most diversification within the marsupials may have occurred on the Gondwanan supercontinent.

The oldest known South American marsupials date back to ~63 mya, from Argentina and Bolivia. Most marsupials from this age were carnivorous didelphimorphians, which later became spectacularly diverse. There followed a period when the shrew-opossums became diverse before dwindling to the seven species found today. The same was true to a lesser extent of the microbiotherians, which are now reduced to three living species – the monito del montes. The sparassodonts (sabre-tooths and their kin) became very diverse early on, then also dwindled and became extinct. The other two South American orders about which we know little, groeberidans and argyrolagidans, appear in the fossil record 20–30 mya. Then they abruptly disappear.

In Australia, the oldest marsupials appear in south-eastern Queensland and are dated at some 55 mya. The fossils from this location are broken and fragmentary, but include some that appear similar to South American didelphimorphians and others that appear like the monito del montes. Most likely, sometime between 45 and 70 mya the monitos, shrew-opossums and opossum orders moved into west Antarctica. At least two of these, the monitos and opossums, continued into Australia, arriving before 55 mya. Of the Australian marsupial orders, the only beast present at this age is something similar to a bandicoot. Probably some of the other Australian marsupial orders had begun to appear by then, but none are present in the earliest deposits. Indeed, we know nothing about them until 24 mya, when all orders had appeared and already diverged spectacularly into nearly all modern families. Conditions then were warm and wet. By 15 mya, drier and cooler conditions had set in, bringing about the decline of nearly half of the older groups. But these changes also saw a rise in diversity for those moving into drier niches, such as diprotodontians (e.g. kangaroos and wombats). These were relative latecomers on the marsupial scene. There was also a spectacular radiation of carnivorous marsupials. Although their relatives had been around for many millions of years, the dasyurids blossomed in the drying continent.

Fig. 5.8. Mulgaras (*Dasycercus* spp.) are part of a suite of arid-adapted carnivorous marsupials that radiated in response to gradual drying of the Australian continent ~5–10 million years ago. Source: Jiri Lochman.

Early relationships among the Australian carnivorous marsupials are uncertain. Part of the problem in resolution is due to the absence of numbats in fossils older than the Pleistocene (1.8 mya). Also, dasyurids are rare in early–middle Miocene fossils, some 5–24 mya. The oldest dasyurid, *Barinya wangala*, is from the early Miocene of Riversleigh, north-west Queensland. Thylacines, in contrast, dominate the earlier fossil records of 23–34 mya.

The 5–10 mya fossil record of dasyuromorphians is unfortunately poor. This is at least in part because the drying climate was harsher on fossil preservation. There is a single dasyurid fossil species recorded from this period; *Ganbulanyi djadjinguli*. However, DNA evidence suggests that the modern dasyurid subfamilies originated 15–20 mya and then underwent an active period of speciation ~5–12 mya. The major dasyurid groups apparently diverged when there was significant global cooling and drying. This radiation of dasyurids perhaps signifies a move into available insectivorous and carnivorous niches left open in the wake of the vanishing thylacinids. Thylacinids may have found the variable climates less favourable, perhaps because they had less access to large prey. In any case, by ~5 mya, there may have been a single thylacinid left: the Thylacine. This animal was originally widely distributed, if only rarely encountered in fossil deposits spanning Australia and New Guinea. Many of the modern dasyurid genera

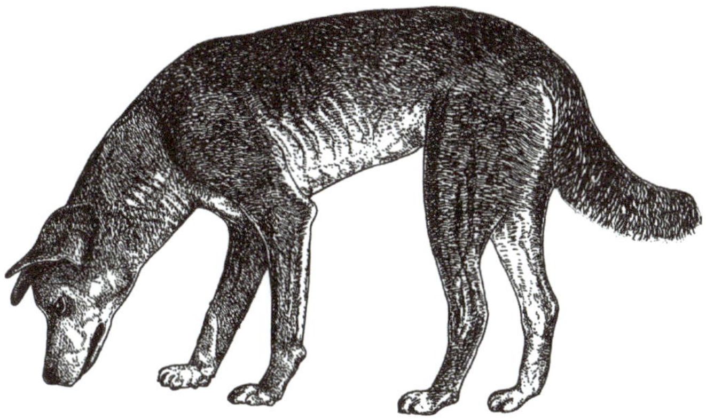

Fig. 5.9. The Dingo (*Canis dingo*). Scientists dispute whether this species should be considered introduced or native to Australia. Dingoes apparently arrived on Australian shores ~4000 years ago. To humans, they are always controversial and, although typically shy, have caused numerous fatalities. Most memorably, they were at the centre of the mystery surrounding the death of baby Azaria Chamberlain in central Australia in 1980. Yet their reintroduction into parts of southern Australia has recently been heralded by some as a way to arrest the continuing decline of Australia's native mammals. Species under threat unfortunately include an array of carnivorous marsupials, which appear to be at the mercy of middle-size predators such as foxes and cats – efficient killers introduced to Australia with European settlement. Dingoes plausibly kill or compete with the smaller cats and foxes, keeping their numbers in check and reducing their devastating effect on Australia's native marsupials. Source: Andrew Baker.

such as *Antechinus*, *Planigale*, *Dasyurus*, mulgaras (*Dasycercus*) and three-striped dasyures (*Myoictis*) had also appeared by this time. The gradual drying of the Australian continent 5–10 mya not only likely caused many extinctions, but also stimulated a general diversification in modern groups. A range of arid-adapted dasyurid features provides testament to this, such as the appearance of 'holey' palates – snout holes on the underside of the skull. The middle ear also became more sensitive to low frequency sounds: a response to environments that were becoming more open and less cluttered as they dried.

In the Pleistocene (from 1.8 mya), there was a further increase in dasyurid diversity. There is fossil evidence of 25 Pleistocene dasyurids. All of these have been assigned to modern genera, including Kowari (*Dasyuroides*), Tasmanian Devils (*Sarcophilus*), dunnarts (*Sminthopsis*), and three extinct species of *Antechinus*. Numbats also first appear in the fossil record during the Pleistocene, with specimens apparently representing the living species. There were periodic land bridges between Australia and New Guinea during the Pleistocene, but interestingly there was probably little movement of dasyurids between the two landmasses. Compare this with the rodents, which underwent successive radiations into Australia from the north. Two dunnart species are shared between Australia and New Guinea, as are three genera: quolls, planigales and dunnarts. Otherwise New Guinean and Australian dasyurid genera have apparently evolved separately over the last 10–12 million years, rendering them unique to those landmasses.

Evolutionary patterns: II. From Pleistocene megafauna extinctions to the present day

Earlier we introduced the lion-sized carnivorous marsupials that stalked the Pleistocene period and alluded to some of the gigantic beasts they probably hunted. Together with large, flightless land birds, giant snakes and other reptiles, these super-sized beasts collectively formed the megafauna. Then, towards the end of the Pleistocene, they mysteriously disappeared. These extinction events took place at different times on different continents and have stimulated much debate about the causes. Because most of the megafauna were mammals, the extinctions saw profound losses of a great range of impressive beasts, particularly in South America and Australasia.

Broadly speaking, the demise of the megafauna was likely driven by hunting by native peoples, coupled with rapid climate change. Climate change caused habitat loss, reduced carrying capacity for herbivores, and increased patchiness and fragmentation of vegetation. These changes in turn may have drastically affected food chains. Constant pressure from human hunting may have pushed many already-teetering mammals over the edge. The megafauna extinction in South America resulted in the loss of 50 genera, representing ~83% of large mammal diversity. Three entire orders disappeared: heavy bodied ungulates (Notoungulata), mammoths/mastodons (Proboscidea) and other ungulates (Litopterna). All large anteater, armadillo and sloth relatives (Xenarthrans) succumbed as well. During this extraordinary event, more species of mega-mammals (over 1000 kg) and large mammals (over 44 kg) were lost than in any other extinction episode that we know of. The extinctions occurred in South America from ~20 000 to 7000 years ago.

In Australia/New Guinea, recent reviews by Steve Wroe, Mike Archer and colleagues suggest that ~88 large vertebrate taxa disappeared during the Pleistocene. The majority of these losses (54 species) occurred within the last 400 000 years. The largest marsupial to go was the 2.8-ton *Diprotodon*, but also lost were the world's most specialised mammalian carnivore, *Thylacoleo*, and the largest lizard the world has known, '*Megalania*', a super-sized (7 m) member of the present-day monitor genus *Varanus*. But there were many other mammal extinctions as well. These include smaller species such as carnivorous dasyurid marsupials, bandicoots and rodents. Other large mammals such as wombats and kangaroos survived the event, some doing so by becoming dwarfed. Mounting evidence points to the loss of many species before the peopling of Australasia (perhaps as early as 60 kya), suggesting a major role for climate change in their disappearance. After people arrived in Australasia, hunting pressure imposed on the background of rapid climate change had major implications for megafauna that had survived to that time.

Much of what we know about such matters has been well synthesised by Chris Johnson in his excellent *Australia's Mammal Extinctions* (published in 2006). Australia's mammals apparently underwent three main phases of extinction: 100 000 to 10 000 years ago; 10 000 to 200 years ago; and 200 years ago to the present. Johnson proposed that predation was the main reason for all the extinctions. There were several contributing factors: the Australian environment had poor soils, unpredictable rain, low mountains and limited river structures. These factors ultimately meant that resources were limited, with the constraining effects of such shortages magnified up the food chain. Coupled with the relatively low base metabolic rate and slow reproduction of Australasian marsupials, this meant there were no very large mammalian predators. Those mammal predators that did exist, the largest of which was the lioness-sized *Thylacoleo*, were relatively small in comparison with cold-blooded creatures that required fewer resources, such as the giant and undoubtedly formidable *Megalania*.

Fig. 5.10. The Tasmanian Devil (*Sarcophilus harrisii*) possibly existed on the Australian mainland as recently as 500 years ago, and certainly until 3000 years ago. Such a recent history on the mainland fuels arguments for the animal to be re-introduced there as part of a re-wilding strategy to help control smaller mammalian super-killers, such as cats and foxes. Source: Jiri Lochman.

As already discussed, hunting by Australian Aborigines (40–60 kya), coupled with rapid climate change over last 100 kya, probably conspired to send most large fauna (over 35 kg) extinct. After Dingoes were introduced ~4 kya, the remaining carnivorous predators on the mainland, the Thylacine (a specialist large mammal hunter) and Tasmanian Devil (a mammal eater and scavenger) disappeared, becoming ultimately limited to Tasmania. Fossil evidence suggests that Devils, at least, existed on mainland Australia until 3000 years ago, possibly even as recently as 500 years ago, just a century before Dutch explorer Jansz visited the Gulf of Carpentaria and northern Queensland. After British settlement, the colonists persecuted Dingoes on the mainland and Thylacines and Devils in Tasmania. They also introduced House Cats (*Felis catus*) and European Red Foxes (*Vulpes vulpes*), both mid-sized predators. Without any larger predators to compete with and keep them in check, these new placental predators ran amok. Both cats and foxes are highly efficient killers with fast metabolisms. They were pitted against slower reproducing marsupials that had not evolved strategies to deal with the alien onslaught. It was a mismatch. The inevitable result: almost one half of the mammal extinctions in the world over the last 200 years are Australian in origin. Consider the contributing effect in this decline spiral of our slow reproducers: the longer our natives have been here, the slower their reproduction seems to be. Marsupials are very slow; placentals, such as the rodents that invaded Australia's north in the last 4 mya, are faster. Yet even these placentals are slower than more recently arriving rodents such as Black and Brown rats (*Rattus rattus* and *R. norvegicus*, respectively). Slow reproduction means slow rates of recovery in the face of the two very efficient killers, cats and foxes. Furthermore, there were few larger carnivores, such as Dingoes and Devils (not to mention Thylacines) to keep them in check. This has prompted recent suggestions to reintroduce Dingoes and Devils to areas of mainland Australia. There are even continuing efforts to genetically clone the Thylacine. In addition to conserving these species, the aim of such reintroductions is to try and control the onslaught of cats and foxes via competition with other predators that have evolved with our native fauna. Dingoes and Devils have large home ranges, consuming fewer prey per unit area, and may reduce fox and cat activity through competition. This proposed 're-wilding' of Australia is a controversial idea that attempts to redress Australia's appalling history of mammal extinctions. Unfortunately, there is unprecedented pressure on our mammals as we teeter at the precipice of Earth's sixth, first human-induced, mass extinction.

VIGNETTE: TRANSFORMATIONAL MOMENT OF DISCOVERY IN THE WORLD HERITAGE FOSSIL DEPOSITS OF RIVERSLEIGH

By Mike Archer
PANGEA Research Center, School of Biological, Earth & Environmental Sciences, University of New South Wales, Sydney, Australia

Since 1976 we had been hammering away at Riversleigh's ancient, incredibly hard limestones. We were intent on finding yet another fossil mammal – one more rare window in time to help us understand how Australia's globally unique mammals came into being millions of years ago. The history of Australia involved evolution in almost total isolation from the rest of the world for the previous 48 million years – a vast dark ages that had so far produced only 73 different kinds of pre-Pleistocene mammals for the whole continent. Having done my PhD on Australia's carnivorous marsupials, I was particularly keen to find even one more of these particularly rare predators.

In 1983 we had become aware of a vast new potentially fossiliferous limestone plateau that hadn't previously been explored by any European. Ken Grimes and the other geologists who were with us that year hiked up over the edge of the plateau, pushing through walls of prickly spinifex, and strode rapidly off into the distance to see what they could find. Sue Hand was still foraging in the back of the 4WD looking for our geology maps. I stood on the edge of the plateau (Fig. 5.11), waiting, so she would know in which direction the team had gone.

While waiting, I idly looked down at the chest-sized rock I was standing on and nearly passed out. Bristling out from all sides of the block were jaws, skulls, limb bones and teeth: more than I had previously seen anywhere

Fig. 5.11. Suzanne Hand and Mike Archer in 1983, standing on Gag Site after excavation work had begun. Many new carnivorous marsupials came from this locality. Photo: Henk Godthelp.

Fig. 5.12. Using light explosives to 'carve' large blocks of the fossiliferous limestone, which is then dissolved using dilute acetic acid to release thousands of fossils from each. This is Price is Right Site, one of more than 250 fossil sites at Riversleigh subsequently discovered after Gag Site was found in 1983. Photo: Henk Godthelp.

else in Australia – ever. Among the many tantalising fossils visible, was a massive blade-like tooth that suggested it belonged to some monstrously powerful carnivore. We later came to realise that it represented of all strange things a carnivorous kangaroo that we named *Ekaltadeta ima*.

That limestone block was packed solid with mammal fossils. When it was eventually acid-processed, it produced on its own 35 new kinds of ancient mammals – nearly 50% more than we had previously known from the rest of the continent! Sue told me that when she heard my shout and ran up the side of the plateau, she found me on hands and knees crawling over the surface of the area making incoherent sounds of hysterical excitement. As we broadened our search, we realised the whole area was a pavement of thousands of the most extraordinary mammal and other vertebrate fossils that we had ever seen. So exciting! That epiphanic moment was the beginning of realisation of just how rich the fossil deposits at Riversleigh would prove to be.

This first-discovered of the fossil-rich sites at Riversleigh, which we later determined to be Middle Miocene in age and ~15 million years old, earned an unusual name because of what happened next. Ken Aplin and Tim Flannery, who were completing their PhD research on fossil mammals from Riversleigh, had spent the day driving in to Mount Isa to deliver far less rich blocks from Riversleigh's 'D Site' to the rail head where they would be sent south to the University of New South Wales in Sydney. That night, when they returned to the campsite along the Gregory River, we told them what we had found. Ken was ecstatic! He was fascinated in particular by bandicoots, but crippled in his ability to research their history by the total lack of adequate fossils of archaic forms – and I told him I'd seen whole fossil bandicoot skulls.

Came the morning and around the campfire a bit of mischievousness crept in. I said to the barely-able-to-wait PhD students, 'Well, wouldn't it be wonderful if what we told you we'd found yesterday really had happened?' Shocked looks and cereal on spoons poised between bowl and mouth. 'Well, sorry to say, we haven't found anything so far in the new area, but **today** may be the day when something like that **is** found!' Everyone who had been involved in the discovery the previous day bit their tongue and played along. So off we went after breakfast, with two very disappointed students in tow. When we got to the area near the edge of the limestone plateau, I casually suggested they check out that same area and we'd be along shortly. Head down and geology picks on shoulder, off they went. About 5 minutes later I heard Ken scream out at the top of his voice from the edge of the plateau 'Archer … you bastard!' Thereby the new site earned its name as Gag Site.

These and hundreds of other new site discoveries that followed led to Riversleigh being placed on the World Heritage List in 1994 along with Naracoorte as a serial nomination, 'Australia's Fossil Mammal Sites'. In terms of carnivores, since 1983 they have clawed their way out of Riversleigh's fossil deposits in rocks spanning the last 25 million years. Many of these are similar to living groups such as dasyurids, thylacinids, bandicoots, ornithorhynchids, bats, birds, reptiles and frogs. Others were unexpected as members of rainforest communities such as notoryctids (marsupial moles). Still others range from strange (*Whollydooleya*, *Malleodectes*) to bizarre (thylacoleonids, yingabalanarids, yalkaparidontids), with no surviving descendants to enable us to determine with confidence how they formed part of the complex ecosystems that made up Riversleigh's ancient forest communities.

Every year for the last 42 years, new discoveries have been made at Riversleigh by our research team and I have no doubt this remarkable place has more than we have already discovered waiting to be found. But very few, if any, are likely to surpass the adrenalin-draining moments when we first discovered Gag Site. For many of us, the rest of our lives have been shaped by that event.

6

THE MAMMAL HUNTERS: DISCOVERY OF CARNIVOROUS MARSUPIALS

In previous chapters we introduced the wonderful variety of carnivorous marsupials and explored their origins and evolutionary relationships. We provided teasers about the extraordinary range of behaviours that they show and snippets of their quirky biology. Before turning to examine their habits and ecological relationships in more detail – and thus peer more deeply into their secret lives – it is first appropriate to meet the pioneers who discovered the carnivorous marsupials and introduced them to us. This is important: although it is easy to 'Google' information, we do not always appreciate how the information has been gained, who acquired it, or how reliable it may be. This is especially so in the realm of science, where methods are often specialised and results can seem esoteric. Science is often presented as a seamless march towards 'truth'. Sometimes it may approach this ideal, and there is great joy and satisfaction when discoveries are made, but in practice, the **doing** of science can also be a bumpy road with many cul-de-sacs and taxes on those who seek to make the journey. Here, we give due recognition to some of the early pioneers who found the carnivorous marsupials just as fascinating as we do, and show the highs and lows, the insights, tenacity and sometimes just blind luck that often attend the hard-won acquisition of new knowledge. We acknowledge both the explorers in the field and the taxonomists and other scientists who sought to identify, describe and bring order to the wondrous oddities that the adventurers brought back with them.

Fig. 6.1. Charles Dorbigny's 1849 rendering of the bizarre opossum (*Didelphis*) clearly depicts its most unusual feature, which stunned European explorers who encountered it in the late 1400s – a pouch. A specimen, replete with young still in the pouch, was presented to an astonished Queen Isabella and her husband King Ferdinand of Spain. The animal justly became famous, representing the mystery and wonders of the New World. Source: Wikimedia Commons.

Early European discoveries of carnivorous marsupials

In 1499, 37-year-old Vicente Pinzón, who had previously sailed on Christopher Columbus' famous voyage, once more set sail for the New World. On arrival, with the help of his crew, Vicente captured a female opossum replete with pouch young. She was plucked from among the trees, shortly after landing. To Pinzón, more than 500 years ago, this furred creature with a pouch was the stuff of nightmares.

He described it accordingly, as:

'A monster, the foremost part resembling a fox, the hinder a monkey, the feet were like a man's, with ears like an owl. Under the belly hung a great bag, in which it carried the young, which they do not drop or forsake until they can feed themselves.'

The young died in transit on his return to Spain in 1500, but Vicente Pinzón proudly presented the female opossum at court to the Spanish monarchs King Ferdinand and Queen Isabella. Curiosity prompted the Queen to poke her royal fingers into the swollen pouch. To her astonishment, the tiny withered young remained there, attached to the teats.

It would be more than 100 years before Captain John Smith and settlers of the Jamestown colony (Virginia, USA), would write the first English description of this animal. The common name 'opossum' was coined: an English translation of the Native American's name for the animal, 'apasum', meaning 'white animal'. The opossum was to play an important role in human transition to scientific enlightenment, being a regular discussion piece for scientists in Europe. It became a symbol of both the strangeness of the New World and the veritable treasures it contained.

European scientific classifications remained simplistic, even after Carolus Linnaeus had structured our taxonomic naming system in the 1700s. Linnaeus himself classified the opossum with a motley assortment of creatures that included insectivores, armadillos and pigs, in his *Systema Naturae* published in 1735. By the time Linnaeus had published his masterwork, several opossum species were named and distinctive Australian marsupials (a possum and wallaby) had already been discovered. Yet connections between them were not made until after European settlement of Australia in the late 1700s.

The first recorded marsupial encounter by Europeans in Australia was in 1606, when Dutchman

Fig. 6.2. The Northern Quoll (*Dasyurus hallucatus*) would not have been seen by early colonists in Australia's south-east because it is limited to northern Australia, but this image shows the distinctive white-spotted coat, possessed by all members of the genus, which so captivated the colonists. Source: Alex Hartshorne.

Willem Jansz landed at Cape York in Queensland, north-eastern Australia. Based on journal descriptions, the captain and crew sighted a cuscus. Indeed, a cuscus was captured and eaten by another explorer, Torres, in nearby New Guinea just months later. In 1629, Dutchman Francois Pelsaert ran his ship the *Batavia* aground on Morning Reef, off the coast of Western Australia. There, they reportedly encountered 'cats', which were most likely in fact Tammar Wallabies (*Notamacropus eugenii*).

The first undisputed Australian carnivorous marsupial noted by Europeans was observed during the voyage of Captain Cook's ship *Endeavour* in 1770 after landing at Botany Bay. English naturalist and botanist Joseph Banks recorded in his journal that on the first of May, 1770, they saw 'the footprints of an animal … whose feet were like that of a polecat or weasel.'

Then, when they reached the Endeavour River, the animal itself was sighted and described by shocked natural history artist Sydney Parkinson, as:

> *'a spotted [quadruped] of the viverra kind … about the size and something like a polecat of light brown spotted with white on the back and white under the belly'.*

Members of the genus *Viverra* are civets, first described by Linnaeus in 1758. But this animal was no civet. It was called by Aborigines variously Je-Quoll, Taquol, Jaquol or Dekol. The beast was a carnivorous marsupial quoll in the genus *Dasyurus*. The specimen was described after the voyage as *Mustela quoll*, but unfortunately it was lost. This was the fate of many animal specimens collected during the *Endeavour's* voyage. Nevertheless, after the First Fleet arrived at Sydney Cove, more quolls were soon seen.

Captain of the ship *Supply*, David Blackburn, noted:

> *'I am informed some Tyger Cats have been seen … perhaps the same which is slightly described in Captain Cook's first voyage as resembling a polecat, having the back spotted with white; and is there said to be called by the natives Quoll.'*

Quolls also feature in Arthur Phillip's *The Voyage of Governor Phillip to Botany Bay*, published in 1789. This is the earliest formal account of the new colony, and includes Captain Blackburn's description of 'the spotted opossum', known today as the Eastern Quoll (*D. viverrinus*). By then the animal was recognised as marsupial. It was noted that it 'had six teats placed in a circle, within the pouch', clearly a 'small mammal of the opossum kind.'

Fig. 6.3. Arthur Phillip's 'spotted opossum', dated 1789. This is now recognised as an Eastern Quoll (*Dasyurus viverrinus*). Note the misrepresented squirrel-like tail. Source: Wikimedia Commons.

A Spotted-tailed Quoll (*D. maculatus*), in contrast, was unfortunately shown in this publication as a member of the placental Carnivora, and dubbed the 'spotted martin.'

Phillip himself noted:

> *'the species is about the size of a large polecat … the visage is pointed in shape … the fur is black, marked all over with irregular blotches of white, the tail is not excepted, which has an elegant appearance, and tapers gradually to a point.'*

Quolls were described by English zoologist John Latham from patchy notes and simplistic drawings. The 'spotted opossum' was correctly labelled a marsupial, but its tail was compared to that of a squirrel. John White's (1790) journal, which appeared a year after Phillip's voyage, dutifully depicted a squirrel-like tail on the animal in two varieties: plain and spotted. White had erred in combining two carnivorous marsupial species – the Spotted-tailed Quoll and the Brush-tailed Phascogale (*Phascogale tapoatafa*).

Englishman George Shaw formally described the Eastern Quoll in 1800, and he noted the discrepancies in morphology, but showed faith in White by publishing his two illustrations under a single name: viverrine opossum. Unbeknown to Shaw, the Brush-tailed Phascogale had already been described by Friedrich Meyer in 1793.

Fig. 6.4. Quolls (*Dasyurus* spp.) were encountered by settlers almost as soon as they stepped ashore after a long sea voyage. To many colonists they were pests; ferocious predators of livestock, especially poultry. Early representations, such as this one by a painter from Port Jackson (possibly Thomas Watling who died in 1814), are not terribly accurate. This is partly because the strange Australian mammals were often merged taxonomically and artistically with known species from other continents, which they superficially resemble. Source: Wikimedia Commons.

Taxonomic connections among the marsupials found by Europeans in Australasia and the Americas were precipitated by the discovery of the even more curious monotremes – the echidnas and Platypus (*Ornithorhynchus anatinus*). Discovery of the Platypus sent shock waves through the scientific world and beyond, with its odd combination of duck's bill, webbed feet and furry, beaver tail. It was thought by some to be a hoax, baffling the greatest thinkers of Europe. The Platypus was called by early Australian settlers the 'water mole' and the first scientific specimen was speared by a local Aboriginal. This was urgently sent to England in 1798 by Captain John Hunter, then Governor of New South Wales. The task of describing the Platypus fell in 1799 to George Shaw of the British Museum. Shaw examined the specimen in shock, doubting the evidence before his very eyes, carefully examining the bill to ensure it had not been sewn on. George Shaw believed it to be a mammal because of its furry body, but if all mammals produce milk, how could the Platypus suckle young encumbered by its strange bird's bill?

Fig. 6.5. The mysterious Platypus (*Ornithorhynchus anatinus*). Was it bird, mammal or hoax? John Gould's famous rendering from *The Mammals of Australia* was made in the mid-1800s at the height of this controversy. The mystery would not be fully solved until the 1880s. Source: Wikimedia Commons.

Fig. 6.6. A painting by Walter Stoneman of famed mammalian taxonomist Michael Rogers Thomas (1858–1929). Thomas preferred as a Christian name 'Oldfield', which was his mother's maiden surname. By middle-age, Thomas had health problems and disliked travel, relying on his extensive network of collectors to deliver him mammalian wonders from all over the world. Thomas described the new species with speed and precision. His knowledge of global mammalian taxonomy became second to none. Source: Natural History Museum.

Thirty years after Shaw described the Platypus, German biologist Johann Meckel produced an accurate series of Platypus anatomical studies. Meckel reported simple glands beneath the thick fur of the female. He suggested they produced milk, but this was doubted by several other leading European scientists. It was not until 6 years later that a lieutenant of the British Army stationed in New South Wales described the atypical mammary glands in a live animal. The milk simply oozed out, rather than being efficiently delivered via nipples. So how did the Platypus babies manage to suckle?

Richard Owen, a palaeontologist of renown from the British Museum, now entered the debate. Owen ordered baby Platypus sent to him from the colonies. Two were collected and he immediately realised the bills were poorly developed in the young so suckling would not be a problem. But a paradox remained. It was maintained by Australian Aborigines and others that the Platypus laid eggs. How could this be? Mammals by definition had fur and suckled young that were born live. Owen ordered several further Platypus specimens sent to him to solve this puzzle. After examining them he maintained that if they were born from eggs, they must be hatched inside the body rather than laid. Of course, eggs aplenty were sent to Owen with claims of being from Platypus, but these **were** all hoaxes, turning out to be bird and snake eggs.

The mystery was finally resolved by William Caldwell, a Scottish zoologist. Four years after graduating from Cambridge in 1884, Caldwell travelled to Australia under a Royal Society scholarship to solve the Platypus puzzle once and for all. After several months, with the help of Aborigines, he shot and killed a female Platypus on the Burnett River in northern Queensland. She had just laid one tiny egg and was in the process of laying another. Caldwell tersely wired London with his astounding results: 'Monotremes oviparous, ovum meroblastic.' In plain terms, monotremes lay eggs, and the eggs are similar to those of reptiles – only part of the egg divides as it develops. Caldwell's four momentous words signified the end of an 86-year-old mystery. Yet, as early as the 1830s, it had been realised that differences between monotremes and other mammals centred on reproduction. Scientific eyes then naturally turned towards reproductive differences between marsupials and placentals. This ultimately led to their taxonomic separation during the 1870s and 1880s.

Taxonomy of carnivorous marsupials: the pre-eminence of Oldfield Thomas

The European taxonomic history of marsupials is, like that for most fauna, tumultuous. The list of contributors to our taxonomic knowledge of the world's carnivorous marsupials is therefore long, but several names stand out. First and foremost was British taxonomist Michael Rogers 'Oldfield' Thomas (1858–1929). Few doubt that Thomas was the greatest of mammalian taxonomists. His prodigious output encompassed an astonishing range of mammal groups studied during 51 working years at the British Museum. Over this time, Thomas produced an average output of a new species description every week!

Thomas died in 1929 aged 71: he shot himself in the head. A year before, Thomas had lost his beloved wife, Mary Kane, after 38 years of marriage. He was in mourning, ill, listless and a staunch advocate of euthanasia – so he took the matter to hand. Thomas' taxonomic work within the mammals covered the gamut of forms. Yet, despite his prodigious output, Thomas' legacy for mammalogy, even today, is still only partly realised. We can learn much from a careful scrutiny of his past contributions.

Thomas grew up in South Africa, demonstrating a youthful interest in natural history by collecting insects on Table Mountain. His father then returned to England to become local vicar, where the boy received a first-rate education at Haileybury. In 1876, aged just 18, after showing no particular academic talent, Thomas secured a junior clerkship in the Secretariat at the British Museum. An aspiring scientist, he attended Thomas Henry Huxley's lectures at the Royal Institute for 2 years. He was described as an alert observer with most capable hands. His beautiful dissections, and the ease and speed with which he made them, won admiration from his teachers. Reports of his prowess soon reached the hallways of the British Museum. Aged just 20, Thomas was duly appointed an assistant in the British Museum's Zoology Department. Some months before his transfer to Zoology, Thomas had received word of his change in fortune. As the day of his appointment grew closer, Thomas came to understand that he would work in the section studying echinoderms – starfish, sea urchins, and the like. In preparation, he took up research on the group with enthusiasm, but on his first day at work the formidable Keeper, Albert Gunther, sent for him and said sternly: 'Thomas, you will do the mammals.' A stunned young Thomas muttered something in objection about echinoderms to which Gunther rounded on him and thundered, 'you **will** do the mammals!' Thomas was duly put in charge of the mammal collection. As days went by, he became more despondent, remonstrating with Gunther about his lot, having visions of working on cows, sheep and pigs for the rest of his life. But Gunther was about as flexible as a short steel rod. He

Fig. 6.7. One of the many species of carnivorous marsupial from Australia described by Oldfield Thomas: the Rusty Antechinus (*Antechinus adustus*). Source: Jiri Lochman.

Fig. 6.8. One of only two species of New Guinean carnivorous marsupial described by Oldfield Thomas: the Black-tailed Dasyure (*Murexia melanurus*). The species varies incredibly in size, colour and distribution and is likely to contain a suite of cryptic (i.e. hidden) species. Thomas hungered for more New Guinean specimens to describe, but his collectors were mostly unable to overcome the many perils and difficulties presented to any explorer game enough to venture into the island's dark interior. Source: John Hornbuckle.

continued to insist stubbornly that Thomas should simply 'do the mammals'. Eventually, Thomas capitulated and, again at Gunther's behest, began to compile a detailed catalogue of mammals in the rapidly growing collection. To Thomas' relief, he soon discovered the collection was diverse and extended well beyond everyday livestock. He gradually warmed to the task. By 1888, at the age of 30, Thomas finally published his comprehensive and masterly 400-page *Catalogue of the Marsupialia and Monotremata in the Collection of the British Museum (Natural History)*. It was momentous, and remains a benchmark reference text to this day.

Thomas described an impressive range of Australian carnivorous marsupials, including the dunnart genus (*Sminthopsis*) and a species therein, four antechinus species (*Antechinus*), a planigale (*Planigale*), a false antechinus (*Pseudantechinus*), a phascogale (*Phascogale*) and a marsupial mole (*Notoryctes*). He also described two New Guinean dasyurids, including a murexia (*Murexia*) and the Red-bellied Phascogale (*Phascolosorex*). However, Thomas' contribution to science extends well beyond this, to other carnivorous marsupial groups in the Americas, such as the Monito del Monte (*Dromiciops gliroides*) and shrew-opossums (*Caenolestes* and *Lestoros*). In addition, Thomas described various species within the short-tailed opossums (four species of *Monodelphis*), fat-tailed mouse opossums (*Thylamys*) and the Patagonian Opossum (*Lestodelphys halli*). Yet, as well as marsupials, Thomas named a large sequence of placental mammals, ranging from antelope to bats and most groups in between. Truly, his work had global scope, displaying an astonishing breadth of vision coupled with a tenaciously fine focus for detail and pattern. It is fair to say that descriptions of new species were in those days often short, some just half a page in length. Further, Thomas rarely offered ecological information about ways the animals might have lived. In comparison, a single mammal species description today may occupy 10–20 pages and cover all known aspects of its life history. However, in Thomas' day there were no shortcuts: no genetics to point the way, no detailed knowledge of species' distributions and no instantaneous email to contact dozens of potential collectors with relatively easy access to field localities. Taxonomists of Thomas' era depended solely on their knowledge of comparative morphology, which, in Thomas' own case, was obtained from his painstaking studies of a patchy if global set of specimens.

Thomas focused on describing, so others did the collecting. To name just a handful: Sherrin, Tunney, Wilkins, Stalker, Fritsche, the Pratt brothers, Kloss, Loria and Meek. In the end, hundreds of thousands of mammal specimens would be collected for Thomas by these intrepid men and a host of others (sadly there were few or no female collectors in those days). To present-day readers, this may seem a colossal waste of mammalian life, but it was just the opposite. Each specimen was labelled and identified, immortalised as part of a priceless reference collection to be studied by future generations of scientists. The specimens enabled the description of thousands of new species. These beasts would otherwise remain enigmatic, their scientific knowledge hidden from the world, until precisely categorised. Such was the crucial work of Thomas and other taxonomists. Their pioneering studies recognised the uniqueness of each species. It was only then, when species had been categorised, that a new creature's many biological secrets could be revealed and their significance understood.

In 1891, Thomas married Mary Kane, heiress to a small fortune and daughter of Sir Andrew Clark, an eminent physician. With his increasing scientific reputation and new-found wealth, Thomas was able to considerably expand the Museum mammal collection. Thomas and his wife supported mammal collectors across the world, financing expeditions where collecting was the main objective. Thomas kept regular correspondence while these adventurers were out in the wilds collecting on his behalf. He forwarded them money from the British Museum trustee's war chest, or his own pocket, to fund their exploits.

Consider, for example, Thomas Eberhardt. This adventurer travelled the world obtaining specimens for museums, but by 1928 had fallen on hard times. He had spent months collecting mammals throughout remote South America and shipped the booty to Hamburg. There, they were impounded because Eberhardt had insufficient funds to send them on the last leg of the ocean voyage to the British Museum. He wrote Thomas a hasty letter asking for money, explaining that he had 19 skins and three alcohol preserved specimens of various valuable South American mammals. The consignment included a female opossum in spirit, replete with nine pouch young. Thomas didn't hesitate, wiring the 11 English pounds (today about US$1000) from his own funds to a grateful Eberhardt. The invaluable specimens arrived several weeks later, and were eagerly described by Thomas. The British Museum later reimbursed both Thomas (11 pounds) for his initial outlay and Eberhardt (5 pounds) for extra shipping costs. In all, there were four specimen consignments containing hundreds of mammals, earning Eberhardt a total of 60 pounds (about US$5000).

Such specimens represented a huge time investment, not to mention risk of death for intrepid collectors – and Thomas demanded more than his pound of mammalian flesh. For each animal caught, he required accurate details of locality, altitude, date and sex of the captured game. He was necessarily fastidious with his labelling and preserving requirements. Without accurate labelling, a specimen was next to useless, and poor preservation meant the specimen would rot and not last in perpetuity. Many specimens from remote locations were already registered in the museum with vague or one word locality descriptors that made understanding variation within a species almost impossible. Thomas was loath to add dross to the collection.

As well as the Americas, Europe and Africa, Thomas was keen to obtain mammals from Australia, with its rich and diverse marsupial fauna. By 1923, the Australian fauna was already recognised to be disappearing under the combined pressures of feral pests, land clearing, agriculture and the expansion of other human enterprises. Thomas urged the British Museum trustees to finance an expedition to eastern and northern Australia, thus securing new type specimens before they were lost to science forever. They engaged one Captain George Hubert Wilkins, aged 34, to lead the trip. Wilkins was an Australian-born polar explorer, war correspondent, naturalist, climatologist and aviator.

He had excellent pedigree, serving as naturalist and second in command on the 1920–21 British Imperial Antarctic Expedition led by Sir Ernest Shackleton. This well-publicised trek brought him to the attention of Sir Sydney Harmer, Director of the British Museum. Wilkins seemed the perfect choice to lead the Australian expedition.

Wilkins appointed an all-Australian team of enthusiastic collectors and naturalists as assistants. From April 1923 to June 1925 the team ranged across thousands of kilometres in remote wilderness between mid-New South Wales on the Australian east coast to the Top End of the Northern Territory. However, Australia, especially areas west of the Great Dividing Range, was then in the midst of drought. In northern Australia, the heatwaves of 1923–24 were record breaking. Despite the generous assistance of the Queensland (Qld) Museum, free rail travel and freight, conditions for collecting were difficult indeed. The landscape was hot, dry and fire-prone. Once they reached the Top End, it became even hotter and at the onset of the wet season, storms and floods were frequent. The mosquitoes and flies were almost unbearable. These hardships took their collective toll. At many sites, zoological specimens were often scarcely encountered, mammals least of all. Between 8 July and 14 August 1923, four months into the trip and camped in Cape York, north Queensland, Wilkins wrote two despondent letters to his Australian contact, Qld Museum Director, zoologist Albert 'Heber' Longman:

> *'Dear Mr Longman,*
>
> *… the authorities cannot blame us for exterminating the native fauna … I am tired of walking miles and miles a day without seeing anything …. If our luck does not soon change you will begin to think that I am out here on a holiday trip and not making any serious effort to obtain mammals. It seems ridiculous to state that we have been able to secure only one or two specimens of one or two species during our six weeks in this locality.'*

In two and half years of trekking, other than kangaroos, wallabies (some of which were new to science) and rats, mice and bats in generally limited numbers, other mammal groups were encountered only in ones and twos. All told, Wilkins' crew secured just a few carnivorous marsupials: northern quolls (*Dasyurus hallucatus*) and some antechinuses.

Fig. 6.9. A recently collected Mainland Dusky Antechinus (*Antechinus mimetes*) from Ebor. This was the same locality that George Wilkins had visited in 1924 and Oldfield Thomas later described the collected animal as a subspecies of Dusky Antechinus. The subspecies has since been raised to species status, adding yet another described species to the legion ranks of Thomas' carnivorous marsupials. Source: Gary Cranitch.

About the only place Wilkins had decent trapping success for mammals was in the 1.5 km high misty mountains of the Guy Fawkes district near Ebor in north-eastern New South Wales. Here, in April 1924, Wilkins happily reported to Longman that they had 'little difficulty in securing a number of the small mammal species that inhabit the district.' One specimen was to prove particularly important. It was a large antechinus, captured in high-altitude rainforest. The specimen was eventually shipped, via the Qld Museum, back to Oldfield Thomas in England. He recognised it as distinctive, describing it the same year as a new subspecies of Dusky Antechinus (*Antechinus swainsonii mimetes*), but a present-day revision of the Dusky Antechinus complex has recognised Thomas' subspecies as a new species of carnivorous marsupial (*Antechinus mimetes*). Captain Wilkins' 1924 specimen provided a crucial piece in helping to solve this taxonomic puzzle.

Thomas was understandably keen to enlist those who had both first-hand knowledge and appreciation of the nuances of mammalian taxonomy. One such was Captain Guy Shortridge: an Englishman who served in the Boer War (1899–1902). After the war, he remained in South Africa to collect mammals and birds for the South African Museum. Then, in 1904, Shortridge returned to England, from where he conducted a range of expeditions on behalf of the British Museum to Australia (1904–07) and New Guinea (1910). These journeys resulted in the

Fig. 6.10. Oldfield Thomas originally named the Long-tailed Planigale (*Planigale ingrami*) as a species of *Phascogale* in 1906. The unique genus of *Planigale* was not nominated until Curator of Mammals at the Australian Museum Ellis Le Geyt Troughton erected it in 1928 to accommodate the three species of small, flat-headed dasyurids known from Australia at the time. Source: Jiri Lochman.

description of a range of new mammal species, including several carnivorous marsupials.

Shortridge's trip to Australia focused on the south-west. Shortridge, then aged just 24, arrived at Fremantle in Western Australia in November 1904. Over the next two and a half years, he trekked throughout remote south-western Australia, visiting some 15 different sites and covering several thousand kilometres. Fortunately, his London connections allowed him to secure a free rail pass, which became his primary mode of transport. His method was to travel by rail to an interesting area and then use this as a launching point into the wilds, hunting mammal and bird specimens. Shortridge had an array of guns and purchased a two-wheeled carriage. He also had 200 'wooden spring' ('gin') traps of various sizes to capture anything from the size of a mouse to that of a dingo. However, he was not fond of these or the methods of dispatch used by Aborigines, because they often damaged the precious specimens.

In one of his detailed journals, Shortridge noted:

'A lot of people who bring me in animals are very fond of giving them a crack across the head, breaking the skull. I have a great bother to get the natives out of it, as their first idea when they've got an animal is to kill it, no matter how. The smaller ones are often broken by the spring traps that have an awkward way of just catching things below the head … When a lot of traps have been set over a good range of country, no matter how early one gets to them, the sun will have generally got on some of them. By this time they have all been cooked and as such they don't last half an hour in the sun getting rotten in an incredibly short time. Also, you would not believe how many [specimens] I have had to throw away on account of ants. Lots of them work all night and as soon as a small thing gets caught they find it out, and next morning when I go to the traps I find half of them with hair or noses eaten off. Sometimes the whole thing is almost a skeleton … What is badly wanted in a hot country is a small trap, for catching the smaller [species] alive, for several reasons.'

Fed up with the shortcomings of snap traps, an enterprising Shortridge designed and successfully deployed a live-capture copper wire trap, the schematic of which he proudly sent to Thomas.

Although he had only a small number of these live-capture traps, Shortridge found them superior to any other trap for catching small mammals because the captured specimen was undamaged and larger animals tended not to inadvertently set them off. Shortridge also made good use of dogs in his collecting. He was fond of using terriers, which he found could put up bandicoots and wallabies and be relied on to point to trees where various mammals were nesting. He liaised closely with the Director of the Western Australian Museum, Bernard Woodward, and renowned museum collector John Tunney, to ensure he visited the best locations and targeted sought-after species. In return, a share of Shortridge's spoils found their way into the Western Australian Museum mammal collection.

Shortridge had been directed by Thomas to secure a complete collection of mammals, 'at least one specimen of everything' – and he made a good job of it. In the end, Shortridge amassed thousands of mammal specimens, including 31 species, 26 of which were marsupials. These represent ~60% of the mammal species known to occur in Shortridge's collecting areas today. There were five species of carnivorous marsupials in Shortridge's collection, including two species of quoll, a phascogale, dunnart (which were scarce and found only under tree bark) and the Numbat (*Myrmecobius fasciatus*) (also rarely encountered). The specimens were sent home in batches to Thomas, who promptly described them in a series of papers published between 1906 and 1909. This included the description of a new subspecies of Western Quoll (*Dasyurus geoffroii*), which is now presumed Extinct in the Wild. Of his 3 years in Australia at the beginning of the 20th century, Shortridge noted with uncanny foresight, 'Animals are dying out here as fast as aboriginals – both are fading before the products of a tougher civilization.' He appealed to the Western Australian Government to set aside reserves in which fauna of the state could be preserved.

After leaving Australia, Shortridge headed straight to Java, amassing some 1500 specimens that represent the greatest collection ever obtained there. In 1908, he headed to Guatemala to obtain live animals for the Zoological Society. Then in 1910, again at the behest of Thomas and the British Museum, Shortridge (then aged 30) joined a party of explorers on an expedition to the Snow Mountains of New Guinea.

Fig. 6.11. The Red-tailed Phascogale (*Phascogale calura*, pictured) collected by Guy Shortridge in his Western Australian Expedition (1904/5) had already been described by John Gould in 1844. However, while Shortridge was abroad, Oldfield Thomas described a subspecies of Brush-tailed Phascogale (*P. tapoatafa*) from the Northern Territory (*P. t. pirata*), which has since been raised to species status. As well as a plethora of mammal species, Thomas described many subspecies, a large number of which are now being recognised as worthy of full species status with the advent of modern taxonomic and DNA techniques. Source: Jiri Lochman.

The Dutch had already conducted two expeditions attempting to reach the snowline of New Guinea. The first, in 1907, was largely unsuccessful, but the second short trip from October to December 1909 reached the permanent snowline at a height of 4460 m. This trip also yielded some new carnivorous marsupials from high altitude, including the Long-nosed Dasyure (*Murexia naso*) and the striking Speckled Dasyure (*Neophascogale lorentzii*). However, these new species came at great cost – four expedition members died on the treacherous descent from the snow-covered peaks.

After these Dutch exploits, the British Museum became intent on also reaching the Snow Mountains,

Fig. 6.12. The striking Speckled Dasyure (*Neophascogale lorentzii*): one of two species named by Fredericus Jentink (Director of the Leyden Museum) in 1911 after the successful Dutch expedition to New Guinea in 1909. Source: Auscape.

conducting a more intensive field survey and finding other species new to science – Thomas' mouth must have been watering. Shortridge's trip in 1910 was a joint expedition with the Royal Geographic Society to celebrate the British Ornithologist's Union Jubilee. It was truly remote work, in steaming jungles where no white explorers had before ventured. After arriving in New Guinea, and with the help of eager local people, it took the explorers' months to even reach the base of their proposed study site, many kilometres inland. This attested to the difficulty of conditions because members of the expedition party were the stuff of legend. Walter Goodfellow (leader), Wilfred Stalker and Guy Shortridge were renowned collectors, having trekked in remote regions around the globe. Alexander 'Sandy' Wollaston was a medical officer, entomologist and botanist of renown. Surveyor Eric Marshall was famed for his involvement in the first Shackleton-led British Antarctic Expedition from 1907–1909. Yet the experienced group was plagued with difficulties from the outset. Shortly after arriving, Stalker died suddenly and mysteriously, aged just 31. Wollaston noted in his journal:

> *'We buried him under a tree about one hundred yards behind the camp … I read the short burial service. Besides Marshall and Shortridge and myself there were a Dutch soldier, two convicts and about fifty Papuans, who stood quietly in a wide circle about the grave. I think the ninetieth psalm was never read to a more remarkable congregation. The grave was the first of many who left their bones in New Guinea.'*

By June 1910, Shortridge became so ill with malaria that he was forced to leave, travelling south to Sydney, Australia, to recuperate. On arrival, he spoke to the *Sydney Morning Herald* newspaper about the expedition's difficulties: 'the people we were camped amongst have never seen a white man. They have never seen a match struck before. Somebody picked up a gun in the camp for a shot at a pigeon or something. They'd never heard a gun, and they vanished into the forest like smoke. The women did not show up again until the next day … It teems with rain and the forest swarms with mosquitos. We take quinine against malaria, but any time of the day when you get into the forest mosquitos simply cover you, and as long as you are there they remain.'

Despite working months in the foothills and following the winding river 65 km inland, the expedition was destined never to reach the Snow Mountains. In his journal, Wollaston lamented their failure:

> *'It was more than tantalising to see day after day those virgin peaks so comparatively close at hand and yet as unattainable as the mountains in the moon. I looked and longed for 15 months, and at the end, with an aching heart, I had to turn my back and leave those fields of untrodden snow as remote and mysterious as they were when we came out.'*

Safe behind his museum work desk in England, Thomas also lamented Shortridge's failure in New Guinea. In the following years, Thomas tried to get several other collectors back there for detailed surveys, but it didn't happen. New Guinean mammals eventually came in ones and twos to the British Museum from 1928 onwards, thanks to Australian collector Fred Mayer, but this was mostly too late for an aging Thomas, who had long considered it the most important place in the world for the 'long British arm' to reach, lest its mammalian wonders be exploited by someone else. And indeed they were (by the Americans), but not until almost a decade after Thomas' death, as we will see later in this chapter.

Shortridge was not only a superlative collector, but a taxonomic talent in his own right. During a series of African expeditions between 1922 and 1930, his numerous letters to Thomas show he was intimately familiar with the finer points of mammal taxonomy. Shortridge took to the task of describing some of the African specimens himself in a series of papers after

Fig. 6.13. Not all carnivorous marsupials collected from New Guinea required explorers to penetrate the dark interior. Indeed, the first such species described from New Guinea, *Myoictis melas*, was simply handed by local people to S. Müller and his delegation from the Natural Science Commission for the Netherlands Indies Co., who had shipped into port during 1828. The local people revealed to the visiting Europeans that the small marsupial lived in the forest on the ground, foraged mainly at night and was known to them as 'Insinsie'. In 1840, Müller provided a cursory (undiagnostic) description in his report of the expedition and the species' discovery is attributed to him. Source: Auscape.

Thomas' death, and published them in 1931 to great acclaim. In 1934, he published another landmark work, *The Mammals of South-west Africa*. Shortridge was duly appointed member of the South African Association for the Advancement of Science – a prestigious honour.

After all his many adventures and near-death experiences in wars and wild, unexplored places, ironically it was misadventure of a more mundane sort that brought Shortridge undone. As a result of injuries sustained when he was knocked down by a car, he suffered frequent blackouts for 2 years. By 12 January 1949, aged 69, he'd had enough. Shortridge committed suicide by strychnine poisoning. Suicide ran in the family: his two brothers had also taken their own lives. Shortridge was found dead at his desk, among his beloved mammals in the Kaffrarian Museum. During his time as Curator and Director, the institution had focused almost entirely on mammals. And Shortridge, like Thomas, was a monomaniac. At the time of his death, the Kaffrarian Museum held a collection of ~17 000 specimens, most of which had been collected by Shortridge himself.

Some collectors of the time were independently funded by Englishman Walter, 2nd Lord Rothschild: a banker, politician and zoologist. Walter was eldest son and heir of Lord Nathan Rothschild – an immensely wealthy financier and the first Jewish peer in England. Money was no object and Walter knew no restraint. At its largest, Rothschild's self-financed collection included an incredible array of specimens: 300 000 bird skins, 200 000 birds' eggs, 2 250 000 butterflies, 30 000 beetles, and several thousand mammals, reptiles and fishes. Taken together, this formed the largest zoological collection ever amassed by a private individual. As a boy of 7, Rothschild had been an avid animal collector and vowed to his parents that he would make a museum. By age 10, he had converted a garden shed for just this purpose. Before long, his insect and bird collections had grown so large that his parents had to

Fig. 6.14. At 6 feet and 3 inches (190 cm) tall, with a speech impediment, Walter Rothschild was by all reports a shy man, but he wasn't beyond staging for attention. He had his photograph taken riding on a giant tortoise and drove this carriage harnessed to six zebras to Buckingham Palace to prove that the animals could be tamed. Thanks to his family fortune he was wealthy beyond all reason and spent unstintingly to amass the largest collection of animal specimens held by a single individual. Indeed, his collection was so vast, some millions of specimens, that his father built him an entire museum to contain it. Source: Wikimedia Commons.

rent rooms off-site as storage. By 1889, when Rothschild was 21 and his specimens numbered in the millions, his father finally built him a real museum on his Tring Park estate. In 1892, Rothschild opened the Tring Museum doors to the public. He promptly appointed two curators: entomologist Karl Jordan and ornithologist Ernst Hartert. Both Jordan and Hartert assisted Rothschild in studying and maintaining his vast collection for almost 40 years, but in 1931, Rothschild was forced to sell the majority of his bird collection to the American Museum of Natural History, after being blackmailed by a wealthy, aristocratic former mistress. The bird collection, minus Rothschild's sentimental favourites the cassowaries, sold for US$225 000. This converted to about a dollar a specimen. However, it represented just a fraction of the collection's true worth at ~US$1 000 000.

After his death in 1937, the remainder of Rothschild's massive private collection went to the British Museum. Rothschild and Oldfield Thomas were peers. Rothschild regularly invited Thomas from the British Museum to describe the small mammals found on his funded surveys. Between private collectors, the Museum and Rothschild, Thomas had unfettered access to a peerless collection of oddities – and he bent his head to the ceaseless task of describing them, naming several new species after his generous benefactor Rothschild.

Thomas had no love for outside work, and rarely ventured afield. Collecting for him was merely a means to provide more precious specimens for description. His dislike of the outdoors is partly explained by the fact that early in middle age Thomas began to worry about his health. His family had a history of illness and Thomas was troubled by heart palpitations and other symptoms he little understood.

Although he became increasingly deskbound as his career progressed, Thomas was certainly no recluse. Even when working, he preferred the presence of at least one other person. He was single-minded, methodical, pragmatic to a fault and had an almost pathological dislike of solitude. This perhaps provides some insight into Thomas' unravelling following his wife's passing, after their many happy years together. Thomas and Mary had been unable to have children; with the loss of his wife, Thomas' enforced solitude was complete. In the 13 months following her death, Thomas missed her sunny and reassuring presence more and more. Eventually, he ceased to become interested even in his zoological work, showing the unmistakable signs of depression. His closest friends were unsurprised when he took his own life.

At an enquiry following Thomas' death, his cousin and private secretary, Reginald Oldfield, commented that, although Thomas had never threatened to take his life before, he remained deeply depressed over his wife's death. The day before his death, on Saturday 15 June, according to Reginald, Thomas had appeared well enough, attending a croquet match in the afternoon and listening to the wireless into the evening, before retiring at 10:30 p.m. At 8:30 a.m. next morning, one of the maids found Oldfield Thomas lying next to his work desk in the garden tool shed of his Kensington home, shot through the mouth, with a discharged colt service revolver in his lifeless hand. Thomas left a suicide note on the desk, which read: 'Friends, relations and all dear ones. Forgive me, for I have fought as long as I could and can do no more. I cannot get over her [his wife's] loss. Hopeless and too terribly miserable. It returns again and again at the least mental effort. I have tried to leave everything arranged. My ring for Geoffrey; the photograph for Daisy.'

Thomas had been indifferent to social advancement and without conceit. In his will, he made numerous bequests to relatives, museum staff and specialist societies. He left money with the British Museum

Fig. 6.15. Photograph from a 1930 obituary of Oldfield Thomas published by the Royal Society. Thomas had been made a Fellow of this prestigious society in 1901. Source: Wikimedia Commons.

trustees to finance future expeditions to collect yet more mammals. He also funded the installation of a small lift just inside the entrance of the museum, to save the legs of anyone unable to use the stairs when moving between the mammal room and main hall. This was a trek Thomas made several times a day. For a man in self-confessed poor health who didn't even like to leave his working desk, the incessant, short expeditions became a burden. Today, the lift bears a plaque in his honour.

After Thomas' death, with rare specimens from legions of collectors continuing to flood in from all over the world, the job of their description passed to others. Capable were the ranks of taxonomists who followed, but Thomas was a nonpareil. His comparative knowledge of the world's mammals was freakish and unsurpassed then and since. Thanks to Thomas, the mammal collection at the British Museum had become (and remains) the most important type series of mammals held anywhere in the world. Hundreds of international scientists flock there every year to study its many treasures, using them as a critical reference to describe new species from around the globe.

Oldfield Thomas is deservedly remembered as a taxonomic giant of the front rank. As testament to his profound contribution to mammalogy, 13 species and a genus of the world's mammals have been named in his honour.

Strangers in strange lands: other hunter–collectors of carnivorous marsupials

Other famous names feature heavily in the early taxonomic history of Australasian carnivorous marsupials, such as John Gould, John Gray and George Shaw. Both Gray and Shaw were trained in medicine and skilled in anatomy – attributes so crucial for application to the taxonomic arts. These men both became keepers at the British Museum and had direct access to its many biological treasures. Other names of prominence in Australasian mammal taxonomy include William Macleay, Baldwin Spencer and Ellis Troughton. Later would come George Tate (notably in antechinuses and false antechinuses), Mike Archer (in particular, planigales, ningauis and dunnarts), Darryl Kitchener (dunnarts), and Steve Van Dyck (antechinuses, murexias and dunnarts), among others. They each made impressive contributions to

Fig. 6.16. Lithograph of famed British artist and naturalist John Gould by T. H. Maguire, 1849. Although his first passion was birds, on visiting Australia Gould became intrigued by Australia's bizarre assortment of marsupials. With invaluable help from his trusty collector John Gilbert, and artists Elizabeth Coxen (his wife) and Henry Richter, he eventually published the natural history work: *The Mammals of Australia*, to worldwide acclaim. Source: Wikimedia Commons.

Fig. 6.17. An impression of a Kultarr (*Antechinomys laniger*), appearing in the prestigious *Proceedings of the Zoological Society of London, 1880* (Plate XLV). Note the relatively long hind legs – the only carnivorous marsupial that has them. This animal was initially considered to be the Australian equivalent of jerboa – hopping mice from Africa and Asia. Source: Wikimedia Commons.

Fig. 6.18. John Gould's impression of the flesh-loving 'Devil' from *The Mammals of Australia*. The Tasmanian Devil (*Sarcophilus harrisii*) is the largest living carnivorous marsupial and has the strongest bite force of any living marsupial. The powerful jaws can crush bone to powder. Devils are pugnacious and one of the few social carnivorous marsupial species. Social they may be, but their cantankerous nature ensures constant squabbling among their fellows at any gathering. Source: Wikimedia Commons.

our knowledge of Australasian carnivorous marsupial taxonomy and systematics.

Of this group, John Gould, a British artist-naturalist, is perhaps best known. Gould was most notable for his prolific bird descriptions, but he also named numerous mammals from Australasia and donated the spoils to the growing ranks of the British Museum. His early, self-published and self-illustrated natural history works, such as the seminal *The Mammals of Australia*, are the stuff of legend. Gould travelled the Australian continent in the mid-1800s with John Gilbert, his colleague-collector. Together, they amassed mammal specimens, seeking knowledge and assistance from cooperative Aborigines. Gould's first-hand accounts of the appearances and lives of Australia's mammals now serve as a crucial record for the unfortunately long list of now extinct Australian mammals.

Between 1838 and 1840, Gould visited Australia with wife Elizabeth and John Gilbert. This was ostensibly to work on his *The Birds of Australia*. However, in the process, Gould became fascinated by Australia's mammals. He later commented that:

> *'It was not, however, until I arrived in the country, and found myself surrounded by objects as strange as if I had been transported to another planet, that I conceived the idea of devoting a portion of my attention to the mammalian class of its extraordinary fauna.'*

Gould returned to London in 1840 and began a series of scientific papers in which he would swell the ranks of known marsupials. Gould's wife and mother to their six children, died unexpectedly in 1841, aged only 37 years. He was devastated and, aside from the emotional distress, this had a major effect on Gould's output – she was his chief artist. He needed to consign the help of Henry Richter to do much of the remaining art work. Gould's collector Gilbert remained in Australia until 1845, when he was killed by Aborigines on Leichhardt's expedition from Darling Downs to Port Essington. The same fate awaited two of Gould's other principal collectors: Fred Strange and John Drummond. Collecting was a risky business.

Gould's landmark *The Mammals of Australia* is what eventually resulted from the combined efforts of Gilbert, Strange, Drummond, Elizabeth (Gould's wife), Richter and Gould himself. Unfortunately, some of the contributors did not live to see how favourably the work was received. Most of Gould's published work is based heavily on Gilbert's specimens and extensive notes. Gould had no formal zoological training, but was respected by contemporaries and some of his taxonomy has stood the test of time.

Some familiar names also appear in descriptions of the American mammal fauna, such as Gould, Gray

Fig. 6.19. John Gould's Numbat (*Myrmecobius fasciatus*) from *The Mammals of Australia* was painted by his artist Henry Richter in 1845. Richter produced ~3000 lithographic plates and watercolours for Gould. The Numbat first became known to Europeans in 1831. George Moore, a member of a group then exploring the Avon Valley, recounted the discovery: 'Saw a beautiful animal; but, as it escaped into the hollow of a tree, could not ascertain whether it was a species of squirrel, weasel, or wild cat ...' On the following day, Moore noted: 'chased another little animal, such as had escaped from us yesterday, into a hollow tree, where we captured it; from the length of its tongue, and other circumstances, we conjecture that it is an ant-eater.' Moore's observation was close to the mark: Numbats eat termites. Source: Wikimedia Commons.

and George Waterhouse. These men were all powerhouses from the British Museum, with access to global collections. It was no surprise that they flexed the scope of their taxonomic muscle, but other workers of note appeared on the taxonomic stage of American carnivorous marsupials, such as the Frenchmen Anselme Desmarest and Geoffroy Saint-Hilaire and the American Wilfred Osgood. Even famed Swede Carolus Linnaeus, father of modern taxonomy, made a guest appearance, naming a species of four-eyed opossum in 1758. Some of the diverse contributions of these taxonomists to our knowledge of carnivorous marsupials are well worth recounting.

Desmarest was made Professor of Medicine in the veterinary college of Alfort, France, in 1815. Then he occupied chairs of zoology, botany and anatomy. He worked intensively on collections held at the Parisian Museum of Natural History, particularly mammals collected in Australia by Peron, Quoy and Gaimard. These included four genera of Australian mammals. He also named several South American carnivorous marsupials, including a species of lutrine opossum and a fat-tailed mouse opossum.

Saint-Hilaire was a prodigy. At the tender age of just 21, he was appointed Professor of Zoology in the prestigious Parisian Museum of Natural History. He was a prominent member of the scientific commission charged by Emperor Napoleon to study all aspects of Egypt. Decades before Darwin, Hilaire proposed that new species might arise by mutation. Hilaire worked on Australian mammals using the diverse material collected by Peron on trips to Australia by the ships *Naturaliste* and *Geographe*, and named the first species of *Antechinus* in 1804. However, Hilaire's taxonomic

Fig. 6.20. A three-striped dasyure (*Myoictis* sp.) from New Guinea. This exceptional artwork dates to 1858 and was published in the *Proceedings of the Zoological Society of London*. The artist was one Joseph Wolf, a German who specialised in natural history illustration. He moved to the British Museum in 1848 and became a preferred illustrator for explorers and naturalists of renown. These included famed African explorer David Livingston ('Dr Livingstone I presume?') and Alfred Russell Wallace, who independently of Darwin conceived the *Theory of Evolution by Natural Selection*. Wolf depicted animals accurately in life-like postures and is considered a pioneer of wildlife art. Sir Edwin Landseer, himself a prodigious 19th century British artist, once claimed that Wolf was '... without exception, the best all-round animal artist who ever lived.' Source: Wikimedia Commons.

work on mammals had global scope – he also named the American Brown Four-eyed Opossum (*Metachirus nudicaudatus*) in 1803. In naming this and several other species of American mammals, Hilaire benefited greatly from the collections of Alexandre Rodrigues Ferreira. Ferreira was the first Brazilian naturalist. He had received a doctorate in 1779 at University of Coimbra in Portugal, before being appointed as Naturalist at the Lisbon Museum. He returned to Brazil in 1783 on a collection trip commissioned by the museum. The trip would last 9 years, extending deep into the unexplored forest of central-north Brazil. Four naturalists were originally included on the roster, but, due to financial cuts, Ferreira was the only one who went. With him alone lay the responsibility of collecting species, classifying and preparing specimens for the return journey to Lisbon, studying the agriculture and constructing maps of the area. He made a good fist of it. On returning to Lisbon in 1793, Ferreira deposited in the museum some 417 new species of animals, including 65 new species of mammal. However, Napoleon (under Junot) invaded Lisbon in 1807 and confiscated the lot. The invaders returned the diverse collection to Paris into the care of an eager Hilaire, who diligently set to work formally naming the mammalian booty.

Numerous collections of South American mammals were made in and around Brazil in the early 1800s, such as those by the German naturalists von Spix and von Martius in 1817–1820. Everything the young Germans saw was new to them. Yet all they could possibly collect and preserve was easily reached along trails. Danger surely lurked in paths less-travelled; they dared venture no further. The Viennese Johann Natterer was far less cautious. He was commissioned by the Museum of Vienna in 1817 to investigate the

Fig. 6.21. An excellent early print of a Yellow-sided (Short-tailed) Opossum, *Monodelphis*. The original specimens were presented to British naturalist George Waterhouse after a successful expedition by members of the HMS *Beagle*. Waterhouse was Curator of the Zoological Society of London's museum and had been asked to join the voyage but declined. Another young naturalist took part in his stead. This was none other than Charles Darwin, originator of the theory of evolution. Source: Wikimedia Commons.

Brazilian biota. He collected there between 1817 and 1835, venturing well into the dark and uncharted Amazon. Natterer's enormous collection was sent back to Europe. It included some 781 specimens of mammals alone, representing nearly half the mammal species known from Brazil today. Natterer's good friend at the Museum, Andreas Wagner, described most of the mammalian spoils. These included two species of short-tailed opossum (*Monodelphis*) and a fat-tailed mouse opossum (*Thylamys*).

Despite the early discovery of the opossum at the time of Columbus, other South American marsupials were discovered only in ones and twos in the 1700s and early 1800s. Early expeditions tended to focus on larger and easier to catch mammal species, such as monkeys, carnivores, hoofed mammals and other placentals. With such a bounty of new species on offer, smaller and harder to catch species such as marsupials, bats and mice were often easily overlooked. Suitable traps for smaller species were also harder to come by. It was left to collectors and taxonomists of the mid-late 1800s and early 1900s to make a more major contribution to our knowledge of carnivorous marsupials from the Americas.

Wilfred Hudson Osgood (1875–1947) was one of the most acclaimed curators ever to serve at the Field Museum in Chicago. He first came to the museum as Assistant Curator of Mammals and Birds in 1909, after working as Assistant Biologist in the Biological Survey of the USA Department of Agriculture. In 1921, Osgood became Chief Curator of Zoology at the Field Museum. He held the post until retiring in 1941. Osgood made many contributions to mammalogy. Perhaps the greatest of these was an understanding of animal populations. Other taxonomists of his day routinely named individual variants as subspecies, species and even genera, but Osgood recognised the natural variation within populations as something to be understood in terms of an animal's environmental history. This was the raw material for natural selection that ultimately gave rise to species.

Osgood based his studies on extensive fieldwork. During his career, he made in total 30 expeditions that took him to the remote reaches of the globe. He led the Marshall Field Chilean Expedition of 1922–23. This began on Isla Chiloé: a large island some 965 km south of Valparaíso. The biota of this area had last been studied by Charles Darwin during his famous voyage on the *Beagle* of 1831–6. Here, Osgood spent six happy weeks, assembling the first North American collection of Chilean mammals, which included a new genus and several new species. But it was a decade earlier in 1911 that Osgood had made perhaps his most important voyage. Sailing from New York on 31 December 1910, Osgood's collection team reached Maracaibo in Venezuela, via Porto Rico and Curaçao, on 9 January 1911. They then proceeded inland, into

Fig. 6.22. An early print of the Brown Four-eyed Opossum (*Metachirus nudicaudatus*) from Richard Lydekker's 1896 work, *A Hand-book to the Marsupialia and Monotremata* (Plate XXXIII). The genus was not recognised until 1854. This species had been discovered in 1803 and classified as a species of *Didelphis*, but it was later recognised that Brown Four-eyed Opossums are in fact pouch-less, unlike the famous pouched *Didelphis* that was proudly presented by explorer Vicente Pinzón to Queen Isabella of Spain in 1500. All marsupials have a paired uterus and vagina (although all have a single external opening) and the name *Didelphis* literally means 'two wombs'. Source: Wikimedia Commons.

dark and truly uncharted territory. High in the mountains, Osgood discovered the bizarre shrew-opossum (*Caenolestes*) scurrying about the dense moss forest floor at a heady 2440 m. After much hard work, he captured 11 animals. Despite being completely isolated and working under trying conditions, he preserved the bodies of two entire 'wet specimens', using formaldehyde and bichloride of mercury. He also prepared several dry skeletons and skins for scientific posterity. These specimens proved crucial in understanding the taxonomy and biology of one of the most mysterious marsupials of the time; an animal that was ultimately to be recognised as a surviving representative of that ancient order of carnivorous marsupials: the Paucituberculata.

The mystery surrounding shrew-opossums had begun in 1860, when R. F. Tomes published his list of mammals from Ecuador. In this work, he described a small animal about the size of a placental water shrew (*Sorex*), with strange protruding lower incisor teeth. However, it couldn't be a placental, he noted, on account of its small and rudimentary pouch. Three years later, Tomes formally described this animal but strangely gave no reference to it being marsupial. Indeed, the specimen was all but forgotten until 1880, when Alston referred to it as a 'small marsupial from Ecuador'. He noted that it may represent a distinct family of marsupials, but the original description was unsatisfactory. It was also mentioned by Oldfield Thomas in his 1888 *Catalogue of the Marsupialia and Monotremata in the Collection of the British Museum (Natural History)*. Thomas noted at the time that it was so insufficiently described no idea of its proper position could be gained; nor had any further specimens been collected. This was remedied in 1895, when a second skin and skull was received by the British Museum.

Thomas promptly described the species in more detail, noting: 'In vindication of Mr Tomes' paper I should like to say, firstly, that his description … proves to agree, so far as it goes, very closely with the present specimen. Secondly … even now, with infinitely greater material and the best of advice, I am unable to be at all positive about the exact position and relationships of the little marsupial described by Mr Tomes … The discovery of this long-lost genus, whose wide distinction from all other living marsupials its original describer does not appear to have fully appreciated, is one of the most interesting events in mammalogy that has happened for many years … *Caenolestes* represents among the marsupials a family, and, perhaps, a suborder, entirely different from any now living … it is evidently a surviving representative of some fossil marsupials from the Santa Cruz [fossil] beds of Patagonia.'

Thomas' interest was well and truly kindled. In his correspondences of 1901–1902 with American Museum curator Joel Allen, Thomas requested a suite of mammal specimens from the forests of Peru and Venezuela. Among these oddities, he doubtless sought further *Caenolestes*, but was denied. Thomas and Allen regularly exchanged mammal specimens of interest from their collections. Yet, in a letter dated 23 January 1902, Allen noted Thomas' request but instead made up a return exchange with mammals mostly from North America. For the purpose of solving marsupial riddles, these were bound to be less interesting. In these letters, Allen made repeated mention of Thomas' planned trips to the American Museum, which Thomas cancelled on several occasions. Even the lure of the veritable treasures held in the American Museum could not entice Thomas from his work-desk at the British Museum.

Although further shrew-opossum specimens were obtained in the years that followed, the material included only dry skins and skulls, similar to that used by Thomas. Soft tissues from wholly preserved specimens were desperately needed. It fell to Osgood in 1911 to provide them, as described earlier. So crucial were these extra specimens in understanding the anatomy and relationships of the shrew-opossums that Osgood based his PhD thesis upon the material he unearthed in the misty mountains of Venezuela. The results of this work astounded scientists the world over. Osgood also described the sole species within another genus of the shrew-opossums, *Rhyncholestes raphanurus*, as well as species of short-tailed opossum (*Monodelphis*) and four-eyed opossum (*Philander*).

As the tales recounted earlier demonstrate, many early mammal collectors were true explorers. They risked life and limb, venturing into remote and unknown territory in search of scientific wonders. Some of these expeditions are now folk lore. One such is the Horn Expedition of 1894 to central Australia. Expeditions of this kind were massive undertakings: months or even years in the planning. Typically, they resulted in descriptions of suites of new mammal

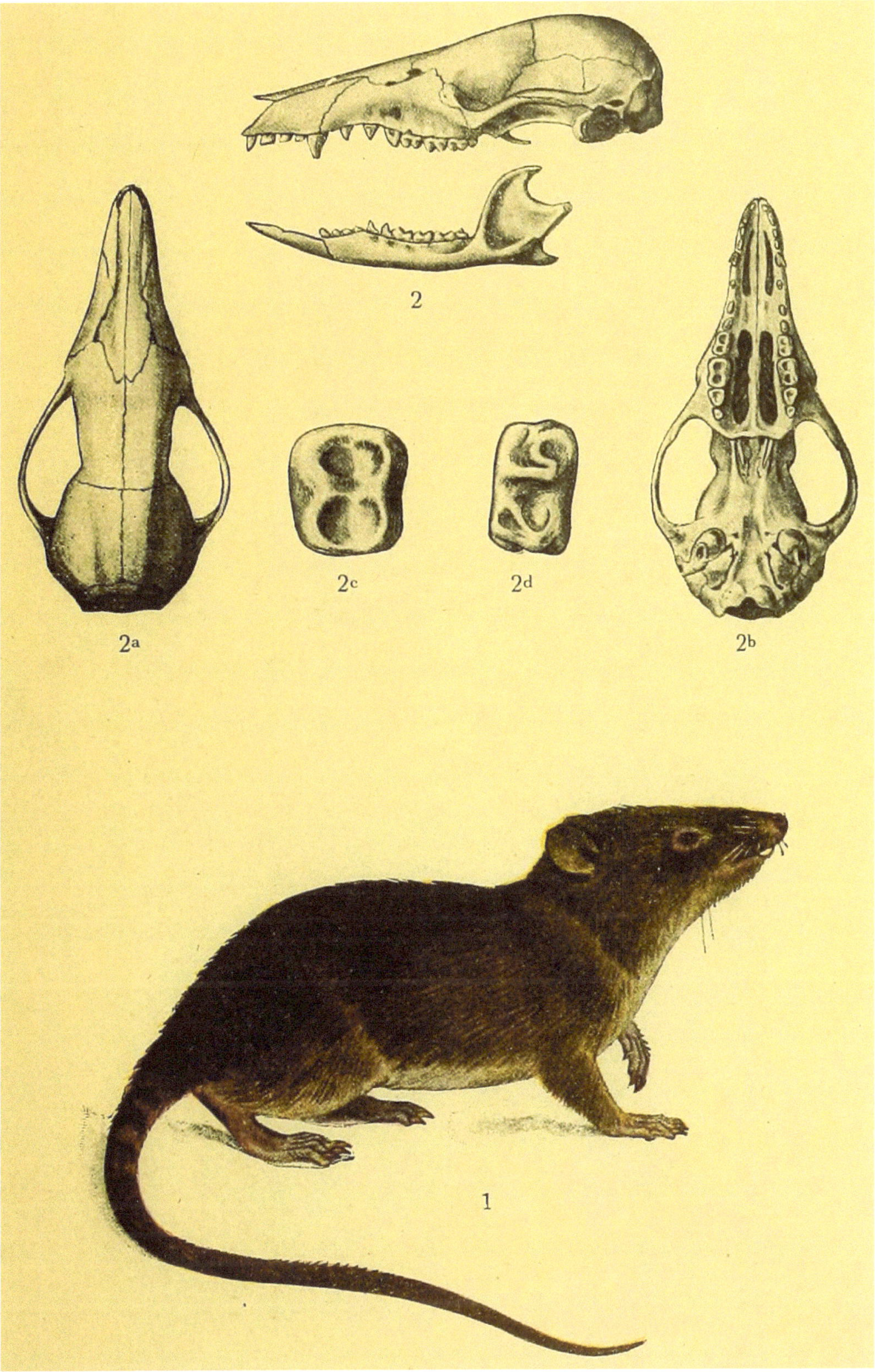

Fig. 6.23. A shrew-opossum (Family Caenolestidae) – another group of 'living fossils' whose relationships with other marsupials date back 80 million years or more, when many of Earth's southern continents were still connected. Shrew-opossums were originally thought to be closely related to the Australian diprotodontids (literally 'two teeth') because of their prominent protruding lower incisors that can be seen in this illustration. This plate is part of the seminal work, *Mammalian Genera: Monotremes and Marsupials* prepared by Spaniard Angel Cabrera in 1919 (plate X). Cabrera was the authority on mammalian studies in Spain during the early 20th century. He learned his craft from the best, spending a month at the British Museum in 1910 under the tutelage of none other than Oldfield Thomas. Source: Wikimedia Commons.

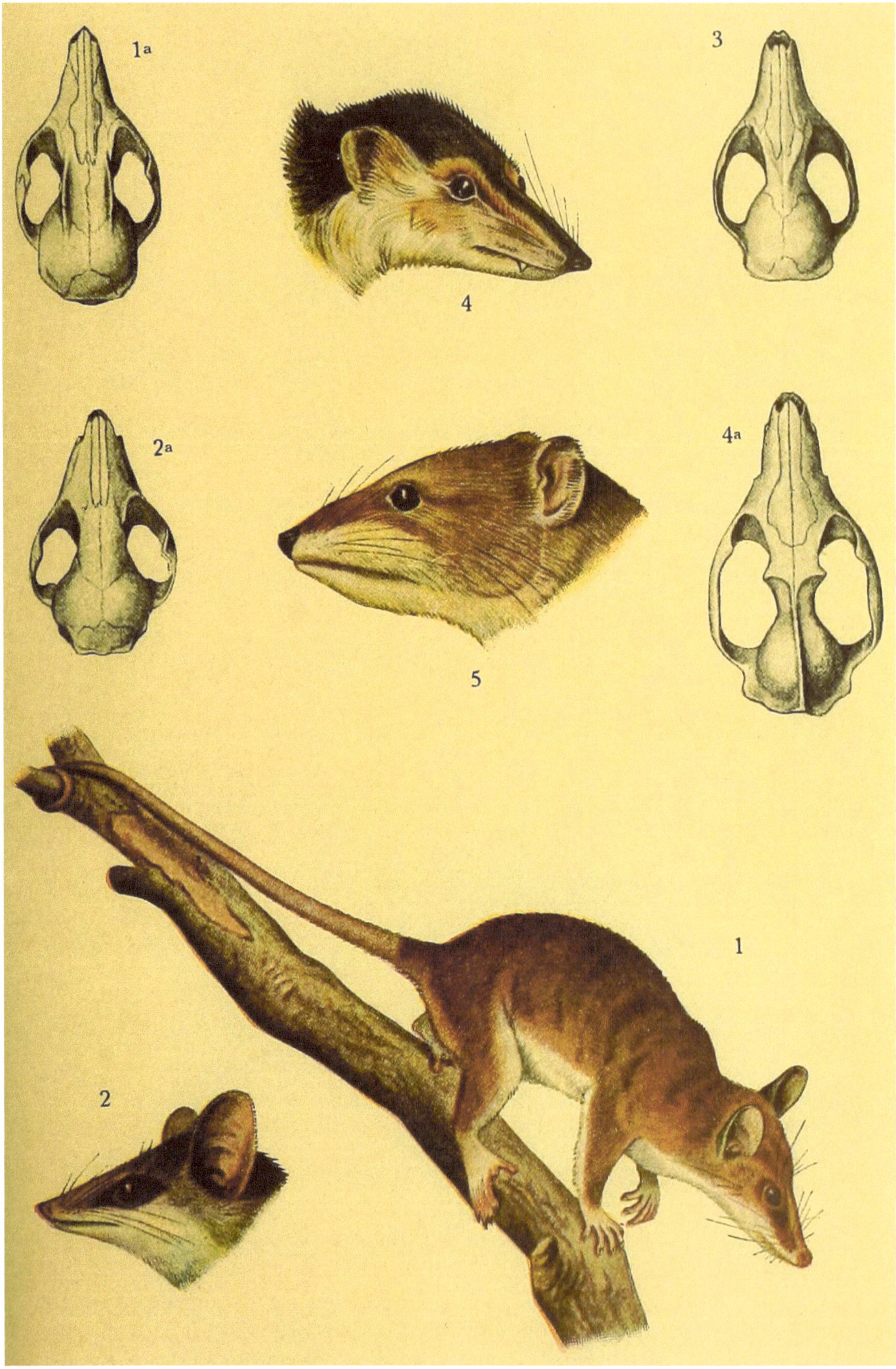

Fig. 6.24. Another of the 19 wonderful colour plates presented in Angel Cabrera's 1919 work, *Mammalian Genera: Monotremes and Marsupials* (plate II). This depicts a range of species, some of which are carnivorous. Number 5, the head shown in centre, is a lutrine opossum (*Lutreolina*); a fat-tailed mouse opossum (*Thylamys*) is shown at lower left (number 2); a short-tailed opossum (*Monodelphis*) is shown top centre (number 4). Source: Wikimedia Commons.

Fig. 6.25. Mr Tomes' Dusky Shrew-opossum (*Caenolestes fulginosus*). There was no easy access to live specimens such as this one in the late 1890s when even scant skin and skull material fuelled so much taxonomic controversy. There is a deep evolutionary foundation underlying their bizarre morphology – shrew-opossums represent an ancient lineage that separated from the other South American marsupials some 80–90 million years ago. Source: Hugh Tyndale-Biscoe.

species. The Horn Expedition was no exception. Here is the tale of discovery for a new mammal unearthed during the expedition – the Sandhill Dunnart (*Sminthopsis psammophila*).

In 1887, Oldfield Thomas named the genus *Sminthopsis* (dunnarts) to account for three unique, small carnivorous marsupials named by Gray (in 1842), Gould (in 1844) and Tarragon (in 1847). In fact, Tarragon's Queensland type specimen of the Red-cheeked Dunnart (*S. virginiae*) had been lost, but other similar specimens were available for Thomas' critical inspection. Decades earlier, British zoologists Gray and Gould had deposited their dunnarts into that well-used taxonomic midden, *Phascogale*. However, during his cataloguing of the British Museum mammal collection, Thomas became convinced the group was unique. As usual, he was right. There had been few detailed zoological surveys of Australia's desert interior. Dunnarts were already known from the region, but there was much speculation as to how many other mammal species and even genera may be hiding in Australia's vast desert expanses.

The mystique surrounding Australia's wild and largely uncharted interior was enhanced by Stirling's 1889 discovery of the bizarre marsupial mole (*Notoryctes*) caught near Charlotte Waters in central Australia. A few years earlier, the manager of Frew River Station (Northern Territory) had been driving cattle when he noticed unfamiliar tracks in the sand. He followed the tracks until he reached a spinifex hummock next to a dry watercourse. A strange, golden creature lay there, huddled in the shade. It appeared blind and mole-like. The beast was sent on a long journey south to Edward Stirling of the South Australian Museum. It was transported in a cartridge box, wrapped in a kerosene-soaked rag. After travelling more than 1000 miles by packhorse and coach, the specimen had begun to decay, but Stirling immediately recognised its importance to science, speculating that the animal might not even be a marsupial. Perhaps it was a monotreme or some other type of very primitive mammal? Yet it was two more years before he reported his findings to the world. It took that long for him to organise a trip north into the desert, travelling by horse and camel during his annual leave. Other scientists were similarly puzzled by the little mammal, noting its astonishing similarity in form to African golden moles, which are placentals. They naturally wondered about possible links between the Australian and African mammals, but the marsupial mole specimens were in ones and twos and poorly preserved. So the stage was set for some intrepid souls to undertake a detailed survey of interior Australia and unearth whatever biological treasures it was shielding from scientific enquiry.

The Horn Expedition was thus devised to investigate in detail the MacDonnell Ranges and surrounding areas of central Australia. The Universities of Adelaide, Melbourne and Sydney supplied the scientific personnel, covering a range of disciplines. The survey was organised and equipped by its namesake, William Austin Horn: a mining magnate, pastoralist and politician from Adelaide. Walter Baldwin Spencer, Professor of Zoology at Melbourne University, was the expedition's zoologist and photographer. Spencer was an Oxford graduate in natural science and had become the first appointed Professor of Biology at the University of Melbourne in

Fig. 6.26. The first *Antechinus* was named in 1803, and the first *Phascogale* in 1793, shortly after European settlement of Australia. Initially, other genera, such as the dunnarts (*Sminthopsis*), were grouped within the same genus with *Antechinus*. We now know that in fact phascogales (*Phascogale*) and antechinuses are sister genera, both exhibiting a spectacular annual male die-off at the end of the breeding period. Dunnarts are less closely related to either of them. In this artist's early impression, the dunnart is at the top (note the fattened tail), the antechinus in the middle, and the phascogale at bottom. This artistic work was one of many made during the voyages of *HMS Erebus* and *Terror*. Source: Wikimedia Commons.

1887. He seemed the perfect choice as expedition naturalist. One Charles Winnecke, whose adventures across central Australia were well known at the time, was chosen as guide. His task was to lead the party safely along a route agreed upon by the scientific members. In May 1894, the expedition set out from

Fig. 6.27. Sir Walter Baldwin Spencer was English-born and an Oxford scholar. In 1886, he applied for, and won, the Foundation Chair of Biology at the University of Melbourne, Australia. He transformed the standards of the university and was an entrepreneur of Australian science. The successful 1894 Horn scientific expedition to central Australia recruited Spencer as zoologist and photographer. Source: Wikimedia Commons.

Oodnadatta, at that time the end of the railway line in South Australia. A small mountain of gear was distributed among the men and stacked on the sturdy backs of 25 camels. In addition to the scientists, the group included two Afghan camel drivers to keep the cantankerous beasts in check. There were also two Aboriginal trackers, two general assistants and a cook.

In his extensive diary of the journey, Spencer noted:

'Our daily programme when on the march was much the same, day after day. We were usually up some time before sunrise. A little after sunrise we had breakfast and the camels were brought in. Each of us loaded his own [camel] and then off we started in single file. After perhaps 10 or 12 miles came the midday halt … we were glad of any shelter afforded by the thin scrub. Mounting again, we travelled on until dusk brought us to our camping place for the night. The camels were unloaded, hobbled and set free to feed. The camp fires were lighted, notes were written up, specimens labelled and packed away, and then, after a final pipe, we lay down on the ground and slept in the open. Often it was so cold that we awoke to find our water bags frozen solid.'

The Horn Expedition lasted three long months, covering some 4800 km. Its findings in all fields of study were widely acclaimed and greatly increased knowledge about central Australia. For Spencer, it was the beginning of a detailed study of the Aboriginal people of central and northern Australia: several tribes had met and assisted the expedition party in capturing mammal specimens. Spencer teamed up with Frank Gillen to eventually produce a detailed work on the culture of the Aranda (Arrernte) Aborigines. This became one of the most celebrated anthropological works of its time.

The type specimen of the Sandhill Dunnart (*S. psammophila*) was collected during the Horn Expedition on 18 June 1894. In his scientific description the following year, Spencer noted that the specimen was first observed by Mounted Trooper Cowle, a member of the survey team in central Australia, near Lake Amadeus. The adult male dunnart was flushed from cover in broad daylight. It led its surprised captors a merry chase, until brought down by an accurately thrown boot belonging to one member of the party. However, the story of this little marsupial does not end with its initial discovery. After Spencer's description, the Sandhill Dunnart then promptly disappeared from enquiring eyes for 75 years. In April 1969, it abruptly reappeared at Mamblyn on the Eyre Peninsula, South Australia. An astonished farmer dropped his hat on a second specimen as it fled in terror from burning spinifex. He nonchalantly retrieved his hat and sent the captive to the South Australian Museum in a battered shoebox loosely tied with string. On arrival, the animal was accurately identified by shocked taxonomists. Just 2 months later, four juveniles were captured as they fled from smouldering debris when confronted by a bulldozer, 80 km south-east of Boonerdoo. Unfortunately, this carnivorous marsupial species has only been encountered sporadically in a small number of sites since, and is now considered one of Australia's most threatened mammals. A larger carnivorous marsupial, the Vulnerable Kowari, was also captured for the first

Fig. 6.28. Members of the Horn Expedition. They boldly entered the uncharted territory of central Australia in 1894. Baldwin Spencer is fourth from the right, standing with the knuckles of his right hand resting on upraised thigh. Source: Francis James Gillen, Museum Victoria.

time on the Horn Expedition. Two years after the expedition, Spencer named the single species of Kowari *Dasyuroides byrnei*, in honour of Patrick Byrne, master of the Charlotte Waters telegraph station, who gave the type specimen to Spencer as the expedition passed through their lands.

Successful capture of mammals on trips such as the Horn Expedition required a combination of science, artistry and seizing the moment. Hats, boots and hands were as useful as any tool to bring down a fleeing animal that might prove invaluable to science. Native peoples of the Americas and Australasia also had a long history of mammal capture, and often served as guides, sharing their many skills and eco-cultural knowledge with European explorers. Indigenous people used deadfalls, snares and nets made from natural materials such as stones, logs or twine. Pitfall traps were dug to capture larger species. Australian Aborigines used returnable throwing sticks (boomerangs) to knock down mammals, fire to flush them out, wooden stakes to stab them, and shallow wooden bowls to dig them out. Larger species were hunted with accurately thrown wooden spears. These people were expert trackers and typically served this crucial role on expeditions, locating animals of interest. Europeans also brought their own traps, adding them to the mix. Steel leg-hold traps appeared in the 1600s and were made for trappers by companies in both England and America. These were modifications of archaic Middle Age designs. There were numerous types, such as spring- and leg-traps. By the 1800s, these were mass produced and distributed to the colonies, driven by high prices for animal pelts. They were effectively used to trap larger species such as the Thylacine (*Thylacinus cynocephalus*), Tasmanian Devil (*Sarcophilus harrisii*) and quolls (*Dasyurus*). The early explorers brought horses and camels with them on expeditions to remote areas. Relay teams of men would chase and run down fleeing mammals until the terrified animal literally ran itself to death. Other large species were spotted at distance and summarily shot. Expedition leaders often pursued interesting species with vigour and determination. In 1798, surgeon-explorer George Bass chased a large wombat on Cape Barren Island that he thought may be new to science. He eventually ran the animal down, grasped it about the belly and carried it a mile or more back to his ship, but not before the squirming beast had gnashed the elbow from his coat sleeve. By the 1900s, metal box traps with spring-operated doors were developed. These could be transported on expeditions in collapsed form, then mass-deployed in an area of interest, which much improved capture success. These devices eventually evolved into the Sherman and Elliott metal traps used today (see Chapter 10).

If Australia's interior was considered remote and daunting, then the mountainous forests of uncharted New Guinea doubtless tested the nerve of even the most intrepid collectors. Enter Richard Archbold, American heir to an oil fortune, who between 1929 and 1939 sponsored three expeditions to the New Guinean highlands in search of new animal and plant species. In the front of Archbold's mind must have been the failure of Shortridge's attempt (recounted earlier) to assail the Snow Mountains nearly three decades before. After two failed attempts, Archbold used a gigantic Catalina flying boat with a 32-m wingspan, which was invaluable in landing the crew in uncharted areas and capturing never-before-seen aerial photographs. In August, Archbold dispatched two exploration teams consisting of Dutch soldiers, convicts and Dyak tribesmen. Each team started at opposite ends of the valley with the goal of meeting in the middle, collecting opportunistically as they went. The ploy worked. The result was massive collections of specimens, with many species new to science. Archbold's plans for a fourth trip were disrupted by the start of the Second World War.

Fig. 6.29. Looking down the barrel of a modern-day metal box (Elliott) trap. To set the trap, the entry door is flipped down and hooked under the 'treadle' step plate (the latter visible here directly in front of the animal). The bait (typically peanut butter and oats) and some bedding material (in this case coconut husk) is placed at the far end of the trap. The animal investigates the smell, walks into the entrance and once it treads on the step plate the hook tilts back, the entry door snaps shut behind the intruder, and it is efficiently trapped. The animal shown captured here is a mosaic-tailed rat (*Melomys*) of the eastern Australian forests. Source: Andrew Baker.

The Archbold Expeditions were coordinated by the American Museum of Natural History. George Tate, an Englishman working in the museum's zoology section, wrote many new mammal species descriptions in collaboration with Archbold based on the expedition specimens. Tate described numerous New Guinean, Australian and South American mammals, including an array of carnivorous marsupials. This included his joint description with Archbold of the New Guinean (antechinus-like) genus *Murexia* and a species of three-striped dasyure (*Myoictis*). But Tate's interests were diverse. He also worked in Ecuador, Bolivia, Brazil and Venezuela, sometimes in truly remote locations. The description of the Patagonian Opossum (*Lestodelphys halli*), sole member of its genus, is a result of his toil.

Indigenous relationships with carnivorous marsupials

As described earlier, in the decades that followed European settlement in Australasia and the Americas the study of carnivorous marsupials (as well as other animal groups) underwent a long and exciting period of discovery and description. These taxa eventually became separated from other marsupials into distinct families, and arrays of new genera and species were described. This was the formal work of western science, but, of course, carnivorous marsupials in the Americas and Australasia were not exactly unknown to people when Europeans first turned up in these regions.

Native peoples of both Australia and New Guinea have lived with carnivorous marsupials as part of their lives and cultures for tens of thousands of years. Most of the modern mammals are well known to them, and larger and more distinctive marsupials are remembered from Australian Aboriginal dreamtime stories. Most mammals, no less various carnivorous marsupials, are used by Australian Aborigines for food. Pelts are sometimes used for clothing or ceremonies; skins are used for instruments. In some cases, sinews are used for tools. Indigenous tribes have close and often insightful knowledge of the distribution, diet, habits, reproduction and trapability of many carnivorous marsupials. This has been used to great effect by Europeans in their searches for various species, in particular many of the elusive dasyurids that are found deep in dark and remote New Guinean forests.

European and indigenous taxonomies, unsurprisingly, often do not match. Sometimes what Europeans recognise as distinct species are known by a single name to Aborigines. However, there is much that each culture can learn from the other to benefit the conservation of dasyurids and other carnivores. Consider the Western Quoll (*Dasyurus geoffroii*), which is known to Aborigines in Australia's western deserts as *Parrtjarta*. When various Indigenous tribes were presented pictures of this quoll, it was recognised by some groups, but not others. Most native people claimed it lived in spinifex; others said it inhabited mulga. All groups agreed that it had disappeared from those places. As we now know, a form of the Western Quoll indeed used to inhabit such areas, but is now apparently extinct. Pictures of Brush-tailed and Red-tailed Phascogales (*Phascogale tapoatafa* and *P. calura*, respectively) shown by researchers to several western region Aboriginal groups met with generally puzzled expressions. Only one group recognised them. The Jameson Range peoples called them 'taling-go-won'. They said the animals lived on the ground in sand hill country. Today, we know this species to be partly arboreal and requiring tall trees. Yet, interestingly, European records from nearly a century ago suggested that Brush-tailed Phascogales occurred in the sandridge deserts of Western Australia, as remembered and predicted by Aboriginal people. Phascogales were also known from the Canning Stock Route in times gone by. This track runs from Halls Creek in the Kimberley to Wiluna in the mid-west of Western Australia: a distance of some 1850 km. Indeed, it is the longest historic stock route in the world, passing through a range of habitats such as salt lakes, rocky ranges, scree slopes, sand dunes and plains, including sections of the Gibson and Great Sandy Deserts.

Fig. 6.30. The Brush-tailed Phascogale (*Phascogale tapoatafa*). Australian Aboriginal knowledge of this and other carnivorous marsupial species has helped ecologists understand to what extent their distributions have changed since European settlement. Source: Jiri Lochman.

A recent study worked with Indigenous peoples in the Northern Territory to determine the timing and causes of marsupial declines in northern Australia. Indigenous information about terrestrial mammals was compiled via interviews conducted across a range of communities. Museum collection skins and photos of the animals were used to facilitate discussion. The earliest memories of these Indigenous peoples indicated that declines, particularly in larger fauna such as Northern Quolls (*Dasyurus hallucatus*) and Northern Brush-tailed Phascogales (*Phascogale pirata*), had increased in recent decades. This matches scientific perceptions of mammal declines in northern Australia. Indigenous participants suggested that several factors were associated with the declines. First and foremost was the effect of changes to Indigenous land management, in particular the use of fire regimes. People emphasised the need to preserve patches of unburned habitat surrounding burns. This was to ensure that animals can move into them and return to feed on regenerating patches later. Several Indigenous elders explicitly highlighted the link between traditional land management and customary social and ceremonial practices. Other participants suggested that the breakdown of traditional controls over hunting and the availability of vehicles and firearms had led to over-exploitation of the fauna. Introduced animals were also implicated by Indigenous peoples as having negatively affected some mammal populations. Feral pigs were blamed for altering the habitat of ground-dwelling species. Feral cats were known predators of small- to medium-sized mammals. Cane Toads were specifically blamed for the decline of Northern Quolls. Much of this insight complements emerging scientific knowledge. The study emphasised the value of Indigenous ecological knowledge and how it can be used to corroborate and inform western-style scientific studies.

New Guinean nationals also have a close relationship with their mammal fauna. For example, the Wola people from the south-central highlands of Papua New Guinea group mammals into two life forms, or families, which they call *sem*. These are categorised according to size: *sab* are larger and *honez* are smaller. Both categories include marsupial and placental mammals. Thus, their taxonomy is profoundly different from ours. Within the *sem* are *semonda* (large families). These are roughly equivalent to our genera and species. Sometimes there are also *semg nk* (small families). These correspond to species and subspecies. For example, the Speckled Dasyure (*Neophascogale lorentzii*) is a carnivorous marsupial known to local people as *Palay*, *Tombil* or *Yogbaen*. It is divided into two small family groups: *suwpalay* (the largest form) and *iyshmondismondis* (the smallest). Other than clear size differences, they are indistinguishable to the natives by shape, colour or behaviour. The local people note that this animal is found in trees and on the ground, but only during the day. This closely matches European scientific knowledge. The species is diurnal (i.e. active during the day) and both arboreal (tree-dwelling) and terrestrial (ground-dwelling). This species is sometimes trapped by Wola, but more often it is killed with a stick or bow and arrow by hunters. It is eaten and the skin is sometimes put to ceremonial use as a drumhead. Intriguingly, perhaps these forms of Speckled Dasyure will also represent different European-based taxa once current DNA studies are complete. Many New Guinean dasyurids are renowned for their large size range, and also are suspected of harbouring cryptic species.

Many Aboriginal people hunted, and in some cases continue to hunt, various dasyurid species. In Australia, Aborigines will ambush basking Fat-tailed False Antechinus (*Pseudantechinus macdonnellensis*) outside burrow entrances. Historically, they hunted Tasmanian Devils. Doubtless, many other dasyurid species were also taken by various tribes.

In New Guinea, the Mianmin, Telefol and Lani people regularly hunt the local dasyurids, especially the large species such as New Guinea Quolls (*D. albopunctatus*), Speckled Dasyures and Narrow-striped Dasyures (*Phascolosorex dorsalis*). They use traditional deadfall traps and bow and arrow with great skill.

Fig. 6.31. Some carnivorous marsupials were known to native peoples as a food source. Aboriginal hunters ambushed Fat-tailed Pseudantechinus (*Pseudantechinus macdonnellensis*) outside burrow entrances, often located in termite mounds as shown here. Source: Jiri Lochman.

They also use hunting dogs, trained for the purpose. New Guinea's mammals include a unique assemblage of relatively small-sized marsupials, but, despite their small size, the hunting of these animals provides a major source of protein for local communities. Impacts of hunting and the influence of the marsupials' life-history strategies on the sustainability of hunting are little known. However, research suggests that cuscuses, bandicoots and ringtail possums (all non-carnivorous) are numerically the most important game animals to New Guinean hunters. Cuscuses provide the main biomass of wild meat, but wallabies, bandicoots and tree kangaroos provide additional protein. Of the dasyurids, only quolls are taken by hunters in appreciable numbers. Hunting success rates for these species are so low (less than 5% of total catch) that they are apparently under no undue hunting pressure. Other dasyurids that are known to be hunted by the New Guinean nationals, such as Speckled Dasyures and *Murexias*, may or may not be affected; impacts have not been assessed.

It is uncertain when humans first arrived in South America. People may have reached Chile 13 000 years ago (kya), or possibly as early as 34 kya. Domestic dogs were brought to South America ~7 kya. The last of the large carnivorous marsupials had likely disappeared 2 million years before this, leaving a cohort of small- to medium-sized carnivorous marsupials in their stead. Southern South America, with its diverse climate and ecozones, probably supported a range of Indian groups. On the pampas and in Patagonia were nomadic hunter gatherers. Central Chile may have supported farmers and pastoralists. Parts of Argentina probably featured village farmers in the tropical forests, and irrigation civilisations may have been a feature of northern Chile. Hunter-gatherers needed to exploit land and sea mammals in areas where farming was impossible or impractical. Llamas (*Lama glama*) and various rodents probably featured regularly on the menu. Carnivorous marsupials were likely captured and eaten opportunistically as well.

Before the arrival of Europeans in 1492, it is fair to say that native South Americans probably had minimal impact on their local mammal faunas. The exception was the farmers of central Chile. Even early Spanish chroniclers noted the absence of wild mammals in this region, due to extensive conversion of forest to crop land. European arrival brought rapid change. Parts of South America were suitable for European agriculture, but bases for extracting natural resources such as gold, silver and gems ruled the day. The mineral riches of Peru were believed to extend far to the south, which prompted settlement in areas such as Argentina. Trading relationships between Europeans and Indians often helped ensure survival of the early settlements. Some places were settled to provide trade routes for transport of valuable gems and minerals from one area to another. Livestock was brought into many areas; horses were regularly traded. By 1699, more than a million head of cattle wandered the pampayan grasslands. Horses were likewise domesticated, and in some areas sheep stretched as far as the eye could see. Couple this multitude of domesticated beasts with feral cats and dogs, and some 200 years after European settlement, the South American fauna, including its unique marsupials, was doubtless much affected. By the late 1700s, Europeans began to turn their eyes from mineral to animal wealth. Fur seal and sea lion pelts were exploited. Elephant seals were hunted for their oil. Whales were harpooned with fervour. On land, deer and chinchillas were dispatched with abandon. But all these mammals were large or abundant. Such is not the case for South America's carnivorous marsupials. They fortunately largely escaped these many types of direct human persecution. But human activity has multi-layered effects on the environment. Many species of carnivorous marsupial do not survive in our much-altered forests, and South American species are no exception. This is especially true of rarer species that may be limited to certain habitat types. The shrew and mouse opossums are sensitive to such human-induced fragmentation. Very few species indeed, such as the Virginia Opossum (*Didelphis virginiana*), have actually expanded their range in the face of the myriad impacts associated with human settlement.

In this chapter, we have shared some of the discovery stories detailing the exploits of intrepid scientists and explorers who have ventured into the realms of Earth's living carnivorous marsupials. The next vignette relates some of the trials facing the modern day mammal hunter – as you will see, such work remains challenging and sometimes dangerous. In the next chapter, we will begin a deeper investigation of how the amazing carnivorous marsupials live their lives.

VIGNETTE: TRIALS AND TRIBULATIONS OF A MODERN-DAY MAMMAL HUNTER-COLLECTOR

By Marissa Parrott

Department of Wildlife Conservation and Science, Zoos Victoria, Victoria, Australia

I first met the amazing Agile Antechinus (*Antechinus agilis*) during my honours year at the University of Melbourne. At the time, I was studying another small marsupial, the Feathertail Glider (*Acrobates pygmaeus*) – a gentle little creature that shared their nest box homes with antechinus, often with dire consequences. I occasionally found empty skins of feathertails and little else. The antechinus eat with surgical precision and can readily kill creatures their own size or larger! When I met my first antechinus, it promptly sunk its canines through the top of my fingernail; a pain I would become familiar with. Yet I fell in love with these voracious predatory marsupials and their big, brave personalities. I began a 4-year PhD studying female mate choice and breeding success in the species, both in captivity and the wild.

The Agile Antechinus inhabits the forests and woodlands of south-eastern Australia, hunting for invertebrates and small vertebrates. They are excellent climbers and have been caught on remote cameras raiding bird nests for eggs and even small adults, which they can dispatch with a lethal bite to the head and neck. In captivity and the wild, I have had large males run at me and try to bite me in an effort to scare me off – an impressive feat given that the 'large' males were less than 40 g! I often said if antechinus were any larger, humans would never have survived in Australia.

Antechinuses are also highly intelligent. I was amazed by their deductive powers and how quickly they learned while working with them in captivity. The females build the most wonderful ball-shaped woven nests

Fig. 6.32. Named for their acrobatic prowess and ability to move nimbly between the ground and tree canopy, the Agile Antechinus (*Antechinus agilis*) is the most commonly encountered small marsupial in the tall forests of southern New South Wales and Victoria. Insects and spiders form the major part of the diet, but small gliders, bats, lizards and fledgling birds are taken down occasionally by this slight, but formidable, predator. Source: Chris Dickman.

for their young – magnificent constructions. In the wild, I had many trap-happy animals that enjoyed my bait (the universal rolled oats, peanut butter and syrup mix) very much. I fondly remember one male who I would catch in multiple traps every day. He would race me to new traps to devour the food (gaining weight at each weighing) before finally leaving with a full bait ball in his mouth to enjoy later!

One of the most fascinating aspects of antechinus biology is their reproductive system. They have a highly synchronised annual 2-week breeding season during which males will attempt to secure as many matings as possible and females may mate with up to seven different males. After this brief, but vigorous, mating season, all males promptly die from stress. This seems to gain great attention when discussed, mostly from women who seem to like the idea. The female antechinus live on to produce and raise their litters, with 98% of litters sired by more than one father. Females are able to store sperm from different males in specially designed isthmic crypts in their oviducts for up to 2 weeks. Thus, they can give birth after the male die-off, meaning that, for a period, there may be no males in the population. I found that female mate choice is important in antechinus, with females choosing males genetically dissimilar to themselves based on scent. Although females do not choose males based on size in captive experiments, in the wild, larger males will sire a higher proportion of young than small males. This suggests that both female mate choice and male competition occurs. Females can have eight or 10 young depending on the number of teats in their pouch and they often produce supernumerary young (an excess number of young higher than the number of teats). This means that most teats are generally occupied; it is survival of the fittest from the start. This is an amazingly complex reproductive system, with many more secrets to unravel.

My work with antechinus in the wild took place in the occasionally well-named Mt Disappointment State Forest. Although it is a beautiful tree fern-filled wet sclerophyll forest, it did frequently conspire to kill my volunteers and me. My volunteers usually consisted of long suffering family members and friends – particularly my poor Mum who luckily is as stubborn and wildlife-loving as I am! The temperatures ranged from boiling hot to below freezing, with weather from torrential blinding snow to baking drought. This meant that large stressed trees frequently pulled out of the ground and could be heard crashing around us. When the winds picked up, the sound of trees crashing and bashing together (or the 'tribal drums', as we called them) signalled that it was time to get out in a hurry. The summer heat also brought the threat of bushfire – in fact, 15 separate fires occurred during my study period. Although some were caused by lightning strikes and an inappropriately named authorised 'controlled' burn (which quickly became completely out of control), the majority were caused by one serial arsonist who was luckily caught before he burnt the entire forest – with us in it.

Due to my field site's proximity to Melbourne, there were many people who also used the forest, from nature-loving campers and legal deer hunters, to illegal hunters. Many of the latter appeared to be moved on by the police in my second year due to possession of illegal semi-automatic weapons, hunting with pig dog packs and their inability to recognise the difference between a large Sambar Deer and a native wombat. Also present were illegal drug cultivators who would regularly sneak into their crops in licence-plate-free cars. I was quite surprised to find one of my field sites trip-wired with plastic explosives to protect an old drug site (which of course I reported to the government). I was also shocked to read in the *Herald Sun* newspaper of a terrorist training cell in the forest, which was busted in my final year by undercover police.

Another entertaining aspect of our fieldwork was leeches. The entire forest was crawling with them. We all had competitions to see who could end the day with the most leeches on their clothes. Mum won with a record 47 leeches on her jacket alone. We both ended up with a leech drinking blood from an eyeball, and my best friend now has the rather unflattering nickname of 'leech-bum', which requires little explanation.

It was all worth it, however, to work with the beautiful wildlife in the area – the Agile and Mainland Dusky Antechinuses, and Bush Rats, which filled my traps. The magnificent parrots and cockatoos flying overhead, and dancing and singing lyrebirds on the forest floor. Biologists **do** suffer for their work, but discovering the secrets of the carnivorous marsupials and the wild lives of the agile antechinus is a life-long passion that continues today … preferably without the leeches!

7

HABITATS AND DISTRIBUTIONS

In previous chapters we saw how carnivorous marsupials came to occur in the Americas and Australasia, and met some of the eccentric and extraordinary characters who brought them to our attention. We turn now to explore which parts of the environment the living carnivorous marsupials occupy, and ask what factors limit these cryptic beasts to the habitats where we now find them. Some of these factors – such as access to food, shelter and mates, or safety from predators – are important at many scales and are recurrent themes in this latter part of the book. In this chapter we focus on how different factors shape where carnivorous marsupials occur at the broad scale of their geographical distributions. In later chapters we examine the resources they use and the constraints they face at finer scales within their distributions.

Species' distributions and habitat preferences

Some collectors and early researchers provided scant information about where their specimens came from, providing locality information as helpful as 'recorded in New South Wales' or 'on the island of New Guinea'. Others were more informative, and certainly more perceptive. These were the kinds of collectors preferred by taxonomists such as Oldfield Thomas, as we saw in Chapter 6, and for good reason. Before his appointment at the Australian Museum, for example, Gerard Krefft took part in an expedition to the lower Murray and Darling Rivers and succeeded in making a very large natural history collection. His notes on the Western Quoll (*Dasyurus geoffroii*) and Red-tailed Phascogale (*Phascogale calura*) provide the only information we now have on where, and how, these species lived in eastern Australia before their untimely demise there in the 19th century. Over time, with repeated observations, patterns begin to appear in the distributions of carnivorous marsupials and we can then begin to define the places where they live.

At a broad scale, carnivorous marsupials occur in all terrestrial environments, and one – the Yapok or Water Opossum (*Chironectes minimus*) – has taken to freshwater habitats throughout its range in the Americas. Generally, more species live both at low rather than high latitudes and altitudes and in areas where the environment is structurally complex. Of course, some species push the limits. The Patagonian Opossum (*Lestodelphys halli*) extends its range to 47°S, whereas the Virginia Opossum (*Didelphis virginiana*) occurs at 50°N in south-western Canada. In Australia, the Mainland Dusky Antechinus (*Antechinus mimetes*) ascends into alpine boulder fields at 2200 m, while altitude records are set in New Guinea at 3900 m by the Speckled Dasyure (*Neophascogale lorentzii*) and in South America at 4500 m by the Pallid Fat-tailed Opossum (*Thylamys pallidior*). As might be expected from ecological theory, more carnivorous marsupials occur in mainland regions than on large islands such as New Guinea or Tasmania; smaller islands in Australasia and the Americas usually support no more than one or two species. Within this broad distributional range, carnivorous marsupials occupy all habitats from dense forest to open desert, and swamps to dry heaths. But what are the factors that dictate where each species will occur?

Fig. 7.1. The Water Opossum (*Chironectes minimus*) is well adapted to thrive in aquatic environments and is the only carnivorous marsupial to be truly semi-aquatic. This allows the species to occupy a large distributional range and exploit the many resources offered by aquatic habitats, such as a broad range of plentiful prey items. Source: Thiago Semedo.

As discussed in Chapter 3, historical and biogeographical events are clearly important determinants of species' broad distributions, but here we will place emphasis on the ecological forces that dictate species' current ranges. This is reasonable because species can only persist in a place because of the prevailing ecological conditions. If the conditions are suitable, they should thrive; if not, they will perish. How long would tree kangaroos have survived on Cape York if they had dispersed there and found no trees? Similarly, how long would the Water Opossum persist if its streams ran dry? We also place emphasis here on patterns of distribution and abundance of carnivorous marsupials that prevailed in the absence of recent human impacts. Human-induced habitat disturbance has distorted the ranges of several species. In Australia in particular, novel predators in the form of the European Red Fox (*Vulpes vulpes*) and feral House Cat (*Felis catus*) have depleted numbers and reduced the ranges of many species. We discuss these effects in detail in Chapter 11.

In taking an ecological perspective, we have divided carnivorous marsupials into two groups: specialists and generalists. Specialists can be viewed as species that excel at exploiting a small range of foods or habitats, often to the extent that they cannot eat alternative foods or use other habitats. This fussiness can severely restrict where such species occur, but it also means they are usually very efficient at exploiting their chosen resources. Generalists, by contrast, employ a 'jack-of-all-trades' strategy and use a variety of food and living resources. They may not exploit all the different foods and habitats that are potentially available to them, but are more flexible in doing so than their specialist counterparts. Indeed, most generalist species show preferences for particular resources, even if they do not use these resources exclusively. Of course, by focusing on carnivorous marsupials, we are talking about a group of animals that already shows some dietary specialisation – there are no vegans here! Specialists and generalists within this group are, nonetheless, quite easy to distinguish.

Specialists

In terms of diet, the Numbat (*Myrmecobius fasciatus*) is perhaps the most specialised carnivorous marsupial. It hunts termites, and turns up its long nose and even more elongate tongue at any other invertebrates it might encounter. Because of its fixation on termites, the distribution of the Numbat is perhaps more easily explained than that of any other carnivorous marsupial. This striking species occurs in the south-west of Western Australia and is closely associated with the distribution of its preferred mound-building termites. In the late 1950s, CSIRO scientist John Calaby recorded 25 species of termites in Numbat droppings. The most frequent prey item was a distinct south-western form of the abundant mound-builder, *Coptotermes acinaciformis*. He also recorded a diverse range of ants, apparently ingested incidentally. Later work has shown *Coptotermes* spp. to be preferred, at least in part, because they are more highly digestible than other termites. If the distribution of the *Coptotermes* is compared with that of the Numbat, the correspondence is good, if not quite perfect. Calaby also reported that fallen hollow logs make critical nesting sites for the Numbat and these are often hollowed out by the termite. Because *Coptotermes* is

Fig. 7.2. Numbats (*Myrmecobius fasciatus*) have a very specialised diet. Their prey, termites, are plentiful, but each adult Numbat must consume up to 20 000 termites a day to survive. Source: Jiri Lochman.

restricted to eucalyptus forests and woodlands, it is no surprise that Numbats attain their highest numbers in these habitats.

There is, however, one exception. Calaby found that Numbats were rare or absent from one widespread forest type – Jarrah (*Eucalyptus marginata*). He thought this was probably because Jarrah wood is hard and resistant to termite attack. Recent harvesting operations have altered the structure of some parts of the Jarrah forest by removing trees and dumping logs and branches on the forest floor. These materials provide nest sites for Numbats and appear to have allowed populations of termites to increase, so allowing the Numbat to extend its range locally into the modified areas. This provides an intriguing, but rare, example of habitat disturbance benefitting a carnivorous marsupial.

When Europeans first arrived, Numbats occupied mulga, scrub and other lightly timbered habitats across much of southern and inland Australia, and they presumably depended upon other, diverse communities of termites. The cause of the subsequent range reduction is uncertain, but the introduced Red Fox is likely to have had a large effect. The scarcity of hollow logs in dry areas perhaps made Numbats conspicuous to the new arrivals, and the open habitats of these areas would have offered poor cover for escaping them.

A different kind of specialisation is shown by the two species of marsupial mole (*Notoryctes* spp.). These bizarre creatures are blind and lack external ears. With a calloused, horny shield to protect the snout, and powerful spade-like claws on the fore-feet, the moles are morphologically specialised for an almost exclusively subterranean way of life. They 'swim' through the soil, back-filling their tunnels as they go. Although widely distributed in central and northern Australia, the lifestyle of the moles essentially confines them to areas with firm, sandy soils and river flats where burrowing is easy. Moles have never been collected in calcareous earth soils, nor from stony desert or red or yellow clay areas, presumably because these soils provide difficult conditions for burrowing.

For reasons unknown, moles also seem to be declining. In *Our Sandhill Country*, published in 1933, Alice Duncan-Kemp noted how colonies of moles, or *ka-ko-ma* as they were called by local Aboriginal people, were found in the red sand dunes of western Queensland. The little burrowers undermined the ground to such an extent that it was dangerous for both horsemen and stock to travel. Now the moles have gone, with no records from Queensland since Duncan-Kemp's observations a century ago. How can we be sure? The lack of records in itself provides a good clue. In addition, moles have not turned up in the diet of expert mole-hunters such as feral House Cats or owls. In parts of the Northern Territory and Western Australia, where moles still occur, a new survey technique has been shown to reliably document their presence. This requires deep trenches to be dug and the sides smoothed to uncover the moles' sand-filled tunnels. Alas, while researchers in Queensland have diligently dug tonnes of sand and squinted hard at the polished trench walls searching for the tell-tale signs of mole activity, none has yet been found.

In the Americas, the Water Opossum shows extraordinary specialisation in both diet and habitat. It is confined largely to tropical and subtropical habitats where freshwater streams and lakes are available. The species depends on fish, frogs and other largely aquatic prey for the bulk of its diet. Because of the extreme dependency of this species on freshwater habitats, it is never found more than a few hundred metres from water. It is able to use slow and fast flowing streams, but prefers those where the waterway is wide and dense vegetation grows on banks. It tends to avoid streams that have heavy loads of sediment, perhaps because these conditions reduce the abundance of aquatic prey or the ability of opossums to detect them. Instead, this species prefers clear water habitats, especially streams with stony beds that provide refuges for prey such as crayfish.

Fig. 7.3. The Lutrine opossums (*Lutreolina* spp.) have a distribution that is closely related to water availability even though they can be found hunting in a range of terrestrial habitats away from water. Source: Gustavo Scrocchi.

Related to the Water Opossum, but not as closely wedded to aquatic habitats, are the Big Lutrine Opossum (*Lutreolina crassicaudata*) and Massoia's Lutrine Opossum (*L. massoia*). These species prefer habitats associated with water, including marshes, swamps and streams, where they hunt prey such as fish and frogs. However, the diets of both species also include terrestrial prey such as small rodents and birds. The Big Lutrine Opossum is even sometimes found in dry pastures or gallery forest in open savanna woodland. Despite forays away from watercourses, both species of lutrine opossum are closely associated with freshwater and, like the Water Opossum, this is the habitat feature that most reliably predicts where they will be found.

Generalists

If food, freshwater and soil type determine where the specialists occur, what are the factors that dictate the ranges and habitats used by the generalist majority of carnivorous marsupials? Insects and small vertebrates occur in all terrestrial habitats in Australasia and the Americas, and so too do the marsupials that hunt them. But, despite this common food source, there is little evidence that the distributions of carnivorous marsupials are generally governed by similar environmental or biotic factors. Let us first consider the smaller predators: the insect-eaters.

For several species, just one or two simple environmental factors seem to dictate the range limits.

Fig. 7.4. The Common Planigale (*Planigale maculata*) is not as restricted by soil or habitat type as its congeners, and occurs in a wide range of habitats in northern and north-eastern Australia. Source: Steve Wilson.

Consider, for example, the planigales. These tiniest of marsupials have an unusual 'squashed' appearance. Their heads are conspicuously flattened and the limbs are attached laterally to the body, giving every impression they have been stepped on. Three species, the Long-tailed Planigale (*Planigale ingrami*), Giles' Planigale (*P. gilesi*) and Narrow-nosed Planigale (*P. tenuirostris*), show a close association with soils that have deep surface cracks, especially grey, brown and red clays, and black earth soils. If these soils are present, the planigales can occupy a variety of habitats from open grassland and savanna to heavily vegetated creek channels. Under certain conditions, such as during prolonged dry periods, cracks will appear even in deep sandy soils. Planigales can exploit these times to move into unusual habitats such as desert sand dunes. Giles' and Narrow-nosed Planigales use soil cracks for foraging, their compressed heads and bodies squeezing into the narrowest of fissures to obtain prey. The deepest reaches of soil cracks also provide secure retreats from larger dunnarts (*Sminthopsis*), snakes and other predators, as well as shelter from the summer heat and winter cold.

Unlike their three diminutive relatives, the slightly larger Common Planigale (*Planigale maculata*) and Papuan Planigale (*P. novaeguineae*) are not restricted by soil type or land form, but are confined mostly to coastal areas receiving more than 1000 mm of rainfall each year. Both species appear to use soil cracks opportunistically, as the high rainfall conditions presumably waterlog the ground and render cracks inaccessible for much of the year. These planigales are exposed to less extreme fluctuations of temperature than their three relatives, which occupy more arid regions, and take shelter in simple nests of bark and grass on the soil surface.

Other small carnivorous marsupials also use soil cracks, albeit not as obsessively as the tiny planigales. These species are largely occupants of dry, inland habitats in Australia, and to a much lesser extent in South America. Thus, the Julia Creek Dunnart (*Sminthopsis douglasi*) is confined to the flat Mitchell Grass Downs country of northern Queensland. This

Fig. 7.5. The Fat-tailed Pseudantechinus (*Pseudantechinus macdonnellensis*) is typically found in rock outcrops, but in some regions the species is closely associated with termite mounds. The pictured animal, not yet fully grown, is pulling an unfortunate dragon lizard from under its shelter to provide its next meal. Source: Jiri Lochman.

species uses crevices in the soil of this treeless and inhospitable region for shelter. Other dunnarts, such as the Fat-tailed Dunnart (*S. crassicaudata*) and Stripe-faced Dunnart (*S. macroura*), occupy a very broad range of habitats, but may use sites with soil cracks as refuges when environmental conditions are poor. Recent doctoral research by Lauren Young and Helen Waudby has effectively quantified the sheltering benefits of soil cracks. Indeed, working in sites with deep soil cracks on the western side of the Simpson Desert, Lauren managed to capture more of these latter two dunnart species in a couple of years than the total number of all dunnart species seen by the authors during their long careers! No South American carnivorous marsupials make exclusive use of soil cracks (or of any particular soil types). However, the Pallid Fat-tailed Opossum uses these landscape features opportunistically if it encounters them.

Rock outcrops, break-aways, jump-ups and other rocky landforms are favoured by several species, often fragmenting their geographical ranges and confining them to sites where these features occur. The Fat-tailed Pseudantechinus (*Pseudantechinus macdonnellensis*) provides a good example. It is restricted to rock outcrops over most of its range, although enterprising animals also make homes in the large termite mounds of the Tanami and western Simpson deserts. This species turned up in Queensland only in 2010 when camera-traps photographed it running about an isolated range of rocky hills on the eastern edge of the Simpson Desert. Four of the five other species of pseudantechinus also show close affinity with rocky landforms. Only Rory's Pseudantechinus (*P. roryi*) dares to be different: venturing from granite outcrops into open woodland and hummock grassland. There is one other rock-dweller of note: the Long-tailed Dunnart (*Sminthopsis longicaudata*). This graceful animal occupies loose scree slopes and rock outcrops throughout its range in central and Western Australia. Small ridges on their hind feet and a remarkable, counter-balancing tail allow this species to clamber nimbly on rock surfaces to access resources and evade would-be predators.

The distributions of several other species of carnivorous marsupials seem to be linked to rainfall. Like the Common and Papuan Planigales, the Brush-tailed Phascogale (*Phascogale tapoatafa*) prefers coastal areas of eastern Australia that receive at least

Fig. 7.6. Clambering easily down a scree slope, this Long-tailed Dunnart (*Sminthopsis longicaudata*) uses ridges on its hind feet to gain grip on the rock surface and its exceptionally long tail to maintain balance. As with other dunnarts, invertebrates form the major part of the diet. Source: Jiri Lochman.

1000 mm of rain a year. The species disappears entirely in areas where annual rainfall is less than half this amount. The distribution of members of the Dusky Antechinus group is similar. Why these species are restricted to wet areas is uncertain. However, Brush-tailed Phascogales and Dusky Antechinuses are among the largest of the primarily insect-eating carnivores and need a lot of food each day to maintain their body sizes. Thus, it is possible that their distributions are limited indirectly by rainfall and more directly by the higher productivity of insects in wetter habitats. Predictive models of the distribution of the Long-nosed Shrew-opossum (*Rhyncholestes raphanurus*) in southern South America support this interpretation by linking the animal's occurrence to high rainfall areas, where prey such as earthworms, and even fungi, are abundant.

For many small carnivorous marsupials, particular habitats, or even components of the habitat, seem to be paramount. In Australia, for example, several species of dunnart (*Sminthopsis* spp.) use grassland habitats dominated by dome-shaped hummocks of spinifex grass (*Triodia* spp.). The spiny, interwoven leaves of these hummocks provide excellent protection from predators and from extreme external temperatures for any creatures brave enough to enter them. The diminutive Ooldea Dunnart (*S. ooldea*) and Lesser Hairy-footed Dunnart (*S. youngsoni*) are small enough to wriggle their way into these hummocks without doing themselves too much damage. By contrast, the stockier Sandhill Dunnart (*S. psammophila*) is too bulky to shimmy through your standard spinifex clump without being impaled. This species seems to prefer old doughnut-shaped hummocks with open centres and gaps between the leaves. While watching these dunnarts gain access to hummocks, wildlife biologist Sue Churchill reported animals springing adeptly from a crouch position over the spiny outer leaves and landing directly in the plant's open centre. Misjudging the jump would presumably be a painful experience!

Many other small carnivorous marsupials also use hummock grasslands, but members of one genus – *Ningaui* – specialise to an impressive degree. These tiny marsupials have small eyes, short-moderate ears and cone-shaped heads that allow them to bore through the closely knit and spiny leaves of the spinifex hummocks without damage. The Pilbara Ningaui (*N. timealeyi*) shows a preference for large, dense hummocks but both the Southern and Wongai Ningauis (*N. yvonneae* and *N. ridei*) use hummocks of any shape or size. Most activity takes place within the protected interior of the hummocks, where all species search for small insects. On occasion, perhaps if disturbed inside the hummock or tempted by the presence of a juicy moth, all species will venture to the outside or to the top of their hummocks. Here, they use their delicate fingers and toes to clasp the spiky leaves and use their semi-prehensile (i.e. capable of grasping) tails to help balance above the pin cushion below them.

Most small carnivorous marsupials in Australia keep their feet firmly on the ground, but there are some notable exceptions. Thus, all species of *Phascogale*, at least nine species of *Antechinus* and the large Spotted-tailed Quoll (*Dasyurus maculatus*) are

Fig. 7.7. Spotted-tailed Quolls (*Dasyurus maculatus*) are excellent climbers and efficient hunters in trees – they dispatch a range of unsuspecting birds and mammals that they extract from hollows or from even higher in the forest canopy. Source: David Nelson.

moderately to highly arboreal (i.e. tree-dwelling). It is likely that several species of *Murexia*, the Speckled Dasyure and perhaps other New Guinea endemics are at least partly arboreal, as are most carnivorous marsupial taxa in Central and South America. For all these species, trees obviously provide crucial breeding sites, nests, hunting surfaces and other resources that animals need to survive. But, stop for just a moment. Hunting above ground is potentially risky. High speed pursuits of would-be prey in the forest canopy increase the chance of a hunter slipping and falling, and probably constrain it to using just its fore-feet to subdue prey; the hind feet would be needed to grip onto the tree. Some species reduce the risk by using gnarly or rough-barked trees that provide an excellent and highly 'grippy' substrate. For example, in research unlikely to be repeated in today's risk-averse environment, 50 years ago PhD student Don Wood clambered 30 m up fig trees and captured Subtropical Antechinus (*A. subtropicus*) in the maze of gnarly branches at that height. More recently, antechinuses have been tracked using radio-transmitters by (university-approved) ground-level researchers. These agile and acrobatic animals were found to use branch complexes at the tops of large old trees; they have also been observed on cameras set just under the canopy. Spotted-tailed Quolls, by contrast, retain their footing above ground by hunting lightly defended bird nests, or by entering and plundering the nest-hollows of hapless possums and gliders.

More generally, at least in the American fauna, there is good evidence that the more time carnivorous marsupials spend above ground, the less carnivorous they are (Fig. 7.8). Thus, tree-dwellers include large amounts of fruit, berries and leaves in their diets: safe types of prey that do not fight or flee when encountered. Highly arboreal marsupials fall into two broad groups. The first comprises relatively large species (100–500 g; e.g. *Caluromys*, *Caluromysiops*, possibly *Glironia*) with diets dominated by fruit and other plant products. Members of the second group tend to be small (less than 100 g; e.g. *Gracilinanus*, *Marmosa*), very agile with long, prehensile tails. These species use lianas to gain access to the upper strata

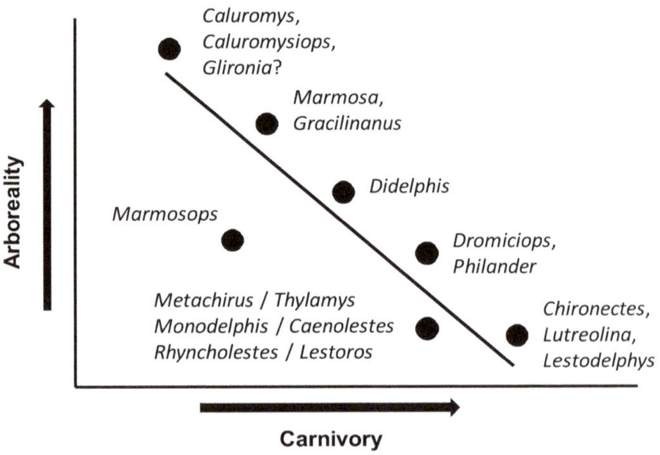

Fig. 7.8. Habitat use, diet and behaviour show strong patterns of association in American marsupials, and together influence where species will occur. In general, the most carnivorous species in genera such as *Chironectes* and *Lestodelphys* are active on the ground, while species in genera such as *Caluromys* that include more fruit and other plant materials in their diets tend to be arboreal. These latter marsupials find their food in trees and hence are restricted to forest and woodland habitats. Source: Figure redrawn, with permission, from Vieira and Astúa de Moraes 2003.

Fig. 7.9. The Northern Quoll (*Dasyurus hallucatus*) has a much reduced distribution since the European settlement of Australia. Today, the species is mostly confined to rocky outcrops that provide protection from feral House Cats and other threats such as frequent fires. Source: Alex Hartshorne.

and canopies of their forest habitats. This group, and also to a lesser extent the larger fruit-eaters, tend to occupy tropical forest habitats where high minimum temperatures and predictable rainfall produce reliable year-round fruit crops. The largely ground-dwelling marsupials likewise fall into two broad groups. The first includes large species (over 300 g; e.g. *Chironectes*, *Lutreolina*, *Metachirus*) that hunt a wide range of small vertebrates and invertebrates, but may also include small amounts of fruit in the diet. The tails of these species are only weakly, or not at all, prehensile. The second group comprises small species (less than 100 g) with either short, weakly prehensile tails (*Monodelphis*) or incrassated (i.e. thickened) tails that can store fat. This group includes *Lestodelphys*, *Rhyncholestes* and most species of *Thylamys* (tail incrassation is slight, at most, in *Caenolestes* and *Lestoros*). The diets of these taxa are dominated by invertebrates or small vertebrates. Because they do not rely on fruit, they are not restricted to forest habitats. Instead, such species occur variably in open shrublands, woodlands and dry, arid environments, often at high altitudes or high latitudes.

Some of the larger American marsupials, such as the Big Lutrine Opossum (weighing up to 900 g), include a lot of vertebrate flesh in their diets, but the largest (six species of *Didelphis*, up to 5000 g) are primarily omnivorous and have some of the broadest diets among the marsupials. Thus, most of the large carnivorous marsupials (over 500 g) occur in Australia and New Guinea. They comprise six species of quoll (*Dasyurus* spp.), the Tasmanian Devil (*Sarcophilus harrisii*) and, until its recent extermination, the Thylacine (*Thylacinus cynocephalus*).

The four species of quoll in Australia have suffered large reductions in their ranges, and now occur patchily around the periphery of continental Australia and in Tasmania. They exploit a variety of timbered habitats, venturing occasionally into lightly wooded country or coastal scrub. There is little evidence that any species is restricted by narrow habitat requirements, food or soil type. However, the Northern Quoll (*D. hallucatus*) shows a marked preference for rock outcrops. Rocky habitats probably provide refuge for this quoll from marauding feral House Cats and other threats. The species used to occur in open savanna habitats too, but has retreated from these areas in recent decades. The most forest-dependent species is the Spotted-tailed Quoll, which is confined largely to tall open and closed forests that receive at least 600 mm of rain each year. These productive forests support a wide range of medium-sized mammals that the quoll includes in its diet, such as possums. This habitat also provides ample opportunity for stalking and pouncing on ground-dwelling prey, such as small wallabies. Old records of this formidable predator suggest it was once common in urban areas. Even now, occasional records turn up on the edge of towns and in low scrub or farmland. Here, animals may be caught raiding chicken coops or harassing domestic animals. Still, most of these records are close to patches of forest, with inland occurrences probably reflecting animals dispersing along river corridors or forested stock routes. Eastern Quolls (*D. viverrinus*) are less reliant on heavy forest, and occur in highest densities in Tasmania where mixed forest and pasture is available.

Alone among the meat-eaters, the Western Quoll (*D. geoffroii*) was once widespread across the arid inland in open grassland and shrubland habitats. We do not know how this species was able to conquer Australia's harshest climatic region when other quolls could not, but the Western Quoll's unusual tolerance of temperature extremes may have played a role. Animals can maintain a normal body temperature when environmental temperatures plummet to 0°C and are able to greatly limit water loss at high temperatures. Like other quolls, Western Quolls do not dig burrows. However, they probably limited their exposure to temperature extremes by using the below-ground warrens of Burrowing Bettongs (*Bettongia lesueur*) and other diggers before these species declined to near extinction.

In New Guinea, the Bronze Quoll (*D. spartacus*) is known from a handful of records in mixed savanna habitat in the Trans-Fly plains, whereas the New Guinea Quoll (*D. albopunctatus*) occurs patchily in a variety of forest habitats over much of the northern part of the country. The ranges of both species occur within zones of high rainfall (over 1500 mm) that probably support abundant small- and mid-sized vertebrate prey. However, whether environmental productivity or other factors influence the distributions of these species remains unclear.

The Thylacine (*Thylacinus cynocephalus*) was likely confined to the open forests and woodlands of northern and eastern Tasmania because these habitats provided the densest populations of wallabies and other prey. Although the wet forests of the south-west are sometimes mooted as the species' last redoubt, especially by hopeful film-makers and the odd crypto-zoologist, its numbers there were probably always low. The Tasmanian Devil (*Sarcophilus harrisii*), with its broad-ranging diet of fresh meat and carrion, occurs throughout the island. Intriguingly, both species were widespread on the Australian mainland until recently, with the Thylacine extending as far north as New Guinea. As we saw in Chapter 5, the extinction of both species on the mainland followed the arrival of the Dingo (*Canis dingo*) ~4000 years ago and coincident improvements in the hunting technology of Indigenous people. The Thylacine was the first to go, ~3000 years ago, followed by the Tasmanian Devil which probably hung on for a little longer. Like the Eastern and Spotted-tailed Quolls, the Devil is in decline in Tasmania due to increasing habitat loss, deaths on roads and, especially, a novel disease that first cripples, and ultimately kills, afflicted animals by the growth of monstrous tumours on the face. Unlike 'Taz', the ravenous and bad tempered Devil who would whirl up a storm in Warner Bros. cartoons (Chapter 4), the tailspin of the real animal is more of an extinction vortex, and is anything but comical (more about this in Chapter 11).

Detective work: how can we deduce habitat preferences?

Building an understanding of the habitats preferred by carnivorous marsupials, and of the factors that shape these preferences, can be difficult and fraught with error. Because most species are small, cryptic and active at night, they are hard to observe in the first place and to keep in view once detected. How then can habitat preferences be uncovered? Researchers can gain an initial idea by setting traps or remotely triggered cameras and recording where animals are detected. Signs of presence can also be used. For example, Spotted-tailed Quolls defecate at 'latrines' that are detectable to the trained eye (and nose), while the faecal remains of Tasmanian Devils resemble short white sausages, unlike those of any animals they occur with. If certain habitats are found to harbour more animals, or signs of animals, than others, it is tempting to think that these habitats might be preferred. Think again. Habitat preferences may change between seasons and years, and thus remain undiscovered in a short-term study. They will often vary over a species' geographical range, so research in one area will not be informative about habitat preferences somewhere else. Chris Dickman (CD) was reminded of this quite starkly in some recent work on the Brush-tailed Mulgara (*Dasycercus blythi*). Having captured many hundreds of mulgaras in the course of long-term research in the eastern Simpson Desert, CD was convinced that sand dunes were the favoured habitat of this species. This misconception was soon exposed on a visit to the western side of the desert in 2013 with CSIRO researchers Jane Addison and Chris Pavey, and PhD student Lauren Young. Here, Brush-tailed Mulgaras were captured on cracking clay amid barren gibber plains with no sand dunes in sight. This was no surprise to the three members of the group who captured mulgaras in this habitat with regularity, but for the fourth member – who was uncharacteristically quiet for the rest of the field trip – it was back to the drawing board.

At a local scale, more insight can be gained about habitat preferences if animals are tracked to determine how they use their environment. Small radio-transmitters are often used. These can be mounted on collars that are attached to animals after capture, with researchers dutifully following the beeps emitted by the transmitters using a hand-held receiver and antenna. This method might show that a study animal lives in forest habitat but never ascends into the trees. It might show, alternatively, that trees are used for 95% of the species' activity time and for nesting. The key advantage of radio-tracking is that it allows researchers to follow their study animals around in real time and get a 'feel' for what they do and where they spend their time. The main disadvantage is that most carnivorous marsupials are active after dark, so unless you happen to be an insomniac, snooping on them can be quite tiring. For example, after radio-tracking Brown Antechinus (*Antechinus stuartii*) in winter in a particularly cold location, Katie Lazenby-Cohen and Andrew Cockburn memorably thanked their volunteers for 'freezing their backsides to rotting stumps in the name of science' in the august journal *Behavioural Ecology and Sociobiology*. Many other researchers swear to having seen small, brightly coloured animals, unicorns, or even low-flying space craft towards the end of long nights of radio-tracking. Bone-tiredness conjures up the most imaginative of hallucinations! Global positioning tags and satellite tags that record animal locations without investigators having to be present

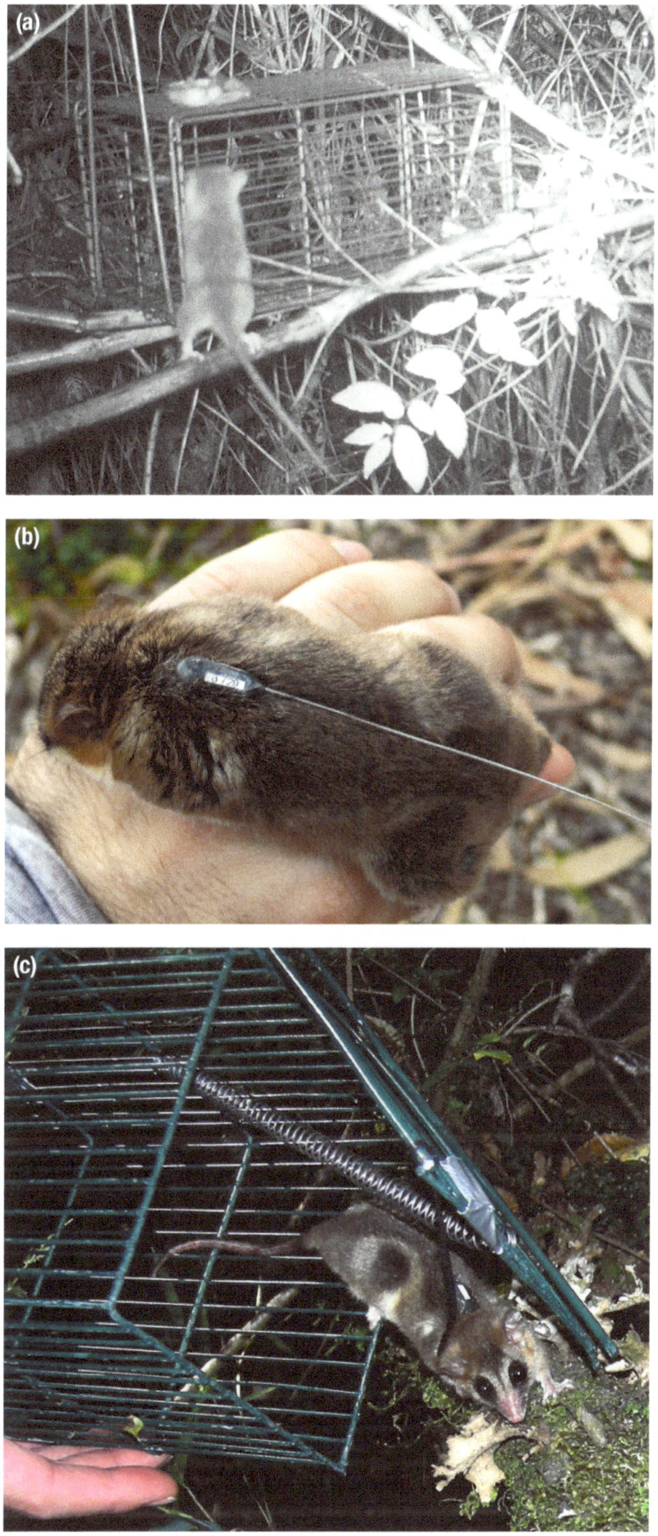

Fig. 7.10. Modern methods of gleaning ecological information have become quite sophisticated. This sequence shows how Monito del Monte (*Dromiciops gliroides*) populations have been studied by: (a) capturing individuals (this animal has been photographed at night investigating a wire cage trap using a remotely triggered camera); (b) attaching radio-tracking transmitters; and (c) releasing them back into the wild, thus allowing animals to be tracked daily for periods up to several weeks. Source: Francisco Fonturbel.

are now becoming available. The smallest are light enough to be carried by large carnivorous marsupials such as quolls and Tasmanian Devils, but have not yet been deployed in any studies.

An array of alternative tracking methods is also available. For example, small bobbins of cotton or nylon thread have been attached to animals using tape or harnesses and, as animals are released after capture, researchers can follow the thread to see where animals have moved. In the first use of this technique, in Brazilian forest, thread-paths of the Brown Four-eyed Opossum (*Metachirus nudicaudatus*) showed this species unexpectedly to be strictly terrestrial. In contrast, thread-paths of the superficially similar Gray Four-eyed Opossum (*Philander opossum*) ascended into the lower forest layers and showed this species to nest in tree cavities 8–10 m above ground. In other tracking techniques, animals have been dusted with fluorescent pigments that fall out of the fur as they walk, allowing researchers to follow their trails later using ultraviolet torchlight, or equipped with small lights (e.g. cyalume light sticks, used as fish lures by anglers) that allow them to be observed directly at night. By revealing those parts of the environment that animals use, and those that they avoid, deeper understanding can be gained of their habitat preferences.

At larger scales, such as the geographical areas that species occupy, all occurrence records can be used to help define their broad habitat use and uncover the factors that underpin this. Dedicated surveys for target species are most useful, because they should reveal where these species occur and where they do not. However, in practice, much insight can be gained

Fig. 7.11. Cotton spool tracking of the mysterious and partly fossorial shrew-opossums (*Caenolestes* spp.) in South America. The cotton spool has been attached to this captured individual which is about to be released. The thread will later be followed by eager researchers to see if the animal stays on the ground, uses burrows or goes in and out of certain types of microhabitat, until the spool runs out or breaks. The data obtained in this way are limited to the length of the spool line (often 50–100 m), but many species' basic movements – especially for some rare New Guinean and South American carnivorous marsupials – are known only as a result of using this method. Source: Reed Ojala-Barbour.

Fig. 7.12. Long known from the western and central sand deserts of Australia, the Hairy-footed Dunnart (*Sminthopsis hirtipes*) was predicted to occur in the Simpson Desert, western Queensland. After intensive sampling, it turned up there in mid-1992. Source: David Nelson.

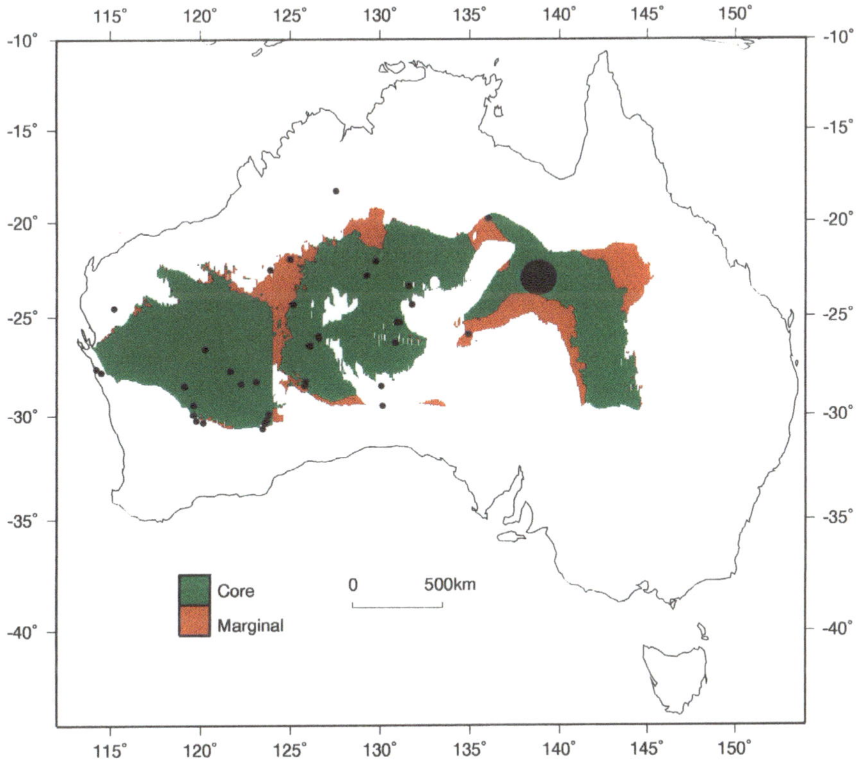

Fig. 7.13. A distribution model for the Hairy-footed Dunnart (*Sminthopsis hirtipes*). In 1991, this species was known from localities in central and western regions of Australia, as shown by the small black dots. The colour shading shows areas with similar temperature and rainfall regimes to those at the species' known localities. Green shading indicates 'core' areas where the climate is most similar to that across the species' known range (using 90% of climate values, and omitting the extremes), and the red shading indicates 'marginal' areas where all climate values are included. The model predicted this dunnart to occur much further east than where it was actually known. It was discovered in 1992 in western Queensland, exactly where it was predicted, as shown by the large black dot. Source: Chris Dickman.

from presence-only records. In the first instance, this is because presence-only records are likely to be available from wide-ranging sources, such as museum records, old reports in the literature, surveys and, for some conspicuous species, reports from members of the public as part of 'citizen science' programs. In several countries, national databases have been constructed to collate all available records, and these can then be interrogated to determine where species occur with some precision.

Such databases provide one further benefit: they can be used to model species' distributions. Why should this be advantageous? So far we have looked at carnivorous marsupials with distributions that can be explained readily by their association with one or two environmental factors. But how reliable is it to assume that an association of this kind is real and not simply a lucky coincidence or correlation? And what can we say about species that have distributions with no obvious associations with environmental factors? One way to resolve these questions is to model the environment experienced by a species in known parts of its range and then prompt the model to predict other areas where the species is likely to be found. If the prediction is correct, and the actual and predicted distributions coincide, we can have some confidence that we have identified the environmental parameters that restrict the species' range. If we are wrong, other parameters can be substituted into the model until we have identified those that provide correct predictions. Sophisticated models allow many different parameters to be entered – such as climate, soils and habitat – and we can evaluate their relative importance. The outputs of these models can serve an important additional function: they predict new places to look for species where they have not yet been recorded.

Let us take an example: the Hairy-footed Dunnart (*Sminthopsis hirtipes*). Survey work over many years revealed this species to occur in sandy habitats in Western Australia and the Northern Territory. However, distribution modelling using rainfall and temperature parameters suggested that it could occur much further east, in sandy habitats of the Simpson Desert in western Queensland where it had never been recorded (Fig. 7.13). The model showed further that this distinctive Hobbit-footed marsupial was known from parts of the continent that receive an average of 246 mm rain a year. Under this regime, it would not be expected to occur in western Queensland all the time, owing to the generally drier conditions there – average rainfall in the Simpson Desert is less than 200 mm a year. However, it might be expected to turn up after years of above-average rain – and this is exactly what happened. Hairy-footed Dunnarts made their first known appearance in the sand dunes of the Simpson Desert in mid-1992 after a year and a half of above-average rainfall (Fig. 7.13). The apparent range expansion – some 500 km to the east of any previous records – provided the first records of this dunnart in Queensland, as well as a compelling validation of the distribution model.

Although models are often used to identify the factors that determine species' distributions, and predict new places where they might occur, they can be useful also at fine, local scales. One local model showed the rare Swamp Antechinus (*Antechinus minimus*) to be associated primarily with treeless heath vegetation at low altitudes in Victoria. Another showed the Monito del Monte (*Dromiciops gliroides*) to be associated with bamboo in Patagonian forest, because bamboo provides animals with nest sites and materials to line the nests. Such fine-scale models can be useful for predicting: how roads, buildings or other development projects will affect the modelled species; which components of the habitat might need to be retained to reduce a development impact; and which habitat attributes should be restored if species populations are to be brought back to an area that has been degraded. Models can also be used to predict where reservoirs of disease occur and the paths that outbreaks might take. For example, some of the more arboreal carnivorous marsupials in South America share their nests with kissing bugs or vampire bugs, and help to sustain high numbers of these blood-sucking insects. The bugs in turn are vectors for the protozoan, *Trypanosoma cruzi*: a parasite that causes the debilitating Chagas disease. This is the affliction that some think caused Charles Darwin's ill health in his later years.

For all their utility, species distribution models sometimes predict local or larger distributions of species that are dramatically wrong. Most models use physical parameters such as climate or soil that can be accessed as readily available data-layers, and thus they predict places where the physical environment is likely to be suitable for their study species. This can be useful. For example, recent work predicted that vast areas of south-eastern Australia are climatically suitable for the Tasmanian Devil. Of course, Devils do not occur on the mainland, but if they were to be liberated there to help recover their failing population in Tasmania, we now have some confidence they would do well. However, the physical environment is not all that species respond to. Carnivorous marsupials, like almost all living things, interact with other species, and models that don't accommodate this can fail horribly. Recognition of the importance of interactions between species was made in a celebrated paper by American ecologist George Evelyn Hutchinson in 1958. For Hutchinson, the potential range – or 'fundamental niche' – of a species was where it could survive and reproduce without interference from other organisms, but its actual range – or 'realised niche' – was usually smaller owing to the presence of other organisms.

What kinds of interactions dictate where species will occur? Predation is an obvious one. Climatically based distribution models suggest that large parts of Australia are suitable for the Western Quoll and Red-tailed Phascogale, and indeed both species used to occur over vast areas of the continental inland (Fig. 7.14). The predatory effects of the larger European Red Fox and feral House Cat, perhaps in concert with loss of habitat, have now reduced both marsupial species to tiny ranges in the south-west of Western Australia (Fig. 7.15). The Northern Quoll also now occurs in a much smaller and more fragmented range than it once occupied. The species is declining due to the advance of the introduced Cane Toad (*Rhinella marina*), which poisons quolls that attempt to eat them, and due to predation from feral House Cats and changes to the fire regime. In these examples, the actual range, or realised niche, may be as little as 10% of the potential range, or fundamental niche, of each species. Novel diseases and pathogens can also reduce species' ranges as effectively as novel predators. Evidence of this can be seen in the massive depletion of Hawaii's endemic birds following the arrival of avian malaria and avian pox in the early 1800s, or the population crashes of Africa's even-toed ungulates following introduction of the rinderpest virus later that century. Such virulent disease organisms are fortunately not known to infect carnivorous marsupials. However, the bizarre, non-viral facial tumour disease that affects Tasmanian Devils is having savage effects on populations of this iconic carnivore.

Other interactions, such as competition, can have more subtle, but still detectable, effects on where species occur. Competition usually occurs between ecologically similar species and is most intense when shared resources are in short supply. In south-eastern New South Wales and Victoria, for example, the Yellow-footed Antechinus (*Antechinus flavipes*) and Agile Antechinus (*A. agilis*) compete for food. Their ranges abut, rather than overlap, probably because neither species gains competitive ascendancy. There is some evidence that their shared distributional boundary ebbs and flows between years, perhaps reflecting a temporary advantage that one species obtains over the other as food availability, or climatic/environmental conditions shift. Other species of *Antechinus* also have abutting ranges suggestive of competition. In south-eastern Queensland, three antechinuses occupy adjacent forested habitats of varying altitude and wetness: the Yellow-footed Antechinus, *A. flavipes* (low altitude, dry), the Buff-footed Antechinus, *A. mysticus* (mid-altitude, wetter) and the Subtropical Antechinus, *A. subtropicus* (high altitude, rainforest). Yet in the absence of the largest of the three species – the Subtropical Antechinus – the Yellow-footed Antechinus extends its range to the coast and even onto offshore islands of south-eastern Queensland, and Buff-footed Antechinuses inhabit the highest altitude rainforest in mid-eastern Queensland.

Many other pairs of carnivorous marsupials have abutting, or 'parapatric', ranges, perhaps indicating that each is preventing the other from expanding into what would be otherwise suitable habitat. Australian examples include the ecologically similar Common Dunnart (*Sminthopsis murina*) and the White-footed Dunnart (*S. leucopus*) in Victoria and southern New South Wales, and Gilbert's Dunnart (*S. gilberti*) with both the Little Long-tailed Dunnart (*S. dolichura*) and Grey-bellied Dunnart (*S. fuliginosus*) in Western Australia. South American examples include the Peruvian Short-tailed Opossum (*Monodelphis peruviana*) and Pygmy Short-tailed Opossum (*M. kunsi*) in southern Peru and north-western Bolivia, and the Chacoan Fat-tailed Opossum (*Thylamys pusillus*) and Dry Chaco Fat-tailed Opossum (*T. pulchellus*) in northern Argentina. Parapatric ranges are not obvious in the carnivorous marsupials of New Guinea. However, better known marsupials in New Guinea, such as the cuscuses (*Phalanger* and

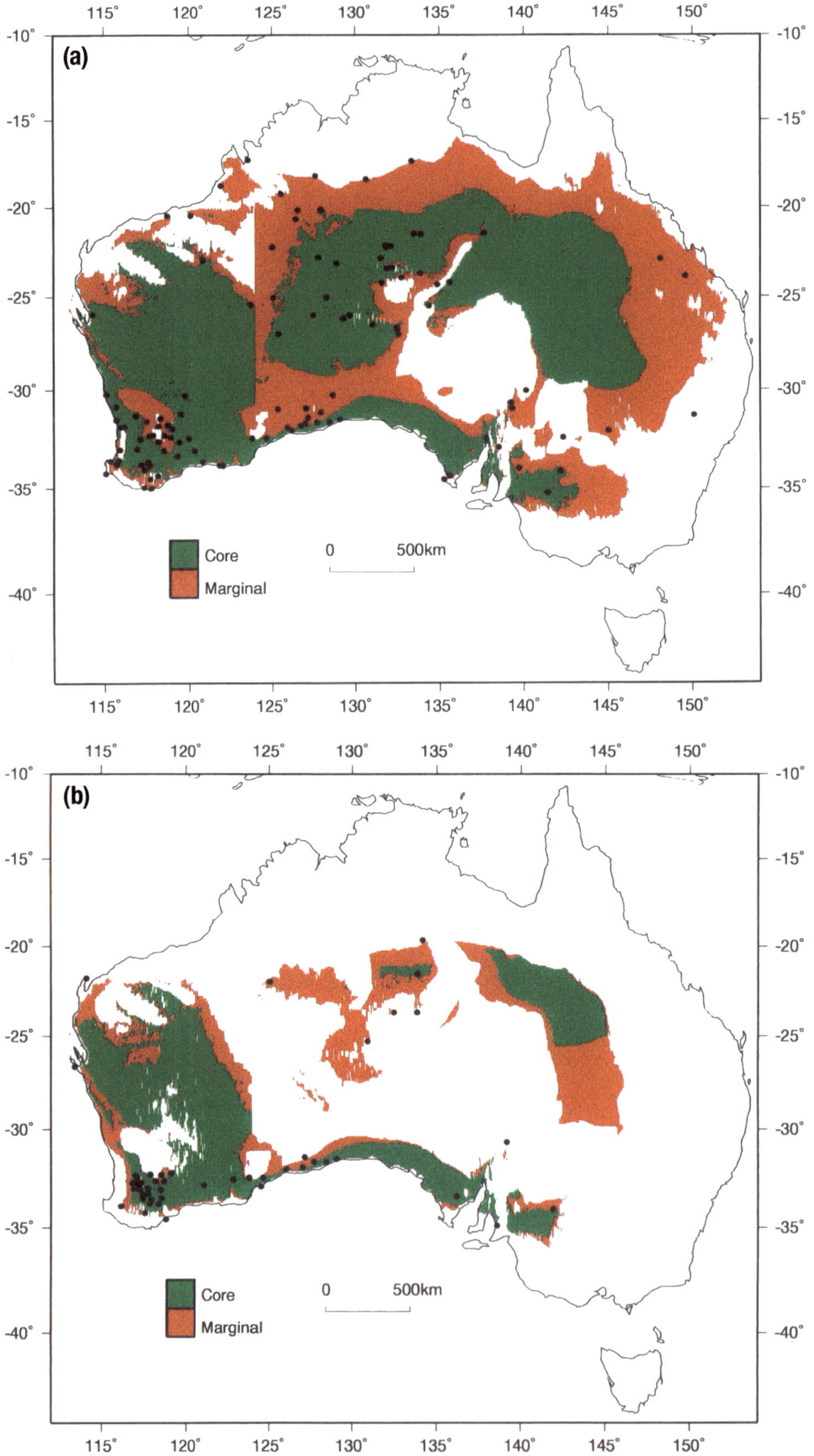

Fig. 7.14. Distribution models for (a) the Western Quoll (*Dasyurus geoffroii*) and (b) Red-tailed Phascogale (*Phascogale calura*) created using records before 1950. Black dots show all known localities from this period, green shading shows 'core' areas where the species were predicted to occur, and red shading shows areas where their occurrence was likely to be 'marginal'. Both species had large distributions that extended from eastern to western regions of Australia.

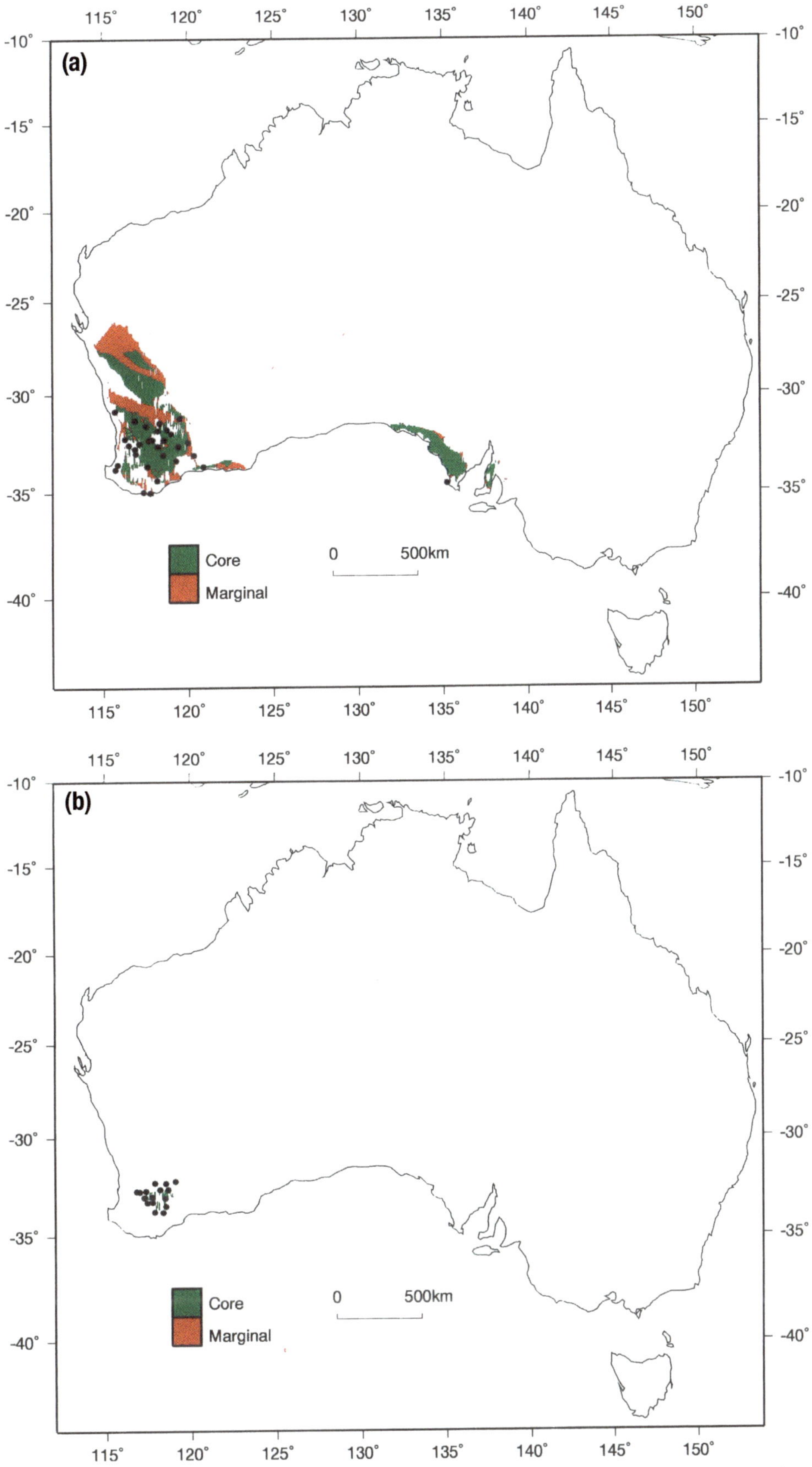

Fig. 7.15. Distribution models for (a) the Western Quoll (*Dasyurus geoffroii*) and (b) Red-tailed Phascogale (*Phascogale calura*) created using records after 1950. Black dots show all known localities from this period, green shading shows 'core' areas where the species were predicted to occur, and red shading shows areas where their occurrence was likely to be 'marginal'. Both species had pan-continental distributions, but are confined now to the south-west of Western Australia.

Fig. 7.16. The Buff-footed Antechinus (*Antechinus mysticus*) (pictured) is apparently squeezed into mid-ecotones due to competition with two larger antechinuses with which it co-occurs: the Yellow-footed Antechinus, *A. flavipes* (in lower altitude dry habitat) and the Subtropical Antechinus, *A. subtropicus* (in higher altitude wet habitat). Source: Gary Cranitch.

Spilocuscus spp.), show marked altitudinal zonation in their ranges that may reflect competition in three dimensions. Some dasyures in the genera *Murexia*, *Myoictis* and *Neophascogale* may turn out to have similarly abutting ranges on the sides of New Guinea's magnificent central mountain range, but this speculation awaits confirmation.

Finding species with abutting ranges is one thing, but confirming that competition causes this pattern is another. To obtain confirmation, one species could be experimentally removed from along parts of its range boundary and, if competition has been operating, the second species should then move in and take over the space occupied previously by the first species. Of course, experiments of this kind are not easy (especially with today's ethical considerations) and, when carried out, they have been on a small scale. Despite this, the results show that competition can have wonderfully subtle effects, as well as the more obvious effects that manifest on the edge of the range.

Consider the Agile Antechinus and Mainland Dusky Antechinus (*Antechinus mimetes*). These two species co-occur in forested habitats of south-eastern Australia, with the larger Dusky Antechinus (65 g) preferring wetter forest and high-altitude heath than its smaller (25 g) relative. Removal of the Agile Antechinus from small-scale experimental plots had very little, if any effect on the larger species, but reciprocal removals produced dramatic results. Thus, Agile Antechinus population sizes increased after the Dusky Antechinus was removed, and animals survived for longer. They descended from the trees where they had previously been most active and started to use complex ground-level habitats that had been the preserve of the Dusky Antechinus. They also shifted their diet, eschewing to some extent the small crunchy beetles that were available to them in the trees and preferring larger and softer-bodied larvae and termites that they extracted from leaf litter on the forest floor. Adding Dusky Antechinus back into the removal plots reversed most of these effects. In other examples, Brown Antechinus (30 g) have been found to banish Common Dunnarts (17 g) from habitats regenerating after fire, and Fat-tailed Dunnarts (12 g) appear to confine smaller planigales (5–8 g) to cracks in the soil where they are safe from encounters with the larger dunnart. In all these cases, as in the great majority of examples of competition between species, the larger species is dominant. The smaller species persists in a realised niche that is considerably smaller than its fundamental niche, whereas the niche space of the larger species is little affected by the smaller competitor.

Interactions between species do not always have such restrictive effects. There are many examples of species that are so dependent on each other they cannot persist alone. Or, if they do, they hang on in smaller areas/lower numbers than they would if their partner organism were present. For example, many plants have special relationships with particular pollinators or seed dispersers, with each partner faring poorly if the other is absent; lichens are formed from tightly symbiotic relationships between fungi and algae or cyanobacteria. Most such interactions involving carnivorous marsupials are loose and opportunistic. Thus, for example, before their disappearance from Australia's arid regions, Western Quolls used to commandeer the excavations of Burrowing Bettongs for shelter. South American mouse opossums often usurp the nests of birds for temporary or longer term shelter. An even tighter relationship has been documented for one cluster of species: the monito del montes. These cryptic marsupials are the only known dispersers of Quintral (*Tristerix corymbosus*) seeds: a mistletoe endemic to the temperate forests of Patagonia. The mistletoe is very fussy about its association with the monito del montes because its seeds also need to pass through the animals' guts before they will germinate. Other research suggests that these marsupials disperse seeds of some 80% of the flora with fleshy fruits within their range. This suggests that monito del montes gain energetic and nutritional benefits that might be expected from their diet, while the flora – especially Quintral – obtain unique and irreplaceable benefits from their interaction with these marsupials. Provided bamboo is present for their shelter, the numbers of monitos are linked strongly with fruit production.

Next generation models are beginning to incorporate both the physical and the biological determinants of species' distributions. As our understanding grows, these new models – ecological niche models, mechanistic niche models, functional trait models, ensemble models, resource selection functions and many other variants – are dramatically improving our ability to predict where species will occur at local and geographical scales, and to incorporate the effects of human disturbance. Importantly, these models are also being used to predict multiple, interacting effects on species such as loss of habitat, climate change, competitors and predators. This information in turn provides more insight into which species are likely to face future peril, and which aspects of the environment we should manage to ensure their continued persistence. Iconic species such as the Koala (*Phascolarctos cinereus*) are currently subject to such research, but, alas, equivalent inquiry on the carnivorous marsupials is lagging. There are excellent opportunities here for enterprising researchers!

Big picture patterns: above the species level

As we get better at identifying the habitats of individual carnivorous marsupials and the factors that define their geographical ranges, we can begin to inquire with more assurance about macroecological, or big picture, patterns in their occurrence. This level of inquiry gets away from the species-by-species

Fig. 7.17. Desert-dwelling carnivorous marsupials in Australia, such as this dunnart (*Sminthopsis*), often have mobile home ranges, matching scant resources that are patchy in both time and space. Source: Steve Wilson.

approach and investigates instead whether there are general, unifying principles that dictate where communities of carnivorous marsupials will occur. As we have seen, there is a broad tendency for more species to occur at low than at high altitudes, at lower rather than higher latitudes, and in environments that provide opportunities for species to segregate in the habitats and resources that they use. However, there are many exceptions to these patterns, and also some counter-intuitive surprises. These large-scale patterns have been studied most extensively in Australia.

A first step in uncovering the big picture is to compile distribution maps for carnivorous marsupials, and then overlay them. This produces a contour, or species density map, that depicts at a broad scale where different numbers of species co-occur. Because the distributions of some species are still incompletely known, this step should ideally use predicted distributions that have been generated using species distribution models, but in practice basic maps drawn from field guides also produce robust results. Species density maps compiled using climatic modelling for both small (less than 500 g) and larger (more than 500 g) species in Australia show very striking and unexpected patterns (Fig. 7.18). Three areas across the Top End of northern Australia are predicted to have no small species, whereas hotspots containing as many as 14–15 species are predicted to occur in parts of arid Australia. Between four and seven species are predicted to co-occur in coastal and subcoastal regions. If the land area of the continent extended no further north than 20° south of the equator, the general decline in species density from 20° to higher latitudes would be similar to big picture patterns for mammals and many other taxa elsewhere in the world. However, because the continental landmass extends north to ~10° 40° south, the decline in numbers of species in the Top End bucks the usual trend. Large carnivorous marsupials also show an unexpected pattern, with three species co-occurring at the highest latitudes, in Tasmania, and none to two on the continental mainland. What is driving these intriguing patterns?

Consider first the extraordinarily high numbers of small, primarily insectivorous marsupials that are predicted to co-occur in Australia's arid regions (Fig. 7.18a). Deserts in other parts of the world usually have only one or two insectivorous mammals, while semi-deserts such as the Kalahari may exceptionally have half a dozen or so such species, perhaps more if ant- or termite-specialists such as the Aardvark (*Orycteropus afer*) and Bat-eared Fox (*Otocyon megalotis*) are included. Yet this is still only half the number of insect-eating species predicted to occur in arid Australia, especially if we add the marsupial moles and specialists such as the Short-beaked Echidna (*Tachyglossus aculeatus*) to the tally.

Several explanations can be advanced to account for these patterns. First, arid habitats may provide excellent opportunities for speciation, so that in effect the deserts act as evolutionary 'hotspots' for the generation of new insectivores. Although arid conditions have prevailed in Australia for only the last few million years, arid cycles have oscillated with more temperate conditions for much of this period. These oscillations would have led to populations of ancestral species fracturing, coalescing and fracturing again, thus providing fertile conditions for speciation. Scincid (skink) and gekkonid (gecko) lizards are also very diverse in arid Australia and perhaps have responded to similar opportunities to radiate in the splendid isolation of the central deserts. If much of the recent evolutionary history of insectivorous marsupials has taken place in arid environments, many species could be expected to be adapted to the harsh conditions. Second, despite their apparent barrenness, resources such as food, shelter, vegetation and other habitat components may be more diverse or

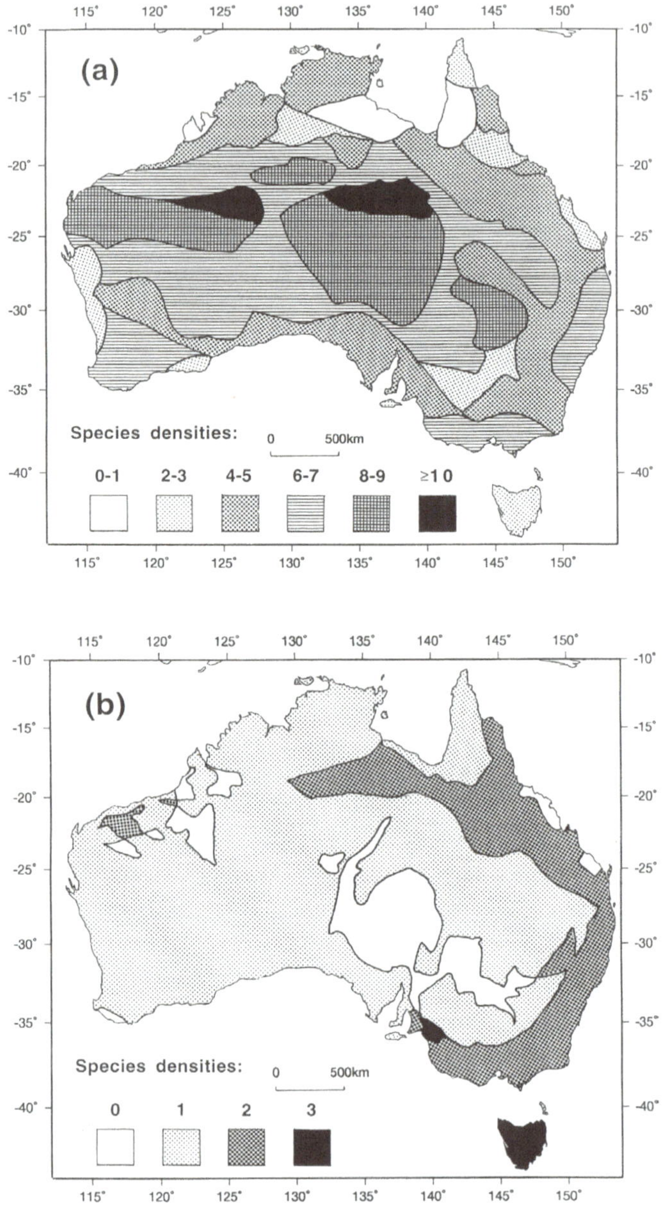

Fig. 7.18. Species density maps of Australian carnivorous marsupials, created by superimposing the predicted distributions of all species except the specialist Numbat (*Myrmecobius fasciatus*) and marsupial moles (*Notoryctes* spp.): (a) The result for species weighing less than 500 g; and (b) the result for species weighing more than 500 g. Source: Chris Dickman.

readily available in arid than in temperate forested regions, allowing more opportunities for niche segregation and hence for more species to co-occur. Spinifex grass, for example, provides homes for ningauis that burrow into mature hummocks, for dunnarts that use the hummock interiors for shelter and hummock edges for foraging, and for mulgaras that dig burrows in the consolidated sand underneath spinifex plants. A third possibility is that, even if resources are not very abundant or diverse in arid Australia, harsh environmental conditions keep populations of resident species at low levels and thus prevent any from becoming dominant. This process (termed 'rarefaction') seems plausible, because populations of desert insectivores seldom achieve densities of more than five animals per hectare – an order of magnitude less than the densities of some of their forest-dwelling relatives. At such low densities, competition is unlikely and no single species would be able to gain ascendancy. Finally, some desert insectivores are very mobile, capable of travelling more than 3 km in a night and over 10 km a week. Such grand movements allow interchange of species across broad inland regions, thus increasing the chance of many species occurring together when local conditions are suitable to support them. By contrast, forest-dwelling insectivores are quite sedentary and seldom move outside their small and prescribed home ranges.

Species density maps provide broad-scale insight into big picture patterns of species occurrence, but they can also be misleading for the unwary. You would be very lucky to go to an area where 10 or 12 species are predicted to co-occur and actually find them all. Indeed, you would be doing well if you encountered even half of the predicted number. The reason is that, within broad areas, different species prefer different habitats and seldom move beyond them. Nonetheless, Australian desert habitats still support more species of small insectivore than temperate habitats, probably because of the diverse opportunities they provide for different species to partition food and shelter. Field surveys show that hummock grasslands sustain an average of 5.3 species of small insectivores, while arid woodland and other desert complex habitats support on average 3.3–3.4 species. By contrast, open forest, rainforest and heathland habitats sustain, on average, less than two insectivore species, with three turning up very rarely.

Counter-intuitive though it appears, forest habitats may be more difficult to partition between species than the apparently simple habitats that characterise arid regions. Why? We speculate that members of the genus *Antechinus* might be involved. Cumulatively, species in the Brown Antechinus and Yellow-footed Antechinus groups are dominant in timbered habitats

Fig. 7.19. Are antechinuses, such as this Yellow-footed Antechinus (*Antechinus flavipes*), part of the 'Goldilocks' group, with just the right body size and habits needed to best exploit forest resources? And, if so, why are otherwise common antechinus species absent from many suitable continuous forest blocks on the east coast of Australia? The animal pictured here is a female with a dry gum leaf in her jaws that she is taking to construct a nest for her young. Source: Chris Dickman.

in eastern, western and southern Australia, are active on the ground and in trees, often present at high densities, and exploit potentially all available microhabitats such as logs, leaf litter, tree surfaces and different strata within the forest. We saw earlier that antechinuses restrict each other's distributions and access to resources via competition, and it seems quite plausible that they could similarly restrict other small insectivores. Indeed, when other species do co-occur with these antechinuses, they are either much larger and probably dominant over their smaller counterparts (e.g. Brush-tailed Phascogale, members of the Dusky Antechinus group), or smaller and more tolerant of habitats disturbed by fire or other disturbances than are the antechinuses (e.g. dunnarts). Why antechinuses might have achieved such ascendancy in Australian forests is open to speculation – perhaps they are the 'Goldilocks' group: they have just the right body size and habits needed to successfully exploit forest insect resources. Both authors have spent many years working on antechinuses, as have many dozens of PhD students and other researchers. After all this effort, we find it curious and perversely satisfying that bold speculation is all we can offer about the functional role of these bolshie, but wonderfully enigmatic beasts in shaping the structure of forest mammal communities!

In contrast to their smaller relatives, large carnivorous marsupials in Australia have never been strongly represented in arid regions. The Western Quoll was widely, but thinly, spread across the continental inland before retracting to its present refuge in the south-west of Western Australia. Only the Northern Quoll now has any presence in arid areas. Wetter forest environments presumably provide the larger carnivores with a more reliable supply of small- and medium-sized vertebrates than arid habitats. If the potential climatic range of the Tasmanian Devil is projected onto the mainland, this species also would be confined to forested areas. Its presence, in addition to that of the Eastern Quoll, would then increase the species density for the south-eastern corner of the continent from two (Fig. 7.18b) to three and, at the same time, resolve to some extent the unexpected trend of more large carnivorous species occurring at higher latitudes.

Although quoll species densities of two or three can be predicted for the Australian mainland, **actual** species densities never exceed one there now. Even before the loss of the Eastern Quoll and the dramatic decline of the Western Quoll, these larger carnivorous marsupials probably overlapped very little. Mainland Eastern Quolls appear to have preferred drier and more open forest than the wetter forest habitats occupied by the Spotted-tailed Quoll, while the northern-most populations of the latter species are restricted to rainforests, which are not used by the Northern Quoll. In addition to this habitat separation, the Spotted-tailed Quoll is more arboreal than either of its smaller relatives, further reducing the chance of the two species encountering each other.

The situation in Tasmania is quite different, with Eastern Quolls, Spotted-tailed Quolls and Tasmanian Devils sometimes occupying the same habitats. Competition appears to be a major force structuring local communities of these species, with complementary separation occurring on the basis of prey size and foraging areas. As with the smaller carnivorous marsupials, competition has probably occurred over long periods of time and has produced regular differences in size, teeth and other trophic characters that serve to minimise overlaps in resource use between the three protagonists. Intriguing recent research further suggests that resource partitioning is maintained by the use of odours. Thus, Spotted-tailed Quolls recognise the odour of Tasmanian Devil faeces, using it as a cue to increase their vigilance and hence reduce their chance of a direct encounter with a Devil. Devils, by contrast, do not avoid and show little response to the odour of either Spotted-tailed Quolls or feral House Cats, suggesting they competitively dominate both of these smaller predators.

The carnivorous marsupials of New Guinea provide a marked contrast to their counterparts in Australia. They are confined largely to lowland or highland tropical forest. Only the endemic Papuan Planigale and Bronze Quoll and two immigrant species from Australia – the Chestnut Dunnart (*Sminthopsis archeri*) and Red-cheeked Dunnart (*S. virginiae*) – occupy savanna and tropical grassland habitats. Overlaying distribution maps suggests that up to 10 species overlap in parts of the rugged Central Cordillera. However, this is almost certainly an overestimate because the distributions of some species are patchy and other species differ in the range of altitudes that they occupy. In practice, no more than three to four species occur at any site. It is uncertain how these species coexist, but differences

Fig. 7.20. Competition among American opossums has rarely been reported, even between large species, such as this Brown Four-eyed Opossum (*Metachirus nudicaudatus*), which often co-occurs with smaller species. Source: Thiago Semedo.

in body size, degree of arboreal activity and possibly differences in diet or time of activity may be important. The high structural complexity of New Guinea's forest environments presumably offers great potential for species to avoid each other. Most species also occur sparsely and there is no indication that one group, such as the antechinuses in Australia, assume dominance. Indeed, some species, such as the Broad-striped Dasyure (*Murexia rothschildi*) and Tate's Three-striped Dasyure (*Myoictis wavicus*) are known from less than a couple of dozen specimens in total.

In the Americas, species densities of opossums increase at low latitudes and reach a maximum in temperate and tropical forests. Although the distributions of carnivorous species are patchy and often incompletely known, the richest latitudinal zone for all opossums lies between 10° and 20° south and contains over 40 species. At least 30–32 species occur in 5° bands to the north and south of this, with diminishing numbers occurring towards the more southerly reaches of South America and also from north of 10° north in Central America. Opossum diversity overall is associated strongly and positively with mean annual temperature, but negatively with the annual temperature range, indicating that opossums occur preferentially in consistently warm forest conditions. Carnivorous marsupials appear to broadly follow these trends, with at least 24–25 species occurring between 10° and 20° south, and 21 species occupying the northernmost parts of South America. Some 26 species occur in the band between 20° and 25° south, but diversity falls rapidly from here as the cooler latitudes are attained.

Species density maps of American carnivorous marsupials show large areas where no species persist and relatively small areas in tropical or lower montane forest where up to eight species co-occur. As in New Guinea, however, many species are rare and poorly known. For example, the Santa Rosa Short-tailed Opossum (*Monodelphis sanctaerosae*) is known from a single recently described specimen from Bolivia. Several others such as Gardner's Short-tailed Opossum (*M. gardneri*), Handley's Short-tailed

Opossum (*M. handleyi*) and the Single-striped Opossum (*M. unistriata*) are known from less than 10 specimens each. In field surveys, it is unusual to find more than four carnivorous species together at a site (see also vignettes present chapter and Chapter 3), although marsupial diversity is often increased by the presence of largely fruit-eating taxa that include insects as a small part of their diet. Where species do co-occur, they usually differ in size and use of different forest strata. Thus, the Brown Four-eyed Opossum (*Metachirus nudicaudatus*) (450 g) is strictly terrestrial, whereas several different species of four-eyed opossum (*Philander* spp.) that commonly co-occur with *Metachirus* may be similar in size but are more active in the lower forest strata. Both *Metachirus* and *Philander* species co-occur with the scansorial (i.e. climbing) and much larger opossums in the genus *Didelphis* (over 1000 g) or the ground-dwelling and much smaller short-tailed opossums (less than 100 g).

Long-term research by Mariana Ferreira and colleagues in Brazil on *Metachirus*, *Philander*, *Didelphis* and two species of small frugivorous or omnivorous opossums found little indication of competition between species, but some evidence of seasonal regulation. Thus, opossum populations were constrained either in the dry season by cool conditions that limited food production or in the wet season by limited food or space. In both seasons, opossums competed for resources within species, their separation in size and habitat use presumably reducing overt conflict between them. In general, competition has been little reported between species of American opossums. It may be that the generally broader diets of these marsupials compared with their Australasian counterparts and access to complex forest environments over much of South America allow them to more effectively avoid each other, and thus reduce competition for food or habitat. Such separation may have arisen due to the action of competition operating in the past over long, evolutionary timeframes. If so, the separation in resource use that we see today would be an example of what American ecologist Joe Connell called the 'ghost of competition past'. Of course, we all know that some ghosts don't lie down. Research by Vanessa Kuhnen in Brazil has recently found evidence of competition between the large (1000 g) Southern Black-eared Opossum (*Didelphis aurita*) and the smaller Brown Four-eyed Opossum, with the latter species increasing in population size if the former is absent. The American carnivorous marsupials are an intriguing and fertile group for further ecological study!

VIGNETTE: A SHORT TALE ABOUT SHORT-TAILED OPOSSUMS

By Emerson M. Vieira
Laboratory of Vertebrate Ecology, Department of Ecology, Universidade de Brasília (UnB), Brazil

During my fieldwork in the altitudinal grasslands of southern Brazil, I was surprised by captures of Southern Short-tailed, or Orange-sided, Opossums (*Monodelphis dimidiata*). My surprise was not due to the fact that we had encountered this didelphid marsupial, but because the captures were so frequent. Southern Short-tailed Opossums had never been commonly captured during my previous field studies on small mammals in central or south-eastern Brazil. I had always thought the main reason for such a pattern was a methodological bias and not actual rarity of these opossums. These animals infrequently enter the kind of live traps that I commonly used: mainly Sherman (metal box) and wire-mesh traps. But, for unknown reasons, Southern Short-tailed Opossums were being captured in southern Brazil with those same traps that did not seem to work well for them in other regions.

Of course, I was excited about capturing tens of these remarkable animals. Despite their small body size, with adults ranging from 25 to ~80 g, they are generalist predators, feeding on several kinds of invertebrates and small vertebrates. But do not be fooled by their size, Southern Short-tailed Opossums are ferocious killers, subjugating even rodents whose size is close to their own by means of a neck bite. Indeed, these marsupials are considered to be the living – albeit down-sized – counterparts of the extinct sabretooth predators, with cranial

Fig. 7.21. A Southern Short-tailed, or Orange-sided, Opossum (*Monodelphis dimidiata*) in southern Brazil. The taxonomy of this large and diverse genus is in urgent need of revision, with this strongly red-flanked individual likely representing a cryptic species that differs from its relatives elsewhere in Brazil and neighbouring countries. Photo: Emerson M. Vieira.

adaptations typical of these predators of the past. These adaptations include a wide gape with canine teeth whose relative size is within the range of those of the sabretooths. Indeed, these canines are relatively the largest among all living marsupial carnivores. I used to tease my students by asking them to imagine the terror felt by insects and small rodents having to live with the possibility of being attacked by a living sabretooth predator suddenly leaping from the grasses at any moment!

Different to most other genera of small Neotropical marsupials, Southern Short-tailed Opossums are basically terrestrial (i.e. ground-dwelling) animals. They are typical inhabitants of the open grasslands of southern South America, in southern Brazil, Argentina and Uruguay. Moreover, also differing from other didelphids, which are mainly nocturnal, these opossums are almost strictly diurnal, being active even during the hottest hours of the day! This is possible because hunting takes place under the protection of the dense cover provided by tall grasses. Besides providing thermal insulation, this cover enables protection against potential predators such as birds of prey and foxes. Furthermore, the 'hot' days in the natural habitat of Southern Short-tailed Opossums are in fact not so hot, with average annual temperatures ranging from just 12 to 20°C. Indeed, during winter the daily lowest temperatures drop to 2.5°C or less. Thus, being diurnal (i.e. day-active) is a useful adaptation displayed also by most small mammals that share open grassy habitats with Southern Short-tailed Opossums.

Reproduction in the Southern Short-tailed Opossum is also quite peculiar. This marsupial was the first mammal for which a 'semelparous' reproductive strategy was suggested. This strategy, also called 'suicidal reproduction', is very rare in mammals, occurring only in a handful of species occurring within two families: the Dasyuridae in Australia and Didelphidae in South America. It consists of a mortality syndrome that comes into effect after only one mating period. Male Southern Short-tailed Opossums reach maturity in spring, when they undergo a sudden increase in body size, and die during the subsequent summer, just after passing through a short seasonal reproductive period. Adult females, in their turn, disappear ~2 months later during early autumn, after rearing their litters of 8–14 young. The semelparous strategy is related to an insectivorous diet, occurrence in higher latitudes, seasonal predictability in abundance of arthropod prey, and a short annual breeding season. All of these conditions are met in Southern Short-tailed Opossum populations.

These small animals emit clicking vocalisations whose function in the wild is still not clear. It is noteworthy, though, that the use of playbacks with these vocalisations in the field promptly attracts birds of prey. This suggests that these marsupials vocalise frequently in the wild and also that birds of prey are able to recognise such sounds, using them as cues for preying upon these carnivorous marsupials.

Short-tailed opossums, as a genus, are quite widespread, occupying a variety of habitats from open grasslands to tropical forests in Central and South America. *Monodelphis* is one of the least studied genera of the Family Didelphidae, notwithstanding its diversity and widespread occurrence. Basic data on ecology, geographic distribution, and taxonomic status of several short-tailed opossum species are still lacking. The systematics of the short-tailed opossums are also poorly known and some species have yet to be fully described. At present, these opossums form the most speciose living genus within the New World marsupials, with at least 23 species currently recognised.

The conservation status of the Southern Short-tailed Opossum (*M. dimidiata*) has recently been changed from Near Threatened to Least Concern in the IUCN Red List, but the conversion of grassland habitat to agriculture is a potential threat for the species. At present, however, there is not enough available information for a satisfactory evaluation of its conservation status. Some ecological aspects of the species, such as its low densities, marked reproductive seasonality, and possible dependence on native grasslands, suggest that this marsupial may be under some level of extinction threat due to expanded human disturbance of their original habitat. Besides conserving its original habitat, one of the priorities for the conservation of Southern Short-tailed Opossums is to increase our scientific knowledge on several biological aspects of these animals. The population of short-tailed opossums that I studied in southern Brazil, for instance, was composed of individuals with reddish belly fur and smaller body size than the typical Southern Short-tailed Opossum. These and other differences lead some mammalogists (including myself) to believe that these animals in fact belong to a distinct species, named *Monodelphis brevicaudis*!

8

HABITS: SECRET LIVES

As we have seen, carnivorous marsupials clearly vary in the habitats they use. Some occur in a single type of habitat, either because they are extraordinarily fussy or because they have been evicted from other habitats by predators or competitors. Others occupy a broad range of habitats and often also have large geographical ranges that encompass these habitats. Despite this variation, one golden rule holds for all species: persistence is possible only if the benefits gained from using a habitat exceed the costs of being there. Benefits include access to food, mates and shelter, as well as all the resources that are needed for animals to maintain their social structures and repertoire of behaviours. Costs include coping with the vicissitudes of weather and longer term climate, the presence of predators, competitors, pathogens and other enemies. We defer discussion of the foraging and reproductive habits of carnivorous marsupials to Chapters 9 and 10, and focus here on the other ways that these animals lead their lives – secret lives – in their chosen habitats. As in Chapter 7, we consider the specialists and generalists separately.

Specialists

As discussed in Chapter 7, because of its dependence on termites, the Numbat (*Myrmecobius fasciatus*) is largely restricted to eucalyptus forests and woodlands that support reliable populations of these prey. Although mound-building termites occur in treeless areas such as the Nullarbor and Tanami, Numbats seldom breach these mounds and instead search for termites that are active in shallow, sub-surface galleries near their nest. Sticks and small branches are all that need to be turned over to expose the termites. Numbats track the activity of their prey, being most active around the middle of the day in winter when termites have started to move, and more active after dawn and before dusk in summer when the midday heat drives termites deeper underground.

Daytime activity is unusual in Australian carnivorous marsupials, and the unique banding pattern of the Numbat almost certainly provides a measure of camouflage from avian predators such as the Brown Goshawk (*Accipiter fasciatus*) and Little Eagle (*Hieraaetus morphnoides*). Numbats have highly attuned senses, allowing them to detect not just their subterranean prey, but also the appearance of predators and other native enemies. If escape is needed, Numbats will flee to a log or low tree hollow. The animals use these shelters at night when at rest. They can also dig shallow burrows in the soil, but

Fig. 8.1. We know that the marsupial moles (*Notoryctes* spp.) possess anatomical specialisations to aid their hearing and sense of smell, that they have a low metabolic rate and are very likely to be solitary. However, most other aspects of the lives of these carnivorous marsupials remain enigmatic. Source: Jiri Lochman.

ready-made nests within logs are preferred. Diligent radio-tracking by ecologist Tony Friend has shown that many shelters are used within each Numbat's home range, but one or two nests are frequented most often. During cool conditions, nests are ~5°C warmer than the external environment, allowing animals to save energy on keeping warm. However, if food is scarce, Numbats conserve energy using a further mechanism: torpor. This is a physiological state in which movement and other energetic activities are reduced. It is relatively short in duration and shallow compared with hibernation, which is a more profound state of dormancy. No carnivorous marsupials are known to hibernate; indeed, among the marsupials as a whole, only one – the nearly carnivorous Mountain Pygmy Possum (*Burramys parvus*) – enters hibernation on a regular, seasonal basis. During torpor, Numbats reduce their metabolic rate and drop their body temperature to as low as 19.1°C, keeping it at this level for up to 15 hours. Careful observations by ecophysiologists Christine Cooper and Phil Withers of wild Numbats carrying temperature loggers indicated that torpor could reduce the daily energy expenditure of an animal in a burrow by an impressive 13–42% over 12.5 hours.

In contrast to the relative wealth of information about Numbats, the secret lives of the most enigmatic of all marsupials – the marsupial moles (*Notoryctes* spp.) – remain just that: utterly secret. Anatomical studies show that the moles have large olfactory bulbs, indicating a good sense of smell. Animals also have a well-developed vomeronasal organ in the roof of the mouth. In other mammals, the vomeronasal organ detects pheromones that are used in communication between individuals, and we might speculate that it serves a similar function in marsupial moles. Both mole species also possess anatomical specialisations of the middle ear that may allow detection of low frequency sound. It is possible that subterranean rumbles allow individuals to contact each other but, contrarily, the only sounds that animals have been recorded to make are high pitched squeaks. As moles

Fig. 8.2. The mottled black, grey and white fur of the Water Opossum (*Chironectes minimus*) likely provides effective protection from aerial predators while it is foraging in the aquatic reaches of its habitat. Source: Thiago Semedo.

have no external ears, the significance of these vocalisations is unclear. Observations of a captive individual Kakarratul (*Notoryctes caurinus*) by Phil Withers and colleagues showed the animal to have a low metabolic rate and low, but highly variable, body temperature (range 22.7–30.8°C), suggestive of torpor. These adaptations likely allow energy to be conserved when animals are inactive, during periods when food is scarce, or conditions are otherwise unfavourable. Limited field observations suggest that marsupial moles are solitary, do not construct nests, and do not have defined home ranges. Scant though they are, these snippets of information are tantalising and have been hard won over decades of exploration. Uncovering the secret lives of marsupial moles remains an elusive goal, if not **the** Holy Grail, for many marsupial biologists!

The largely aquatic diet of the Water Opossum (*Chironectes minimus*) and, to a lesser extent, the two species of lutrine opossum (*Lutreolina* spp.), confines these species to flowing or standing bodies of fresh water and associated habitats. All species are proficient swimmers, hunting for prey in shallow water and in the dense vegetation fringing the banks of watercourses. The Water Opossum appears to be unwary while in the water, perhaps reflecting the lack of predators that can pursue it there, or perhaps the effectiveness of the camouflage provided by its strikingly mottled coat. It can be readily detected by researchers if it is present, often by the bright reflection from its eyes when a torch is shone towards it. On land, both the Water Opossum and lutrine opossums are sensitive to noise and will either run for dense cover or take to the water if this refuge is close. All species swim rapidly if fleeing from danger, and even dive briefly under the water surface if pursued. The Water Opossum can achieve a swimming speed of 2.6 km/h, and the Big Lutrine Opossum (*Lutreolina crassicaudata*) 1.4 km/h. Cornered animals will hiss at attackers or, in the case of the Water Opossum, utter sharp staccato barks. All three species are largely solitary in the wild, and construct nests lined with dry grass and leaves. These are used for rest by both sexes during the day and by females when leaving their young at the end of pouch life but before the young become independent. Although the Water Opossum and both lutrine opossums are good climbers, their nests are most often constructed at or near ground level, with the entrances concealed to reduce the risk of them being discovered by larger predators. Unlike the Numbat and the marsupial moles, these three New World opossums do not appear to use torpor to conserve energy. This may reflect the relatively large size of these opossums (adults exceed 500 g), or perhaps the reliable and energy-rich sources of food that they exploit.

Generalists

On the benefit side of the ledger, habitats provide carnivorous marsupials with nests, burrows or other shelters that provide short or longer term sanctuary from external threats. These retreats are extraordinarily diverse and, in the generalist species, range from simple pre-formed structures, such as tree hollows, to more elaborately constructed nests. Because the shelters and associated habits of desert-dwelling carnivorous marsupials differ in many respects from those of their forest and woodland dwelling counterparts, we will discuss them separately.

Desert-dwellers

In arid environments, carnivorous marsupials typically use cracks in the soil, spaces in living or dead vegetation, fissures in rocks or even structures such as termite mounds. One species, the Red-tailed Phascogale (*Phascogale calura*), is unusual in nesting above ground in trees in its current range in the south-west of Western Australia. It most likely nested above ground in its former range in central arid areas too. With a few exceptions, all of the arid-dwelling species are solitary and share their shelters only during breeding or when females are nursing dependent young. Perhaps because these animals forage and den alone, and also because they often move widely during their activities, the shelters are simple and unadorned and may be used only once or twice before animals desert them. Consider the Lesser Hairy-footed Dunnart (*Sminthopsis youngsoni*) in central Australia. Using tiny radio-tags (less than 0.5 g) glued to the nape, these diminutive (9 g) beasts have been recorded moving 2–3 km a night and using up to four temporary shelters while in transit – perhaps to catch their breath. The animals sometimes return to the same daytime shelter over consecutive days, but more usually shift to new shelters each time they come to rest near the end of the night. These dunnarts never dig burrows for themselves, but usurp the burrows of rodents, spiders or scorpions, or time-share with day-active animals such as dragon lizards. Only when females have young will they return consistently to a shelter. They usually modify a pre-existing hole in the ground, such as a dragon burrow, and redecorate it with dry leaves or grass before moving in.

Although radio-tracking has revealed extraordinary insights into the private lives of dunnarts and other carnivorous marsupials, the tiny radio-tags that must be used, as well as having short battery lives (often just a few days) can also sometimes give researchers false signals due to 'signal-bounce'. This can make it appear that an animal is in a different place than where it really is. On more than one occasion, researchers have excavated an animal from a burrow while thinking that it was elsewhere, leading to great surprise and some consternation by both parties (Fig. 8.3)! Snooper-scopes, fibre-optic scopes and other reverse periscope devices are available to peek into animal burrows, but have been little-used to spy upon carnivorous marsupials. Our own, admittedly limited, attempts to do this have been frustrated by animals waking up and biting the ends of the scopes, but there is little doubt that intimate insights could be gleaned about the private lives of many carnivorous marsupials if the technology was deployed more widely.

Fig. 8.3. This Lesser Hairy-footed Dunnart (*Sminthopsis youngsoni*) was excavated accidentally at the bottom of a scorpion burrow in the Simpson Desert, central Australia. The burrows of large desert scorpions descend in a characteristic corkscrew shape to depths of ~10 cm. The spiral pattern of the burrow can be seen above the dunnart. The tiny radio-tag, with its long antenna made from piano wire, can be seen attached between the dunnart's shoulders. Enterprising dunnarts find scorpion burrows to provide a snug fit and probably, on occasion at least, a ready meal of the scorpion itself. Source: Chris Dickman.

A small number of arid-dwelling carnivorous marsupials dig their own burrows, and these can be large and impressive structures. Occurring in the same areas as the Lesser Hairy-footed Dunnart, the Brush-tailed Mulgara (*Dasycercus blythi*) constructs burrow systems with up to six entrance and exit holes that can cover an area on the ground surface of ~3 m^2. The holes are connected by tunnels that descend to 0.5–0.6 m. Some of the tunnels end blindly, and contain one or more enlarged chambers that animals line with dry grass or acacia leaves. It is not clear whether these burrow systems are excavated by one or both sexes, but males and females have been found in them living solitarily or, occasionally, in pairs. Surprisingly, radio-tracking by ecologist Gerhard Körtner and colleagues showed that burrows were used on average for only 3.2 days before animals moved on, with up to 15 burrows used over the course of 54 days. This turnover in burrow use begs the question: do animals do their own digging every time they move, which would be energetically costly, or do they take over the burrows of other desert denizens? There is some precedent for the latter strategy. The Kowari (*Dasyuroides byrnei*) exploits vacated burrows of the Bilby (*Macrotis lagotis*) in gibber soil country in western Queensland, and the Patagonian Opossum (*Lestodelphis halli*) in southern Argentina readily takes over the burrows of tuco-tuco rodents (*Ctenomys* spp.). The Pallid Fat-tailed Opossum (*Thylamys pallidior*) in the Monte Desert occupies the nests of birds and burrows of lizards after it has first eaten the eggs or young of the owner. Brush-tailed Mulgaras also commandeer the burrows of fellow diggers, but they are undoubtedly good diggers in their own right. While excavating the burrow system of a Brush-tailed Mulgara, Pat Woolley – a world authority on dasyurid marsupials – memorably described the sound of an animal tunnelling frantically ahead of her.

Arid environments are usually characterised by extremes of temperature, prolonged periods of water shortage and occasional deluges when droughts break. They also have sparse vegetation cover, thus exposing small animals to the risk of predation from larger predators. Shelters provide carnivorous marsupials a means to cope with these varied threats to their existence. Working in open cracking clay landscapes for her doctoral research, Helen Waudby showed that both the Fat-tailed Dunnart (*Sminthopsis crassicaudata*) and Stripe-faced Dunnart (*S. macroura*) made extensive use of soil cracks in both the summer and winter, with animals descending to depths of 12–20 cm when at rest by day. During summer, the cracks were up to 19°C cooler by day than the ground surface, where temperatures were often above 50°C for several hours. In winter, the cracks were 13–14°C cooler than the ground surface by day, but 6.0–7.9°C warmer during the coldest part of the night. In addition to providing a welcome temperature buffer for the dunnarts in both seasons, the relative humidity of soil cracks was higher than on the ground surface during summer and relatively stable between seasons, thus helping animals maintain their water balance. These physiological benefits to animals increase with depth. The deep burrows of Brush-tailed Mulgaras and Kowaris provide highly stable conditions year round, as do the deep crevices used by rock-dwellers such as the Fat-tailed Pseudantechinus (*Pseudantechinus macdonnellensis*).

Despite their obvious importance, burrows and other types of shelter are not always used passively by carnivorous marsupials, and they are not the only means of escaping climatic extremes. For example, although almost all the generalist arid-dwelling carnivorous marsupials are nocturnal (only the Patagonian Opossum and Kaluta (*Dasykaluta rosamondae*) are active by day as well as by night), many will lie about in or near the entrance of a shelter and bask in the early morning or late afternoon sun to warm up. Such basking behaviour is common in larger species such as the mulgaras and the Fat-tailed Pseudantechinus. Although it has been little reported in smaller species, this behaviour may simply have been overlooked. Burrows that are used for longer periods, such as when animals are making nests for their young, will often be insulated by dry grass and other vegetation. Enterprising dunnarts and fat-tailed opossums have even been caught lining their nest chambers with feathers they have removed from bird carcasses. There is nothing like feather-down to keep the cold at bay! Animals may also snuggle together to maintain warmth. The record is held by the Pallid Fat-tailed Opossum, where up to 19 animals have been found huddling together (see vignette, Chapter 9). In his pioneering studies on the Fat-tailed Dunnart, ecologist Steve Morton found that groups of up to eight animals would huddle together. Up to 70% of the animals in his study population shared nests in winter compared with less than 40% at other times.

Two further intriguing observations arise from Morton's research. First, aggregations of nest-sharing animals were found to be so random that they even included erstwhile prey items such as the occasional House Mouse (*Mus musculus*). Nest sharing between species is very rare. Although mice in this study ran the risk of having a bed and being breakfast, the mutual need for warmth was presumably great enough for both the dunnarts and mice to huddle harmoniously. Second, Steve's methods of catching animals were inventive and unusual: he overturned rocks under which animals were sheltering or detected them by spotlighting, in both situations catching animals by hand. As virtually all other studies of carnivorous marsupials have used traps to capture animals that are out and about, rather than catching them while at rest, is it possible that we have under-estimated the extent to which they socialise? At least among arid-dwelling carnivorous marsupials, we have little evidence that nest-sharing is widespread. Outside of breeding periods, it has not been recorded in 10 species of opossums, dunnarts, planigales or ningauis that have been subject to intensive radio-tracking. The Brush-tailed Mulgara has been recorded on occasion to nest-share, and both sexes will bunk down together amicably in captivity. The Kowari is equally tolerant in captivity, as is the relatively placid Fat-tailed Dunnart. However, most other desert-dwellers fight ferociously if placed together in a cage, suggesting that nest-sharing among these feisty beasts is indeed the exception rather than the rule.

As well as extremes in temperature, arid environments typically also experience large

Fig. 8.4. Unlike most carnivorous marsupials of the arid regions in Australia, the Kaluta (*Dasykaluta rosamondae*) is active at different times both day and night. The animal depicted is devouring a scorpion. Source: Jiri Lochman.

fluctuations in the availability of free water. Long, dry periods can be punctuated by flooding rains that abruptly transform previously barren and wind-blown landscapes into lush green meadows. Paradoxically, it is not the enduring droughts that spell trouble for many carnivorous marsupials, but rather the flooding rains. Insects, many other invertebrates and small vertebrates are rich in protein but also have a high content of free water (60–80%). This means that water intake and turnover are relatively high in carnivorous marsupials, especially in comparison with similar-sized seed-eating mammals, even if no free water is available for drinking. If conditions are unusually dry, animals can increase their water intake by searching for especially juicy prey such as spiders or cockroaches and eschewing prey such as beetles that contain less water. In addition to their high intake of water in food, most carnivorous marsupials have highly efficient kidneys and can produce concentrated urine that assists further with water conservation. This ability is most pronounced in the arid-dwelling Chacoan Fat-tailed Opossum (*Thylamys pusillus*), which produces the most concentrated urine known among the marsupials. Indeed, this South American creature's urine is so concentrated that a spilled drop can burn through the clothing of an unwary field researcher! Surviving dry periods, then, is not an issue. Surviving floods is the problem, especially in Australia's arid regions.

Arid Australian landscapes are flat to undulating over large areas. Drainage is poorly developed due to the gentle topography, allowing water to pool over vast areas for periods of days to months after heavy rainfall. The low lying parts of the landscape are also places where soils tend to be more consolidated and thus where most digging animals prefer to excavate their burrows. Because these are the burrows that many carnivorous marsupials use for daytime shelters, the animals may drown in flood waters. Alternatively, deprived of effective shelter, they may have poor survival prospects due to increased exposure to extreme temperatures, predators and other threats. In support of this scenario, several authors have reported seeing sad and soggy planigales and dunnarts clinging to vegetation in higher and drier parts of the landscape after heavy rains. Moreover, depleted populations of these species and others, such as Kultarr (*Antechinomys laniger*), have been observed in the months and years following a heavy rainfall event. However, not all carnivorous marsupials are similarly affected. Proficient diggers such as mulgaras and the Kowari stopper their burrow entrances to reduce the inflow of water, or simply dig new burrows in drier places if they do get flooded. These problems are generally much less concerning for American carnivorous marsupials. The arid landscapes where they occur, such as the Monte Desert, tend to have greater topographic relief that are less prone to local flooding than the undulating landscapes of arid Australia.

Before we leave the subject of burrows and other shelters, it is worth asking why arid-dwelling marsupials show such little fidelity to their shelter sites and occupy what have been termed either 'drifting home ranges' or 'serially nomadic ranges'. Several studies suggest that invertebrate food resources are unreliable and shift across the desert landscape over time. Locusts provide a good example of this. Beetles, cockroaches, termites and other potential prey also show spatial patchiness over time as they respond, in turn, to changing environmental conditions. The movements of carnivorous marsupials might then arise if they track these ever-shifting resources. There is some evidence supporting this idea. On the one hand, species that can access relatively stable sources of food, such as the rock-dwelling Fat-tailed Pseudantechinus, tend to be more sedentary than their relatives in open desert landscapes, and occupy discrete home ranges with more regularly used shelters. On the other, some species of carnivorous marsupials are known to move several kilometres towards sites that have received recent local rain, presumably to gain access to newly regenerating food resources. They must be hungry to undertake such long and potentially risky journeys, if not at the beginning, then certainly when they arrive. How do animals 'know' that food is likely to be found over the next sand dune or at even more remote locations? We have no clue, but speculate that wind-blown odours, differences in barometric pressure, or gradients in humidity may play a role (see vignette, present chapter). Such directed dispersal remains a marvellous mystery that begs resolution!

Although animals may respond to spatial pulses in food resources, there are other reasons why vagrancy might arise. Because small carnivorous marsupials are constantly borrowing the burrows of other desert creatures, stays of short tenure may reduce the risks of running into the original burrow builders. Such risks can be severe: we have excavated the burrow of a desert bird-eating spider (*Selenocosmia* sp.) and found its nest chamber lined with the desiccated corpse of an unfortunate dunnart! However, animals remain highly mobile even if they encounter burrows that have been long-disused and the risk of a run-in with the original owner is negligible. An explanation for this must therefore acknowledge some other cost to staying long term in a shelter. One possibility is that long stays allow time for parasites, such as fleas, to build up and complete their life cycles. Another is that repeated use of a single shelter allows odours or other cues of an animal's presence to accumulate, which then act as signposts for larger predators. There is, in fact, compelling evidence for this last possibility: voracious reptilian predators such as Sand Goannas (*Varanus gouldii*) frequently dig up the burrows of spiders and scorpions and consume the owners, and detect prey using their excellent sense of smell. Research by doctoral student Nicole Hills showed that goannas quickly dig up artificial burrows that have been laced with small mammal odour, but show no interest in burrows containing either no odour or the odours of non-prey species. More research is needed to resolve the relative importance of these competing ideas.

Despite the critical importance of burrows and other shelters, arid-dwelling carnivorous marsupials cope with the extreme conditions of their desert habitats in a variety of other ways. Physiologically, many species have well-developed abilities to tolerate extremes of temperature. In cold conditions, maintenance of the body temperature in Australian carnivorous marsupials depends on access to a supply of food and on an extraordinary ability to produce heat by elevating the metabolic rate eight to nine times above normal resting level. Early experiments showed the Kowari using this ability to maintain its normal body temperature for at least four hours at temperatures as low as –10°C. At high ambient temperatures, all species can lose heat by panting; dunnarts can increase heat loss still further by licking their limbs and effectively using their saliva as 'sweat' to cool down.

Shortage of food is another risk of desert life, and one likely to be faced by most species at some time or another. As we saw earlier, many desert-dwellers are

Fig. 8.5. Like many of the arid-dwelling carnivorous marsupials, some forest dwellers also store fat in the tail. In this case, the Monito del Monte (*Dromiciops gliroides*) relies on fat storage to sustain it when food resources are limited, or during intermittent bouts of torpor. Source: Agustin Iriarte.

exceptionally mobile and can move long distances to exploit remote food sources. However, this strategy carries risks, and is most likely to be used if conditions are dire. Two other strategies are available. In the first instance, many species store fat in the tail as well as in the body. In the appropriately, if prosaically, named Fat-tailed Dunnart up to a third of animals' fat reserves can be located in the tail. Because this fuels only ~20% of an individual's daily energy requirement, fat storage in the tail is likely to provide only a short-term solution to an acute food shortage. Despite this, ~20 species of arid-dwelling dasyurid marsupials store fat in the tail, as do arid-dwelling fat-tailed opossums, the Patagonian Opossum and Long-nosed Shrew-opossum (*Rhyncholestes raphanurus*).

Although tail fat-storage on its own may play only a modest role in helping animals through an acute energy crisis, its value is boosted when used together with a second strategy: torpor. Almost all arid-dwelling carnivorous marsupials that have been studied to date use this strategy, suggesting that it is of fundamental importance in conquering unpredictable desert environments. As we saw earlier, torpor is typically used daily, and is characterised by controlled reductions of metabolic rate, body temperature and other physiological functions. Consider the Stripe-faced Dunnart. This species can rapidly reduce its body temperature from 35°C to just 11.3°C, and is most likely to do so in the second half of the night after a bout of foraging. Torpor can last over a day, but more usually concludes after several hours, with animals re-warming when temperatures outside their retreats begin to heat up during the day. Stripe-faced Dunnarts, like many other arid-dwelling species, can re-warm by burning energy or passively basking in the sun near the entrance to their burrow or crack in the soil. Entering torpor reduces the metabolic rate of Stripe-faced Dunnarts to just 30% of their basal metabolic rate when they are at their normal body temperature, and also reduces water loss by 50%. Long-term research by Fritz Geiser and colleagues has shown that smaller desert-dwellers can drop their body temperatures to lower levels than their larger relatives, and are also more likely to enter deeper and more frequent bouts of torpor. Take the Kowari as an example. This species co-occurs with Stripe-faced Dunnarts, but does not always enter torpor; the body temperature falls to 25.3°C and lasts for a maximum of only 4 hours.

Torpor has often been seen as an emergency measure that is taken only when animals have no recourse but to switch the thermostat to a lower level. In reality, however, it seems more like a regular part of life for small marsupials and is elicited under a wide range of conditions. The Pallid Fat-tailed Opossum enters torpor if the ambient temperature falls to 15°C or if its body temperature falls to 29°C. By contrast, Fat-tailed Dunnarts increase their use of torpor if the supply of food becomes unpredictable. Stripe-faced Dunnarts enter torpor during cold conditions, when hungry, or if the arrival of the next meal is uncertain. Despite its prevalence, however, torpor may not be inevitable. In the Patagonian Opossum, torpor is expressed when cold or when food is short. Animals drop their body temperature to a remarkably low 7.7°C and can remain torpid for over 42 hours. Yet, the Long-nosed Shrew-opossum – which occurs almost as far south as the Patagonian Opossum – has not been recorded to use torpor. This enigmatic beast is active year round and evidently cannot arouse if it becomes hypothermic, but has nonetheless been recorded running about actively on top of the winter snowpack. It remains unclear how it maintains its high energy budget or copes with shortfalls of food if they occur.

Despite the magnificent array of adaptations that carnivorous marsupials display to cope with the extremes of weather and fluctuations in the food supply in arid environments, they remain at risk from one further threat: other species. Disease organisms, pathogens and parasites have been little studied in desert-dwelling marsupials and, apart from occasional bites from fleas and ticks, they cause few obvious problems. Competitors also do not appear to have strongly negative effects. Rather, the major threats come from predators. Hot desert environments globally contain rich and varied communities of predators, including scorpions, spiders, snakes, birds of prey and carnivorous placental mammals. Large carnivorous lizards also may be present – Australian varanids (goannas) are good examples.

The generally open habitats of arid areas mean that deadly encounters between carnivorous marsupials and larger predators cannot be easily avoided. However, there are several ways that such risky encounters can be minimised, and survival prospects enhanced if they do occur. With the exception of the sun-baskers already noted, the Kaluta and the strikingly marked and day-active Numbat, carnivorous marsupials in arid areas are nocturnal, thus precluding encounters with day-active predators. All have uniform brown, grey or tan coloured coats that likely make them less conspicuous to other visual hunters of the night. All species studied remain close to the cover provided by shrubs or cracks in the soil while active at night, and move swiftly into cover if disturbed. On several occasions in the Simpson Desert, we have attached small lights to dunnarts and mulgaras to keep sight of them as they run around. These lights – tiny fishing lures that emit a luminous green glow – fade and fall off after a few hours, providing observers with a superb opportunity to watch the animals directly. The slightest noise, a hasty movement or the approach of another animal is

Fig. 8.6. In some cases of deep torpor, the creature can appear dead to the unwary ecologist, such as this Monito del Monte (*Dromiciops gliroides*). The inert creatures have even been collected by researchers to submit as museum voucher specimens, only to become a squirming, lively mass intent on escape just moments later. Compared with the animal in Fig. 8.5, this torpid monito has used most of the fat stores in its tail. Source: Marcela Franco.

enough to spook the experimental animals and make them disappear almost instantly from view. Careful examination of the sites where the lights go out reveals tiny holes, pockets of leaf litter or piles of debris that conceal the animals. These bolt-holes and other micro-refuges are not obvious to the casual observer, but the speed with which animals access them suggests they have intimate knowledge of safe locations within the landscape.

The risk of predation can be reduced even further if animals are attentive to signs that indicate the presence of predators. For example, bright moonlit nights make it easier for hunters with good night vision, such as owls, to detect and swoop on their prey. Several species of carnivorous marsupials respond to these risky conditions by entering torpor, reducing their activity, or by using only those parts of the landscape that provide overhead cover or ready access to bolt-holes. Windy nights evoke similar responses. Ground-active predators, such as the feral House Cat, can stalk unwary prey animals more closely because wind-noise masks their approach. Wet and stormy conditions also cause animals to stay under cover, but on these occasions additional risks may accrue to animals from drowning or hypothermia. Of course, acquiring this information also poses some threat to researchers. It is quite unnerving to be wandering about on top of a desert sand dune listening for the 'beep-beep' of a radio-tagged animal in the swale below, with a metal radio-receiver and antenna in hand, when the first bolts of lightning pierce the sky. But that is a story for another day!

Carnivorous marsupials also use a third sensory system to detect the presence of their predators: olfaction (sense of smell). Anyone who has been up close and personal with a large varanid (as desert biologists often are) can attest to the strong odour of their faeces, especially if the animal has sprayed you as you have tried – very nervously – to handle or photograph it. The faeces and urine of many predators are similarly pungent, although each has its own bouquet. Because these odours are obvious to the human nose, it could be expected that they would be at least equally obvious to other mammals with more acute senses of smell, such as the carnivorous marsupials. Recognition of the odour of a potential predator might indicate, in turn, that the predator is not far away, and hence that evasive manoeuvres are needed. Such avoidance behaviour does occur, albeit with some caveats. When presented with faecal odours of the predatory Sand Goanna and domestic cattle, together with cologne and water as control odours, both Hairy-footed Dunnarts and Lesser Hairy-footed Dunnarts avoid the smell of goanna but show no response to the other odours. Lesser Hairy-footed Dunnarts also assiduously avoid faecal pellets produced by the larger Brush-tailed Mulgara. However, similar experiments with dunnarts using the faecal or urinary odours of the feral House Cat and Red Fox have produced inconsistent results. Although these introduced predators consume many small mammals, it is possible that their arrival in Australia's central deserts is too recent for dunnarts to have learnt about the risks that these new species pose. Nonetheless, evidence is mounting that many small vertebrates in open desert environments are sensitive to a wide range of direct and indirect cues of predation risk, and must make moment-to-moment decisions about when and where to carry out their daily activities. The recently coined term 'landscapes of fear' well describes the gamut of threats that influence the daily lives of many carnivorous marsupials.

Forest and woodland dwellers

Trees provide excellent opportunities for exploitation by animals, both on the ground and in all strata up to the canopy. Many carnivorous marsupials use trees as foraging substrates, as arenas for attracting mates, for nesting or for temporary shelter, and for detecting and evading predators. Others use fallen trees for the same purposes. Almost all use forest products, such as leaves and bark, as constituents of their nests. Even the semi-aquatic Water Opossum will drag dry leaves using its tail to line its burrow in a riverbank, or amid tangles of riparian tree roots.

Forest-dwelling carnivorous marsupials are more sedentary and tend to have more restricted home ranges than their desert-dwelling counterparts. This presumably reflects the relatively high and reliable productivity of wooded environments. Because of this stability, and because suitable sites are often limited, forest dwellers usually return to their nests over long periods and invest some effort in constructing and maintaining them. All species use forest components such as loose bark or piles of leaf litter for temporary shelter if disturbed when active, but they are considerably choosier about where to set up their

Fig. 8.7. Brush-tailed Phascogales (*Phascogale tapoatafa*) are denizens of the trees. They rarely descend to the ground, and will only nest in tree hollows that are just the right size and height above ground. Source: Jiri Lochman.

longer term nests, especially if they depend on a single nest rather than several.

Consider first the ground-dwelling species. In Australia, the Mainland Dusky Antechinus (*Antechinus mimetes*) prefers large logs (over 40 cm diameter) in the early stages of decay for its nest sites. This species will make nests in decaying logs, holes in the ground or in tangles of long grass only if suitable logs are not available. Entrance and exit holes range from 3–7 cm, just large enough for animals to move in and out with a squeeze. Nest site preferences appear similar in the Little Long-tailed Dunnart (*Sminthopsis dolichura*), despite the much drier woodland habitats that this species uses compared with the Mainland Dusky Antechinus. In heathland, where trees are sparse, these dunnarts nest under tangles of sticks and leaf litter, and very occasionally exploit burrows made by other animals. Another inhabitant of heathland, the Dibbler (*Parantechinus apicalis*), digs short burrows under dense mats of low growing plants, constructing nest chambers at the terminus at a depth of 10–15 cm. In New Guinea, the Habbema Dasyure (*Murexia habbema*) also uses holes in the ground, either digging or extending pre-existing burrows on the floor of its forest habitat. Nest chambers may be 80–100 cm deep, and are neatly lined with the interwoven leaves of ferns and other plants.

In contrast to the ground-dwellers, arboreal species nest in tree hollows, spaces under loose and decorticating bark, or tangles of epiphytic plants on the upper trunks and sub-canopies of tall trees. The Brush-tailed Phascogale (*Phascogale tapoatafa*) is one of the fussiest nest-selectors, using only carefully chosen tree hollows. It prefers large living or dead trees with a diameter at breast height of 70 cm or more, and often constructs nests in hollows 10 m or more above ground. The smaller Red-tailed Phascogale nests 3–7 m above ground in mature Wandoo (*Eucalyptus wandoo*) in its remnant range in Western Australia. This species chooses hollows with entrance holes that are 4–5 cm wide. The Agile Antechinus (*Antechinus agilis*) and Yellow-footed Antechinus (*A. flavipes*) also show distinct preferences for tree hollows, constructing nests in suitable hollows 2–13 m and over 1.5 m above ground, respectively. Other arboreal species are less fussy. Nests of the Black-tailed Dasyure (*Murexia melanurus*) are made in hollows, under the palm-like fronds of *Pandanus* trees, and even near the ground in clumps of tall Pitpit grass (*Saccharum edule*). None of the carnivorous South American opossums is known to be strictly arboreal, unlike many of their more frugivorous and omnivorous relatives such as the mouse opossums (e.g. *Gracilinanus* spp., *Marmosa* spp.), woolly opossums (*Caluromys* spp.), Bushy-tailed Opossum (*Glironia venusta*) and Black-shouldered Opossum (*Caluromysiops irrupta*).

Most carnivorous marsupials use forest resources opportunistically for den and nest construction, exploiting shelters below, on or above ground level where conditions allow. This holds for all scansorial (climbing) antechinuses (i.e. all species except the ground-dwelling dusky antechinuses and Swamp Antechinus), quolls (*Dasyurus* spp.), Tasmanian Devil (*Sarcophilus harrisii*) and some dunnarts (*Sminthopsis* spp.) among the Australian fauna. This is also true for the Speckled Dasyure and probably some species of *Murexia* in New Guinea, and most of the South American species that have been studied. The Tasmanian Devil provides a good example of flexible nesting habits. Both sexes make nests of bark, dry

Fig. 8.8. A litter of Western Quoll (*Dasyurus geoffroii*) young, huddled together for warmth. Note the dry grass and leaves that line the nest and provide insulation from temperature extremes. Source: Jiri Lochman.

grass and leaf litter under dense vegetation, in hollow logs, caves or underground in burrows of the Common Wombat (*Vombatus ursinus*). Females prefer to den in rocky habitats. This is perhaps because of the extra security that these retreats afford when young are small and vulnerable. In Mexico, the Tlacuatzin (*Tlacuatzin canescens*) constructs nests in the usual places such as in tree hollows, under logs and rocks, but it also takes over the abandoned hanging nests of birds and makes its own nests up to 5 m above ground in shrubs and even cacti. In Chile, both the Monito del Monte (*Dromiciops gliroides*) and Panchos's Monito del Monte (*D. bozinovici*) make spherical or ovoid-shaped nests ~200 mm in diameter using sticks and leaves, especially the leaves of bamboo (*Chusquea coleou*). They situate the nests in tree hollows or in thickets of bamboo, on lianas or branches several metres above ground, or in logs and under rocks at ground level.

Wherever they are located, nest sites should provide animals with effective shelter from extreme weather conditions and secure retreats for dependent young. If animals select sites with care, they should remain dry during wet weather. If not, nests can sometimes be rendered waterproof by the addition of water-repellent leaves, such as the waxy leaves of *Eucalyptus* or *Chusquea* bamboo. Nests should also provide a stable thermal environment. One study showed that outside winter temperatures in a tall open forest in Victoria fluctuated between a chilly 3°C at night and 13°C by day. In a hollow with a wall thickness of 2 cm and depth of 12 cm, the temperature averaged ~2°C warmer than outside, but in a hollow with a wall thickness of 8 cm and depth of 50 cm the temperature remained at a balmy 10–16°C throughout the 24-hour cycle. Similarly buffered temperatures are provided inside logs and under rocks, with the buffering effect increasing with depth inside the structure.

To make home life even more comfortable, all carnivorous marsupials studied to date in wooded habitats line their nests with leaves, moss, bark or other materials such as feathers or even snakeskin that further insulate them from climatic extremes. Australasian species use their jaws to select and carry these materials back to the nest. They will often carry a single leaf at a time, requiring many trips to complete their interior decor. American species also do this, but more commonly use another method. They select nest materials using their forepaws, then pass the bundle back for transport to the nest by the tail. Although such behaviour is best known in highly arboreal mouse opossums with prehensile tails, even species with weakly prehensile tails such as the Water Opossum, Brown Four-eyed Opossum (*Metachirus nudicaudatus*) and some short-tailed opossums (e.g. Gray Short-tailed Opossum *Monodelphis domestica* and Orange-sided Opossum *M. dimidiata*) drag or carry nest materials with the tail. The Virginia Opossum (*Didelphis virginiana*) uses its tail to carry large amounts of material. This species builds nests that have circular walls 10–25 cm high and 10–35 cm thick. This adaptable opossum shows some preference for suburban areas, and is not above incorporating used socks, plastic bags and sundry other items that it purloins from backyards or rubbish bins.

Although nests with the right characteristics are of great importance to carnivorous marsupials, these denizens of forest and woodland seldom restrict themselves to a single nest site. Unlike their nomadic, desert-dwelling counterparts, however, forest dwellers use multiple nest sites within discrete home ranges that may change little over the lifetime of an individual. There are several reasons why multiple nest sites are used. In the first instance, a single nest is unlikely to provide optimal conditions year round. For example, a site that is warmed by the sun will be beneficial in winter but too hot in summer. A site that is a snug fit for a solitary individual will be too cramped when the family expands. As for the desert-dwelling species, short term use of nest sites by forest dwellers may also be beneficial to avoid build ups of parasites such as fleas and ticks, or to avoid the accumulation of odours that could attract the attention of larger predators. Studies of Agile Antechinus and Brown Antechinus (*Antechinus stuartii*) in artificial nest boxes suggest that animals quit nest boxes that are inspected too frequently. This lends support to the idea that disturbance by potential predators can cause nest abandonment.

Another reason why forest-dwelling carnivorous marsupials use multiple nest sites is that many occupy large home ranges and need access to retreats while moving about in the furthest reaches of these ranges. The Brush-tailed Phascogale provides a good example of this. In low productivity Victorian forests, detailed radio-tracking studies by ecologists Todd Soderquist, Rodney van der Ree and colleagues have shown that females occupy ranges that average 41 ha. In contrast, males cover areas exceeding 100 ha. Both sexes used multiple nest sites. Even in fragmented, but more productive, forest patches where home ranges may be an order of magnitude smaller, multiple nest sites are still used. In this latter situation, female phascogales returned to a single nest tree on 20–30% of occasions and used an average of 11.4 nest trees over tracking periods that ranged from 13 to 35 days. As the nests were scattered widely over about three-quarters of each animal's total home range, it is likely that phascogales simply used whichever nest was closest at the conclusion of each night of foraging. In Western Australia, exhaustive studies by Susan Rhind for her doctoral thesis indicated similarly that a single male phascogale would use on average 27 different nest sites, and a female 38 nest sites, over the course of a year. Except for females with dependent young, animals used any given nest for only 2–5 days before moving on.

Nest site mobility may also allow direct contact between individuals. Alternatively, it may allow for less direct exchanges of information about other animals' social or reproductive status based on residual scent cues left behind at the nests. Forest-dwelling carnivorous marsupials nest solitarily most of the time. The obvious exception is that of nursing females before the maturation and dispersal of their young. Despite this, there are some reports of communal nesting. During her landmark work on the Brush-tailed Phascogale, for example, Susan Rhind found these usually solitary animals nesting communally in one of her four study years. Over half the population shared nests, with two to four individuals of any sex or age combination occurring together. This extraordinary phenomenon coincided with low temperatures and declining food supplies. The researcher concluded that the phascogales were responding to an 'energy crisis' brought about by a prolonged drought and the onset of harsh winter conditions. Intriguing recent research by Diego Astúa and colleagues in Brazil indicates that opossums may also be more gregarious than previously thought. These researchers summarised information on nest sharing in five species of opossum. Of these, only one – the Brazilian White-

Fig. 8.9. Mating in Northern Quolls (*Dasyurus hallucatus*) is brief. Couples typically remain together for a few days at most, leaving little opportunity for nest sharing. Source: Jiri Lochman.

eared Opossum (*Didelphis albiventris*) – can be considered at least partly carnivorous. Nevertheless, observations of 13 mixed sex and age individuals in a single nest is a spectacular record.

Temporary increases in nest sharing occur in some carnivorous marsupials before the breeding season. In some species of quoll (e.g. Eastern Quoll, *Dasyurus viverrinus* and Spotted-tailed Quoll, *D. maculatus*) liaisons are fleeting and last for hours or days. However, in other species, such as the monito del montes (*Dromiciops*), pairs are reputed to stay together for several weeks during the spring breeding period. In the Brown Antechinus (*Antechinus stuartii*), both sexes spend increasing amounts of time in communal nests as winter approaches, with some 15–18 individuals being present at one time. When breeding commences in mid to late winter, males then aggregate in just a few of the communal nest trees and females nest alone. However, females visit and spend considerable time in the male aggregations, where mating takes place. These mating arenas, or 'leks', have not been described in any other carnivorous marsupials. Indeed, in most forest-dwelling carnivorous marsupials, females typically stake out exclusive territories before breeding and males occupy much larger ranges that overlap with those of several females. Transient nest sharing at these times, and increased visitations by males to female nest sites, allow males to monitor oestrus in females. This maximises their chance of mating when females are receptive.

Chemical signposts are also used near nests to assist in maintaining social organisation. Males in most species rub secretions from chest, or sternal, glands onto substrates near their nests or other important communication sites. Some species, such as various antechinuses and the Spotted-tailed Quoll, augment their olfactory charms by defecating at regular latrine sites. The precise functions of such signposts are unknown. However, there is little doubt that their strategic positioning facilitates communication within and between the sexes. Carnivorous marsupials generally have well-developed olfactory systems and thus a keen sense of smell. The few species that have been studied also have a well-developed accessory olfactory system, or vomeronasal system. Located either side of the septum, or dividing wall in the snout, the paired vomeronasal organ allows detection of pheromones. These chemicals carry information between animals of the same species. In the Subtropical Antechinus (*Antechinus subtropicus*), males and sometimes females rub pheromonal secretions from their sternal glands onto tree branches and other substrates before and during the breeding season. Their vomeronasal organs enlarge before breeding, perhaps allowing animals to better detect the odours of their fellows and to gauge their reproductive state. Such chemical communication is particularly evident in the Gray Short-tailed Opossum. Females check the odours of new male visitors to their nests and achieve puberty and initial oestrus only after exposure to pheromones that are present in the urine and faeces of the newcomers.

Deployment of chemical signals certainly helps maintain social networks and promote opportunities for liaisons within species. However, these same signals can also be used by 'eavesdropping' interlopers, especially predators. This is most obvious for faecal and urinary odours, but even pheromones can be detected on occasion by other species. For example, if the sternal gland secretions of male antechinuses can be detected by the human nose (and we attest that they can be!), it is not hard to envisage that wily predators might also detect them and exploit the information to track their prey. Forest-dwelling carnivorous marsupials may be safe if they nest in tree hollows, logs or other structures that are difficult for predators to access. But risks can be reduced in other ways too. First, most species either use the cover of night for activity, or are active in the twilight hours around dawn and dusk. Of course, nocturnal activity provides benefits other than just a reduced risk of encountering predators. For example, these species have access to nocturnal insect prey and cooler temperatures. In fact, it is not clear what the major evolutionary driver of nocturnal activity may have been. But, whatever its origin, night activity allows small carnivorous marsupials to avoid large predatory lizards, snakes and a broad array of day-active hawks, eagles and other birds of prey. During his PhD studies, Luke Leung witnessed Black Butcherbirds (*Cracticus quoyi*) and Grey Goshawks (*Accipiter novaehollandiae*) attacking hapless Cinnamon Antechinus (*Antechinus leo*) that had ventured out by day to find mates during the breeding season. We have seen other love-starved antechinuses being pursued during the day by other avian predators such as the Pied Currawong (*Strepera graculina*), Kookaburra (*Dacelo novaeguineae*) and Pied Butcherbird (*Cracticus nigrogularis*). On one occasion, we even observed a Common Dunnart running frantically to escape a Red-bellied Black Snake (*Pseudechis porphyriacus*).

Nocturnal activity may reduce predation risk for carnivorous marsupials, but does not remove it entirely. Owls, larger mammalian predators and, in tropical regions, snakes, readily include unwary animals in their diets. Hence, most small marsupials are attentive to unusual sounds, odours, movements or new objects while they are moving around. If a potential threat is detected, most species adopt an 'indecision-alert' posture. Here, animals raise one forefoot off the ground, hold the ears erect and keep the eyes wide open, while maintaining tension in the body. This posture heralds the classic 'fight or flight' response. If the threat is perceived to be acute or severe, animals will flee quickly to bolt holes or other shelters. Headlong flight can be risky in itself. Animals occasionally lose their footing or misjudge a leap between tree branches. However, like their desert-dwelling cousins, there is good evidence that many forest-dwelling carnivorous marsupials know the locations of 'safe houses' within their home ranges and seldom stray very far from them. But flight is only one response to a threat. If a disturbance is not judged to be serious, animals may respond with curiosity, feigned indifference, aggression or a range of intriguing stereotyped behaviours. Consider the Atherton Antechinus (*Antechinus godmani*). This small, but feisty, creature will flee quickly if a threat appears serious, but a perceived lesser threat – such as a researcher attempting to extract it from a trap or even get close enough to take a photograph after release – will be met with stern hissing sounds and attempts by the animal to lunge and extract as much blood as possible from its tormentor. Or, take the Brush-tailed Phascogale. This highly arboreal species repeatedly drums its forefeet on a tree branch if disturbed. This sends a loud signal to a would-be predator that it has been spotted and hence that an attack would not succeed. This species, like other members of the genus and the scansorial antechinuses, ascends trees by spiralling upwards, especially if startled or confused. This takes individuals in and out of view of potential predators such as owls, thus disrupting

Fig. 8.10. Sometimes the hunter becomes the hunted. This Olive Python (*Liasis olivaceus*) is swallowing an unfortunate Northern Quoll (*Dasyurus hallucatus*). Source: Jiri Lochman.

the search image of the predator and reducing the chance of a successful predatory attack. There is some evidence that these responses to predators are calibrated depending on the conditions. Animals are often more flighty on still, bright moonlit nights than under the cover of a new moon. This is especially the case if the forest structure is simple and has an open or broken canopy.

One behavioural response of carnivorous marsupials to predators is particularly well known. This is the habit of 'playing possum', or – more technically – 'thanathosis'. Best studied in the Virginia Opossum, this behaviour is elicited if an animal is threatened or handled and no escape route is available. Then, the animal feigns death – lying on its side, tail curled, with its eyes and mouth open and its tongue hanging out. The animal may also salivate, urinate and defecate as it assumes this position. It can remain in this state from just a few minutes up to 6 hours. As a consequence of this performance, the animal appears, and smells, as if it is dead, rendering it unattractive to many predators that prefer to kill and eat fresh prey. Astonishingly, animals reduce their heart rate by about half and their body temperature and breathing rate by almost a third, while playing possum. Yet they retain brain activity and electrocardiogram patterns at normal levels. This suggests they remain conscious and able to continuously monitor the behaviour of their would-be predator. Indeed, if the performance has been successful and the predator walks away from the supposed corpse, the opossum will spring back to life, zombie-like, and escape at the first opportunity. Full recovery is rapid, and complete within 20 minutes. Thanathosis occurs also in the Brazilian White-eared Opossum (*Didelphis albiventris*) and probably in Arlindo's Short-tailed Opossum (*Monodelphis arlindoi*). Surprisingly, thanathosis has not been confirmed in other marsupial species. It is suspected to occur in two or three other species of short-tailed opossums, but does not appear be part of the behavioural repertoire of any Australian marsupials.

Independent animals have varied abilities to detect and avoid predators, but young animals are

Fig. 8.11. These young Mainland Dusky Antechinus (*Antechinus mimetes*) have just been deposited by their mother in a ground-level nest made of dry tree leaves and grass. Aged only 6–7 weeks, these young wriggle vigorously to the bottom of the nest if they are disturbed, causing the rest of the nest to collapse and cover them with dry leaf debris. This reduces the risk that they will be detected and eaten by larger predators. Source: Chris Dickman.

potentially at much greater risk. Females with young attached to their nipples are slowed increasingly by their growing brood. They often restrict their movements at this time to parts of their home range that are safe and provide reliable access to food. They also increase vigilance. The increased wariness of females at this time makes them suspicious of new objects such as traps, and thus difficult to capture. When the young are large enough to be left on their own in a nest, females remain vigilant when active away from their brood. However, the young have few defences. Species that preferentially construct nests in tree hollows, such as scansorial antechinuses and phascogales, usually deposit their young in the nest at an early age (5–6 weeks) when the young are still pink and lack mobility. In these species, the nestling young are both conspicuous and defenceless. However, the risk of predation is minimised if the nest entrance is too small to be accessed by predators. In contrast, log- or ground-nesting species such as dusky antechinuses, the Swamp Antechinus and the Dibbler deposit young in nests at a more advanced age (6–8 weeks) when they are slightly hairy, darkly pigmented on their backs, and able to wriggle vigorously. Nests in these species are constructed using dry leaves that collapse inwards if disturbed. This potentially allows the nestlings to escape by burrowing down amid the nest debris. Because ground-level nests may be difficult to hide or exclude access by predators, risk of predation is probably minimised by the cryptic camouflage and active evasive movements of the nestlings themselves. In other ground-dwelling species, such as the Brown Four-eyed Opossum, nests are well hidden and protected from view by leaves and twigs placed across the nest entrance.

Although most carnivorous marsupials are nocturnal, the complex structure of some wooded habitats provides scope for adventurous animals to emerge and forage safely by day. The most diurnally active carnivorous marsupials include some of the dusky antechinuses (*Antechinus mimetes*, *A. swainsonii*, probably *A. vandycki*, occasionally *A. arktos*); the Swamp Antechinus (*A. minimus*); Atherton Antechinus (*A. godmani*); Rusty Antechinus (*A. adustus*); Speckled Dasyure (*Neophascogale lorentzii*); most species within *Myoictis* and *Phascolosorex*; and at least five species of short-tailed opossum (Northern Three-striped Opossum *Monodelphis americana*, Orange-sided Opossum *M. dimidiata*, Pygmy Short-tailed Opossum *M. kunsi*, Tawny-headed Opossum *M. scalops* and Touan Short-tailed Opossum *M. touan*). Most of these species emerge by day to forage at or just above ground level where dense cover or camouflage is available. For example, radio-tracking has shown the Touan Short-tailed Opossum to rustle about beneath leaf litter when active by day in French Guiana. The Atherton Antechinus similarly hunts invertebrates under logs and in the deep, accumulated litter between the buttress roots of rainforest trees in its haunts in northern Queensland. The dark coats of the dusky antechinuses render them very hard to detect against the deep, dank soils where they are typically most active.

Not all the day-active carnivorous marsupials are strictly terrestrial. At least four species are partly arboreal. Intriguingly, and in marked contrast to the dull brown or grey coats of their terrestrial relatives, two of these exceptional species have conspicuously marked coats. The black stripes of the Northern Three-striped Opossum provide a strong contrast to the animal's otherwise tawny brown fur, while the coat of the Speckled Dasyure from New Guinea is characterised by a frosting of distinctive silver hairs. It is tempting to suggest that these coat markings help to break up the owners' shapes when they are potentially exposed and active while climbing in trees. The other two scansorial species that are at least partly active by day are the Atherton Antechinus and Rusty Antechinus. Neither has distinctive markings. These species co-occur with the Yellow-footed Antechinus and very likely compete for food or

Fig. 8.12. Almost nothing is known of the ecology of the recently discovered Tasman Peninsula Dusky Antechinus (*Antechinus vandycki*). However, we might assume it to be at least partly active during the day, since this is the case for all other members of the Dusky Antechinus group. Source: Gary Cranitch.

habitat resources. It is quite possible, albeit unproven, that these species reduce the chances of encountering each other by partitioning the forest strata and the time when they are active.

Several more species probably show some diurnal activity. This is most likely during winter when daytime temperatures are more benign than those at night, or when animals are hungry and need to extend their foraging time to top up on food. For example, daytime activity has been recorded in the usually nocturnal Agile Antechinus, but only on cold winter days when night temperatures fall below 0°C, and in small individuals that are in poor condition. In captivity, the Patagonian Opossum (*Lestodelphys halli*) shows peaks of activity after sunset and before dawn, but also emerges by day. Although its activity pattern has been little studied in the wild, the ability of this species to hunt prey by day in its arid, but chilly, steppe and shrubland habitats may be highly advantageous, as may its pronounced ability to use torpor.

In a similar manner to their desert-dwelling counterparts, a few forest-dwelling carnivorous marsupials – principally fat-tailed opossums – store fat in their tails to help cope with short-term food deficits. Species within most genera can also enter torpor. Only the largest species, including the Tasmanian Devil and Spotted-tailed Quoll, are not known to use torpor. Torpor in many species is expressed on a daily basis, but increases in frequency, depth or duration depending on both external conditions and food availability. This is similar to the expression of torpor in desert species. However, recent studies suggest that forest dwellers use torpor in other conditions too; namely, in the wake of fire. Thus, Brown Antechinuses enter torpor more often and for longer periods in the post-fire compared with the pre-fire environment, or when compared with animals in unburnt forest. They also allow their body temperature to fall to a low of 19.8°C compared with 27.6°C for animals in unburnt forest, and are active for only half as long. Increasing the use of torpor after fire has two benefits. It allows animals to reduce their energy needs at a time when food resources are likely to be temporarily depleted. It also reduces their exposure to predators at a time when vegetation cover has been reduced and foraging would be more risky.

One forest-dwelling species group has taken the idea of energy conservation to an extreme: the monito del montes. The strongly seasonal conditions in the cold temperate rainforests where these animals live prompt them to store fat under the skin and at the base of their tails in summer so that this can be metabolised in winter when conditions are poor. Under laboratory conditions, animals spend increasingly long periods in torpor as conditions cool, from 10 hours at 20°C to 5–6 days at 12.5°C. Body temperature drops by 20°C from the usual 34.7°C to just 14.5°C, and the heart rate falls from 230 beats per minute to just 30 beats. These measures allow animals to reduce their metabolic rate during torpor to less than 4% of their basal metabolic rate, and to levels even less than this if compared with when animals are actively moving. In these respects, monitos resemble small diprotodont marsupials such as pygmy possums (*Cercartetus* spp.) more than other carnivorous marsupials, and can be regarded almost as short-term hibernators. The impressive adaptations exhibited by monitos exemplify the astonishing range of ways in which carnivorous marsupials as a group exploit their chosen habitats, whether these are cold temperate rainforests, equatorial moss forests or hot continental deserts.

VIGNETTE: TREKKING THE DESERT

By Chris Dickman
University of Sydney, Sydney, Australia

Conventional wisdom holds that small, desert-dwelling mammals will have fixed home ranges of no more than a few hectares. This makes sense: provided that food, shelter, mates and other essential resources can be found, no benefits should accrue to animals that go trekking across the harsh and open landscape away from the comforts of home. I have had the good fortune to radio-track small mammals in both the Negev (Israel) and Kalahari (southern Africa) deserts, and there the animals abide strictly by the rules, occupying small and stable ranges. In the Kalahari, for example, elephant shrews (Family Macroscelididae) use the same runways so regularly within their home ranges that these movement corridors are etched indelibly into the topsoil. The

Fig. 8.13. A Lesser Hairy-footed Dunnart (*Sminthopsis youngsoni*) devouring a bush cricket. These feisty pocket dynamos trek long distances each night in search of food and, like many other small desert mammals, are itinerant or occupy home ranges that shift with the prevailing conditions. Source: Bobby Tamayo.

animals are also assiduous in clearing the runways of debris, so their movements are never impeded by unexpected obstructions.

Armed with such knowledge, I began my studies in Australia's deserts thinking the small mammals there would behave like those in other arid regions. First in the Goldfields of Western Australia and then in the Simpson Desert in Queensland, I set up grids of traps to capture my study animals. The aim was to study food resources, habitats and interactions of small carnivorous marsupials – all dasyurids – and illuminate as many aspects of their secret lives as we could uncover. Having separate, independent sites is an important aspect of sampling design, so I set up the grids half a kilometre or more apart thinking that few animals would move between them. After some pilot tests, pitfall traps (buckets dug into the ground) turned out to be the most effective means of catching animals. Not only did these traps capture a full range of carnivorous marsupials, because of their depth (60 cm), they also provided excellent protection for captured animals from temperature extremes on the ground surface – very useful in a desert climate! Lids on the traps ensured that no animals could fall in between field trips.

The fun begins around dawn when the traps are checked: will there be a ningaui, a mulgara or a completely new animal waiting for you when you peer into the trap? Whatever awaits, the animal is removed, identified, and its vital statistics (sex, weight, reproductive condition) recorded. Then, if you plan to return to the grids at later dates, the animals are marked for later recognition. Very small marsupials can be marked by clipping small notches in the outer edge of the ears, and larger ones by applying small numbered tags to the base of the ear, much like ear-rings. In many studies, small passive integrated transponder (PIT) tags are injected under the skin. These tags, a bit larger than a grain of rice, contain unique bar codes. PIT tags allow many animals to be marked, but their identification numbers have to be read by bar code scanners. Because these devices are quite delicate, and as desert fieldwork necessitates long drives over corrugated roads, we gave up using PIT-tag marking after our first two bar code readers fell to bits while in transit. Whatever method is used, giving animals unique marks means one can gain detailed knowledge of their growth, condition, longevity, movements and reproductive seasonality when they are recaptured.

So far, so good. What did the recapture record of our desert denizens reveal? In the first place, we recaptured very few animals. This was a shock. If you work on many carnivorous marsupials, such as antechinuses or short-tailed opossums, the problem is often not how to recapture them, but rather how to stop them getting back into the traps all the time to raid the bait. Were our animals so traumatised by their first capture experience that they were wary about traps thereafter? Did they wander off to their burrows to sulk until our field trips had concluded? These possibilities seemed unlikely because we could often see marked animals running around on the open sand a night or two after they had been first captured. The next shock came when we compiled the records of the few animals that we did manage to recapture. Many of these had moved between our supposedly independent grids, with some animals – especially Lesser Hairy-footed Dunnarts (*Sminthopsis youngsoni*) – moving distances up to 12 km! If you are a dunnart that has fallen down a hole and then been poked and prodded in the most undignified ways, you may be less than impressed at falling into another hole some kilometres distant and afforded the same treatment. But for us, these records of long-distance movements were astonishing and revelatory: Australia's small desert mammals didn't seem to bother with home ranges, and were regularly making the longest movements ever recorded by any small desert mammals. What was driving this extraordinary behaviour?

We suspected that rainfall might be playing a role. In 1975, two visionary ecologists working in central Australia, Alan Newsome and Laurie Corbett, proposed that patchy rainfall events could create brief and local increases in productivity and food, and thereby provide shifting refuges for any animals able to detect and access them. This novel idea had never been tested. Could our dasyurids provide the first test subjects? Using rainfall records drawn from the study regions, we were first able to show that animals were moving about twice as far during or just after rainfall compared with dry periods. Second, and very importantly, these long movements were not aimless: three-quarters of them were directed **towards** where rain had fallen. Eureka! This is exactly what would be expected if animals were tracking and exploiting shifts in conditions across the desert landscape. Our results therefore provided the first support for the shifting refuge concept.

Of course, as soon as one question is answered, others open up. And here we had unleashed a veritable Pandora's Box of very tricky questions. Do animals move to wet areas to get a drink, or to access food resources that are generated by plant growth following rain? How do they even know that it has rained somewhere on the horizon and then know which way to head? How do they find shelter while in transit? Radio-tracking and some simple experiments provided a few answers to these questions, with that miniature marathon runner, the Lesser Hairy-footed Dunnart, acting as our test subject. Working in the magnificent sand dune system of the Simpson Desert, we captured adult dunnarts weighing at least 9 g and attached tiny radio-tags to them that had been developed initially for those most diminutive of birds, the wrens. Then, assisted by teams of insomniac volunteers, we began our investigations. Long bouts of tracking showed that these dunnarts spent the day asleep in burrows in valleys between the sand dunes, and that just after dusk they sprint 200–300 m up to the dune crests. Here they would forage, seeking out large beetles and spiders that could be seen readily on the open sand. After 4–5 hours, they would return to a burrow at the base of the dune, before rising again an hour or two later to begin another bout of feeding on the dune crest. Males frequently travelled up to 3 km each night and seldom returned to the same burrows they had used previously. Females ranged up to 2.5 km each night, but showed slightly more inclination to use the same burrows, doing so on about a quarter of the occasions when they were tracked.

Direct observations of tracked animals suggested that the main benefit obtained from their nightly commutes to the dune crests was access to food. So, we reasoned that if we provided the dunnarts with supplementary food in the dune valleys before they woke, this should reduce their need to run to the crests. And presto, this is exactly what happened! In a series of experiments using shredded beef and mealworms (beetle larvae) placed in the dune valleys, dunnarts stopped to eat the added food after leaving their burrows and went back to bed early. Their time on the dune crests fell by over 50%. Further radio-tracking by PhD student Adele Haythornthwaite showed that dunnarts took convoluted routes while out foraging, staying close to the edges of spinifex hummocks to maximise their chance of escaping larger predators. Our more recent work suggests that dunnarts don't worry about drinking free water, instead getting all that they need from their food. If times are really tough and there is no rain in the landscape, they seek out the juiciest prey they can find, such as spiders, to ensure that their energetic, nutritional and water needs are met.

The famed explorer Charles Sturt was the first European to set eyes upon the sand dunes of the Simpson, and was in no doubt about the harshness of the environment that it presents. In a letter to his wife Charlotte, written in September 1845, he described his first impressions of the desert: 'A kind of dread, and I am not subject to such feelings, came over me as I gazed upon it. It looked like the entrance into hell.' Yet dunnarts and many other carnivorous marsupials thrive in this fierce environment, eschewing fixed home ranges, evading predators and other hazards, and tracking pulses of food wherever they appear. This vagrancy, or serial nomadism, appears unique to the small mammals of arid Australia and is undoubtedly a response to the extremely unpredictable environmental conditions of the continental interior. At a time when science is becoming more and more formulaic and driven by the products it can deliver, it is refreshing to think that chance and follow-up curiosity still have roles to play. Without it, the secret lives of our little desert trekkers may have remained uncovered still.

9

DIETS AND EATING HABITS

Early observers of carnivorous marsupials were often struck by their ferocity and the savage ease with which they made their kills. In 1866, for example, Gerard Krefft commented that the Yellow-footed Antechinus (*Antechinus flavipes*) would kill any number of mice if placed in a box with them. Thirty years later, Richard Lyddeker noted that 'Like their allies, they [Fat-tailed Dunnarts, *Sminthopsis crassicaudata*] are inveterate enemies to Mice, which they fall upon at once with great ferocity, killing more than they can possibly devour.' Mice fare poorly in many accounts. New Guinean dasyures (*Murexia* spp.), mulgaras (*Dasycercus* spp.) and the Kowari (*Dasyuroides byrnei*) have been reported to dispatch mice with swift bites to the head and neck. Zoologist Oliver Pearson reported that captive Patagonian Opossums (*Lestodelphys halli*) killed mice at 'lightning speed, eating everything – bones, teeth, and fur'. Larger carnivorous marsupials have long been reputed to tackle bigger prey. The Brush-tailed Phascogale (*Phascogale tapoatafa*), for example, was known not just as a killer of poultry, but – in the words of zoologist Frederick Wood Jones – to be a '… determined slayer of chickens, and one which killed not merely to appease its appetite'. The largest of the extant carnivorous marsupials, the Tasmanian Devil (*Sarcophilus harrisii*), had the worst reputation for ferocity. Writing in the mid-19th century, famed naturalist John Gould regarded the Devil to be 'untameable and savage in the extreme, [it] is not only destructive to the smaller kangaroos and other native quadrupeds, but assails the sheep-folds and hen-roosts whenever an opportunity occurs for its entering upon its destructive errand.' These and other accounts of carnivorous marsupials as vampires and blood-thirsty killers have given the group a fearsome reputation. But is this view justified? Here we examine the diets of carnivorous marsupials and describe the range of prey that they hunt. We review the strategies these carnivores use to detect their prey and the weapons they employ to subdue and process them. With carnivorous marsupials stalking landscapes throughout the Americas and Australasia, we finally consider the broader effects of these miniature killers on other species that share their habitats.

Fig. 9.1. No doubt formidable, but is the reputation for untamed ferocity in some carnivorous marsupials, such as this Tasmanian Devil (*Sarcophilus harrisii*), justly deserved? Source: Jiri Lochman.

Diets: moving feasts

In Chapter 1, we defined carnivorous marsupials as those with diets comprising more than 50%

vertebrates or invertebrates. This broad definition allows scope for enormous dietary variation. For example, Numbats (*Myrmecobius fasciatus*) feed almost exclusively on termites, with dietary variation arising only from the different termite species that animals can access in different seasons or locations. The Water Opossum (*Chironectes minimus*) also has a relatively restricted diet, relying principally on fish, frogs and aquatic invertebrates. The diets of a few other species appear to be dominated by particular prey. Thus, insect larvae comprise a large part of the diet of the Gray-bellied Shrew-opossum (*Caenolestes caniventer*), while lycosid spiders are over-represented in the diets of several species of small dasyurids. It remains uncertain whether these invertebrates are preferred or simply reflect prey that are easily accessible within the species' respective habitats. In general, invertebrates such as ants, scorpions, slaters and slugs appear more rarely in the diets of carnivorous marsupials. These prey may provide poor energetic and nutritional returns and thus are not favoured, or are avoided simply because they can fight back or are distasteful. Other invertebrates, such as earthworms, also appear to form a trivial part of the diet of most species. However, they are eaten regularly by South American shrew-opossums, and it is possible that remains of these soft-bodied animals have been overlooked in dietary studies of other species. Shrew-opossums, like many other South American carnivorous marsupials but unlike their Australasian counterparts, also eat fruits, fungi and other non-animal matter. We now consider the diets of these two broad groups separately.

Dietary patterns in the Australasian region

The natural diets of some 32 species of dasyurids have been examined in greater or lesser detail, with most investigations focusing on Australian, rather than New Guinean, representatives. On occasion, researchers have identified dietary items by stealthily following and observing their study animals as they hunt prey, but dietary detective work more usually involves painstaking microscopic analysis of the faecal remains of live-trapped animals or the stomach contents of dead specimens. The results show that all the smaller species are broadly insectivorous, with invertebrates forming most or all of the diet. They eat a wide range of hard-bodied insects such as beetles and cockroaches, soft-bodied insects including caterpillars and moths, and representatives from many other invertebrate groups such as spiders, scorpions, centipedes and amphipods. Some insects, such as termites, are rarely eaten, but can be consumed in large numbers if they are encountered. Termites are usually active underground or within dead wood and are then accessible only to specialist hunters such as the Numbat that can break into their galleries. The Southern Marsupial Mole (*Notoryctes typhlops*) also includes many termites in its diet, presumably encountering them while moving underground. However, on warm, humid nights termites can emerge *en masse* to forage on the open soil surface. Under these conditions, they may be gobbled up quickly by any dasyurids that come across them. Other invertebrates, such as snails, are also eaten rarely. The hard shells of these molluscs probably make it difficult for hunters to eat them and, when snails do show up in the diet, it is usually in winter or after rain when other invertebrates are less available.

Many of the smaller (less than 200 g) dasyurid species also include other vertebrates as part of their diet, but the contribution is usually less than 10% by frequency of occurrence in faeces or stomachs (Fig. 9.2a), and even less on a volumetric basis. Larger dasyurids (more than 200 g) tend to eat vertebrates more often than their smaller counterparts, with the Spotted-tailed Quoll (*Dasyurus maculatus*) and Tasmanian Devil being especially partial to flesh (Fig. 9.2b). Representatives of all classes of terrestrial vertebrate have been recorded in the diets of one or more dasyurid species, including amphibians, reptiles (mostly lizards, but very occasionally small snakes), birds and mammals, with lizards and small rodents being recorded most frequently. Except for the Tasmanian Devil, which readily consumes carrion, dasyurids prefer to catch and kill their own prey. Plant parts, especially pollen and floral structures, also occur in the diets of some of the smaller dasyurids. Although it probably contributes little to the diet, there is compelling evidence that several species transfer pollen between flowers and may even contribute to pollination.

The weak, but positive, relationship between the size of small dasyurids and the amount of vertebrate material in the diet shown in Fig. 9.2a probably arises for at least two reasons. First, the smallest dasyurids would be little bigger than most of the potential vertebrate prey they might encounter. This increases

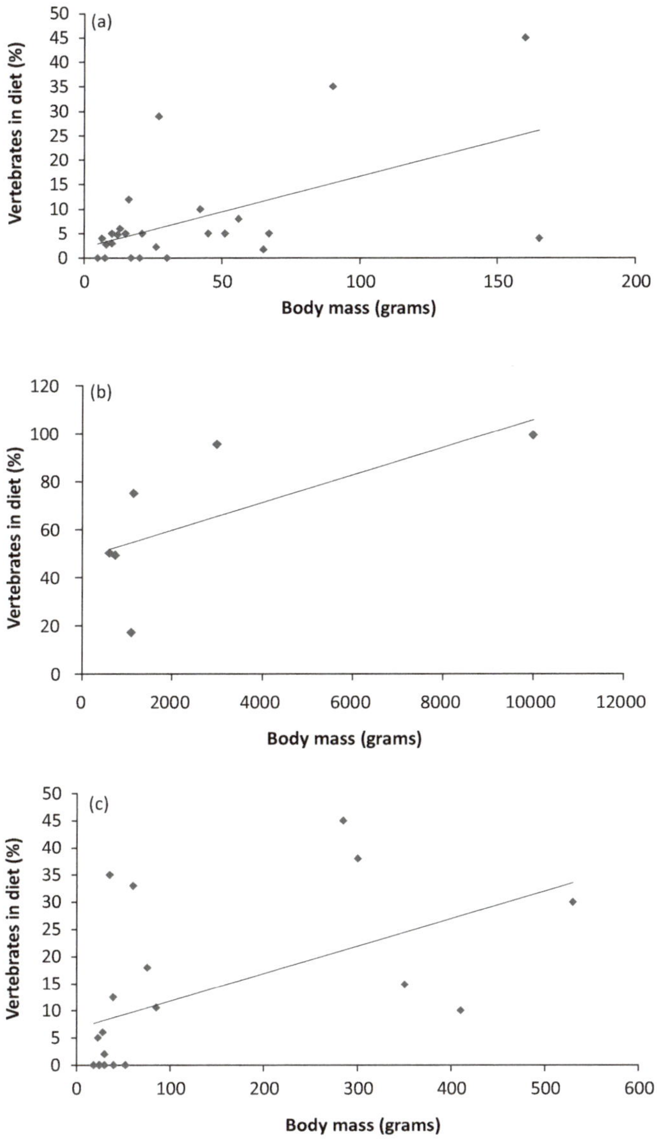

Fig. 9.2. The relationship between body mass and percentage representation of vertebrates in the diets of carnivorous marsupials. Each point represents a different species. If two or more studies described the diet of a particular species, the values were averaged. The lines in each figure represent the best fit between the points, and show a generally positive relationship between the size of carnivorous marsupials and the amount of vertebrate flesh in their diets. Relationships are shown for Australian carnivorous marsupials weighing: (a) less than 200 g; (b) more than 200 g; and (c) American carnivorous marsupials. Source: Chris Dickman.

the costs of pursuing these prey and the risks of subduing them once captured. Such costs would be minimal for larger dasyurids, which would also reap greater energetic, and perhaps nutritional, benefits from hunting larger vertebrates. Second, when foraging for invertebrates, there is considerable evidence that larger dasyurids explore foraging paths and microhabitats that increase their chances of encountering large beetles and other insects. It is quite plausible, albeit untested, that these larger species use a similar strategy to increase their chances of encountering potential vertebrate prey. Similar explanations probably underlie the relationship for larger dasyurids depicted in Fig. 9.2b. In addition to including more vertebrates in their diets than their smaller relatives, the larger quolls and the Tasmanian Devil also hunt larger prey. For example, the formidable Spotted-tailed Quoll has been seen climbing trees to extract hapless possums and gliders from hollows and pouncing upon small wallabies from low-hanging branches. Tasmanian Devils sometimes lie in wait before ambushing prey as large as kangaroos, and can pursue them for long distances until their quarry tires, after which it is quickly killed and eaten.

Despite the general tendency for larger dasyurids to consume more vertebrates, the scatter of points in Figs 9.2a and 9.2b suggests there is much variation in this relationship. At one extreme, small species such as the White-footed Dunnart (*Sminthopsis leucopus*, ~27 g) are more avid hunters of vertebrates than might be expected for their size. Zoologist Dan Lunney and colleagues recorded the remains of five skinks (*Lampropholis* sp.) from just 17 faecal samples of this small predator. At the other extreme, some of the larger species consume relatively few vertebrates. Consider the Brush-tailed Phascogale. Although often exceeding 200 g, detailed studies from Victoria have shown this species to be apparently averse to vertebrate prey. One study recovered the remains of a single vertebrate – a House Mouse (*Mus musculus*) – from over 200 faecal samples produced by this phascogale. Another showed the diet to comprise mainly arthropods (i.e. insects and other invertebrates with hard external bodies) gleaned from the surface of tree boles and branches, leavened with small amounts of nectar from eucalyptus flowers. The Eastern Quoll (*Dasyurus viverrinus*), although often weighing more than 1000 g, sometimes also eats few vertebrates. Its diet in agricultural areas comprises mostly insects (97% of samples), with small mammals and birds eaten infrequently (17% of samples). Most unusually, this species often eats a lot of grass and other plant material, which it perhaps ingests while digging in the soil for caterpillars and beetle larvae.

In addition to the varied amounts of vertebrate material eaten by different-sized dasyurids, diets can

Fig. 9.3. Dunnarts, such as this Little Long-tailed Dunnart (*Sminthopsis dolichura*), can take down vertebrates, but insects and spiders are their more typical fare. Source: Jiri Lochman.

often vary within the same species between times, places and animals of different age. In one example, post-doctoral fellow Xiaolin Chen and colleagues found the remains of rodents and reptiles in 47% of faecal samples of the Brush-tailed Mulgara (*Dasycercus blythi*) during spring. However, rodents were recorded in only 8% and reptiles in 15% of samples collected during autumn in the same area. There were also strong seasonal differences in the frequency of occurrence of centipedes, spiders and small marsupials in the diet of the Brush-tailed Mulgara. In another example, a two-and-a-half year study of the Fat-tailed Pseudantechinus (*Pseudantechinus macdonnellensis*) by doctoral student Sandra Gilfillan showed that animals ate spiders and 10 different orders of insects. A later study of this species in the same locality found insects still comprised a large part of the diet, but rodent remains were also found in half the faecal samples analysed. Interestingly, the Brush-tailed Phascogale, used as an exemplar of almost pure insectivory earlier, includes more vertebrates in its diet in the western part of its distribution. Here, Fiona Scarff and colleagues recorded mammal or bird remains from six of 56 faecal and stomach samples, and observed animals feeding on small reptiles, as well as on a desiccated bird carcass. Finally, consider the Spotted-tailed Quoll. Despite the lust for flesh that is shown by adults, juveniles are more likely to feast upon invertebrates. Age-related dietary shifts are probably common across the carnivorous marsupials because juveniles can only acquire hunting skills as they grow. Yet little research has been carried out on this aspect of diet.

Although the diets of 32 dasyurid species have been studied, those of the remaining 42 species have not. Nonetheless, casual observations, anecdotal accounts and broad similarities in body plans provide no reason to think that prey other than vertebrates and – especially, invertebrates – form important dietary components for any members of the family. Taking all the dietary results together, it seems reasonable to conclude that the vilification and historically blood-thirsty descriptions attached to the small dasyurids were largely undeserved. Although efficient and capable predators, these species are primarily insectivorous. Many of the small dasyurids can undoubtedly kill large prey, but observations of them shredding defenceless House Mice probably reflect fear responses to being trapped or placed within the confines of small boxes or wire cages. In contrast, there is no doubt that the largest dasyurids kill and

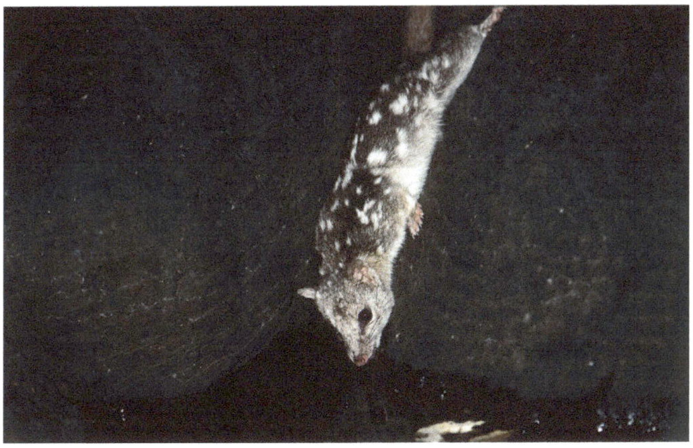

Fig. 9.4. Many carnivorous marsupials in desert regions obtain all their water from food items. For other species, water obtained from food is supplemented by taking free-standing water where it is available. This Northern Quoll (*Dasyurus hallucatus*) has paused for a drink at a pool in its rock-pile habitat. Source: Jiri Lochman.

eat large vertebrates, and that poultry or even lambs will be hunted if the opportunity presents. Spotted-tailed Quolls in particular have been making incursions in recent years into outlying areas of Sydney and the Blue Mountains in New South Wales. Tempted into suburbia by easy access to chickens and ducks in backyard runs, these enterprising carnivores remind us why they were so feared by the first European settlers.

Dietary patterns in the American region

Studies in the wild and captivity show that the diets of South and Central American carnivorous marsupials are much broader than those of their Australasian counterparts. All of them consume plant parts in addition to animals. In captivity, several species show clear preferences for live animals such as beetles, or even mice or lizards, and evince little interest in fruit or other plant materials. The Patagonian Opossum (*Lestodelphys halli*) and lutrine opossums (*Lutreolina crassicaudata* and *L. massoia*) are good examples and, with the Water Opossum, are often considered to be the most carnivorous marsupials in the Americas. However, inspection of the faecal or stomach contents of wild-caught animals usually also reveals the presence of fruits and seeds. These can occur at low frequency (e.g. 2% in the Patagonian Opossum) or be present in most diet samples (85% in the Big Lutrine Opossum, *L. crassicaudata*). In the latter species, as many as 6500 seeds have been found in a single faecal sample. Far from indicating accidental ingestion, this represents very targeted foraging for what must be a favoured food source.

Microscopic analysis of faecal or stomach samples has characterised the natural diets of ~20 species of American marsupials. The invertebrate component varies from ~50% in species such as the lutrine opossums that include vertebrates in the diet, to almost 100%. Most insect groups are represented, as well as a broad range of additional taxa such as flatworms, slugs, crabs and other crustaceans that do not appear in the diets of Australasian species. Beetles, larvae and spiders are among the most frequently hunted invertebrates in the species studied to date, but other taxa – including potentially quite formidable predators in their own right – are also eaten. The Gray Short-tailed Opossum (*Monodelphis domestica*) is particularly adept at capturing scorpions, while about three-quarters of dietary samples from wild Gray-bellied Shrew-opossums (*Caenolestes caniventer*) contain the remains of centipedes.

Vertebrates are included in the diets of about two-thirds of the American carnivorous marsupials, with larger species generally consuming most flesh (Fig. 9.2c). All vertebrate classes are hunted. The two species of lutrine opossum are unusual in frequently including fish and frogs as part of their diets, with the Big Lutrine Opossum feasting also upon birds, small mammals and reptiles. The Water Opossum also relies on aquatic vertebrates, but the relative importance in the diet of these types of prey compared with crabs, crayfish, molluscs and other

Fig. 9.5. Often considered to be one of the most carnivorous of the American marsupials, this Patagonian Opossum (*Lestodelphys halli*) is shown here devouring a large lizard. Source: Dario Podesta.

Fig. 9.6. One of a small group of American carnivorous marsupials to actively hunt snakes, this Gray Four-eyed Opossum (*Philander opossum*) is shown here grappling with a Mexican Parrot Snake (*Leptophis mexicanus*). Source: Chico Sanchez.

invertebrates is not known. Away from streams and lakes, terrestrial didelphids and shrew-opossums hunt rodents, small marsupials, birds and their eggs, iguanas and other small lizards. Incredibly, several species also hunt venomous snakes such as lanceheads (*Bothrops* spp.), rattlesnakes (*Crotalus* spp.) and other members of the pit viper subfamily, as well as coral snakes (*Micrurus* spp.) from the Family Elapidae. Termed 'ophidiophagous', these snake-eaters include the Big Lutrine Opossum, the Gray Four-eyed Opossum (*Philander opossum*) and at least three species of large American opossums (*Didelphis* spp.). Although we have classified *Didelphis* spp. as 'nearly carnivorous' (see Chapter 2), we think that any marsupials capable of dispatching such formidable prey are at least worthy of mention! The Gray Four-eyed Opossum has been observed pulling a coral snake from its burrow and killing it without being bitten. However, even without the dexterity needed to safely subdue such venomous prey, these didelphids have serum proteins that allow them to rapidly neutralise snake toxins if they do get bitten. Interestingly, venom resistance appears to have arisen only once in opossums, probably in the late Miocene in the common ancestor of the *Didelphis* + *Philander* + *Lutreolina* clade. It would have provided twin benefits in allowing the ancestral didelphids to avoid death by snakebite, and in allowing the marsupials to place the snakes themselves on the menu.

The positive relationship between the size of American carnivorous marsupials and the amount of vertebrate prey in the diet is similar to that seen with dasyurid marsupials (Fig. 9.2), and probably arose for similar reasons. That is, larger carnivorous marsupials gain greater benefits and experience smaller risks from hunting vertebrates, but, as with dasyurids, several American species buck the general trend. The small, but feisty, Orange-sided Opossum (*Monodelphis dimidiata*, ~60 g) consumes a diversity of invertebrates, but 30–35% of stomach samples have been found to contain rodent or bird remains. Conversely, the much larger Brown Four-eyed Opossum (*Metachirus nudicaudatus*, ~350 g) is primarily insectivorous, with samples from different parts of the species' range containing anywhere from no vertebrate material to 26%, by frequency.

Vegetation is the other major component in the diets of all American marsupials. It can comprise a minimal part of the diet (1–2%) in carnivorous species, such as some short-tailed opossums (*Monodelphis* spp.), but occurs in 50–100% of diet samples in fat-tailed opossums (*Thylamys* spp.) and in the ever-contrary Brown Four-eyed Opossum. Perhaps not surprisingly, there is no clear relationship between body size and the consumption of vegetation (Fig. 9.7). All consumers should gain similar nutritional and energetic benefits from eating fruit, and experience little risk in doing so. Fruit, seeds, leaves of grasses and herbs are the most commonly recorded plant materials in the diets of American carnivorous marsupials, but mosses and rhizomes have been recorded on occasion (e.g. in the newly described Sangay Shrew-opossum, *Caenolestes sangay*). Species inhabiting high rainfall areas, such as the Monito del Monte (*Dromiciops gliroides*) and Long-nosed Shrew-opossum (*Rhyncholestes raphanurus*), are also known to eat small amounts of fern tissue and fungi. Despite the prevalence of fruit in the diets of many species, several authors suggest that this food type may be underestimated in importance. Direct observations indicate that animals often eat fruit pulp and reject the seed, especially if the seed is large. Because fruit pulp is difficult to identify in stomach or faecal samples, frugivory thus may be even more pronounced than we think.

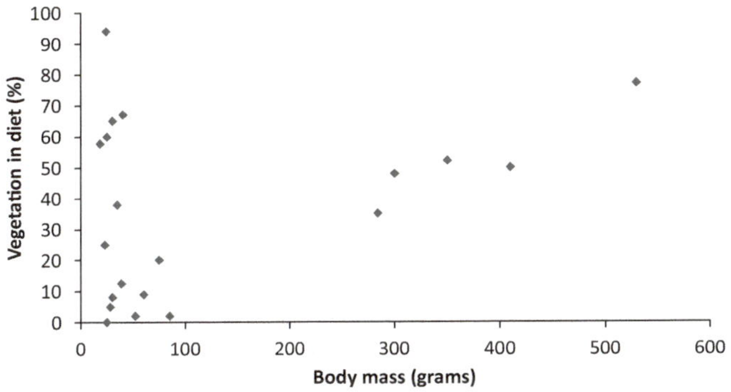

Fig. 9.7. The relationship between body mass and percentage representation of vegetation in the diets of American carnivorous marsupials. Each point represents a different species. If two or more studies described the diet of a particular species, the values were averaged. Unlike the positive relationship between animal size and the consumption of vertebrate prey, no clear relationship is evident between size and vegetation consumption. Plant material – fruit, berries, leaves – appears to be eaten readily by carnivorous marsupials of all sizes, although there is more variation among small species in the amount consumed than there is for larger species. Source: Chris Dickman.

There is emerging evidence that seeds show greater germination success when they have passed through the guts of marsupials, indicating that these animals contribute both to seed dispersal and establishment success. As we saw earlier, the Monito del Monte is the best studied species in this respect, acting as a vector for the seeds of many temperate forest species with fleshy fruits and as the sole disperser of seeds of Quintral Mistletoe (*Tristerix corymbosus*). At least 10 further species of carnivorous marsupials have also been identified as seed dispersers. In addition, the germination rates of seeds often increase after passage through the marsupial gut. For example, following extraction from faecal pellets of the Brown Four-eyed Opossum, the germination success of seeds from two species of Brazilian Cerrado (savanna) tree (*Clidemia urceolata* and *Miconia holosericea*) increased 3.5–4-fold, compared with untreated seeds. Similar increases in germination success for other seed species after ingestion and excretion by other marsupials suggests that this process primes the seeds. It perhaps also provides an additional source of nutrients for the newly growing plants. Unlike their Australasian counterparts, there is little evidence that the American carnivorous marsupials are involved in the transfer of pollen or in pollination. However, their granivorous habits indicate that these marsupials contribute importantly to habitat maintenance by seed dispersal and activation.

Given the wide range of food types exploited by the American carnivorous marsupials, it is not surprising that diets often differ between locations, times and animals of different age, as well as between related species. The South-eastern Four-eyed Opossum (*Philander frenatus*) provides a good example. At different sites in south-eastern Brazil, this species consumes varied quantities of insects, spiders, millipedes, snails and other invertebrates, as well as more than 40 species of fruit. Fruit consumption is greater in drier habitats, especially during periods of rainfall deficit, and perhaps provides a supplementary source of water. Vertebrate remains have been recorded in 0–50% of dietary samples between locations. Rodents are hunted most often, along with lizards and birds, but short-tailed opossums and an unidentified primate also turned up on the menu at one site. The Gray Four-eyed Opossum shows similar variability in its diet between sites, but consumes more fruit than its close relative. Unusually, it also feeds on

Fig. 9.8. Like most South American carnivorous marsupials, the Pallid Fat-tailed Opossum (*Thylamys pallidior*) includes leaves and other plant material as a consistent part of its diet, as well as a wide range of invertebrates. Source: Dario Podesta.

sap, using its teeth to open and reopen wounds on tree trunks in a manner similar to the Yellow-bellied Glider (*Petaurus australis*) in Australia. Dietary flexibility in other species, including the Gray Short-tailed Opossum, Orange-sided Opossum and Amazonian Red-sided Opossum (*Monodelphis glirina*), appears to be largely an opportunistic response to food availability. However, in the last species, juveniles have more restricted diets than those of adults.

One species bucks the general pattern of dietary variability: the Pallid Fat-tailed Opossum (*Thylamys pallidior*). Unlike most other species studied to date, this opossum inhabits arid and semi-arid habitats, and incorporates invertebrates and leaves as the major components of its diet. In detailed analyses of faecal samples from 123 individuals captured in the Monte Desert, doctoral research by Soledad Albanese showed the proportions of leaves and invertebrates to be invariant among habitats, seasons, both sexes, adults and juveniles (see vignette, present chapter). Leaves represent a reliable and abundant type of food in arid regions, as well as a source of dietary water, and for these reasons may contribute to the observed dietary stability. This interpretation is supported by evidence from another *Thylamys*, the Chacoan Fat-tailed Opossum (*T. pusillus*), which also occurs in the Monte Desert. Although only six dietary samples have been obtained for this species, all from the dry season, plant leaves featured almost as prominently as invertebrates in its diet. Whether other desert-dwellers eat leaves as a consistent part of their diet is a question that awaits further research.

By now, one overarching question should be apparent: why do the American carnivorous marsupials include so much plant material in their diets while their Australasian equivalents eschew non-animal food almost entirely? There is no clear answer. However, if we indulge in some speculation, we might base an argument around the composition of the mammalian faunas of the two continental regions. In Australasia, carnivorous marsupials share their habitats with a wide range of other marsupials, most of which are either larger than the carnivores or overlap only slightly with them in their diets.

Bandicoots and echymiperas perhaps show the greatest overlap with dasyurids, but the 18 extant species in the bandicoot family are all more or less omnivorous and exceed most dasyurids in body size (100–5000 g). There is also a modest radiation of ~60 extant species of small (10–1000 g) and mostly herbivorous or omnivorous rodents. Both the bandicoots and rodents consume invertebrates and sometimes small vertebrates as part of much broader diets, but do not rely upon them. In essence, Australasia's carnivorous marsupials occupy their own relatively specialised dietary niche, with little competition from other mammals. Competition from other vertebrate classes such as birds and reptiles is likely also to be limited as these groups are largely active by day whereas most carnivorous marsupials are nocturnal.

Contrast this idyllic and halcyon situation with that in the Americas. Here, carnivorous marsupials share their habitats with many other similar-sized placental mammals that range from specialist carnivores to generalist omnivores. The specialist carnivores include more than two dozen species of shrews (*Cryptotis* and *Sorex* spp., all less than 100 g), as well as several species of small cats and weasels. Less specialist carnivores that consume some plant material include armadillos (20 species in nine genera, 80–60 000 g), marmosets and tamarins (~40 species in six genera, 120–700 g), various other, mostly larger, New World monkeys, and assorted mid-sized species such as coatis, raccoons and olingos (17 species in five families, 5000–8000 g). And then there are the rodents. Around 120 species have been recorded from Central America and Southern Mexico alone, with many hundreds more from South America. Although relatively few have been subject to detailed study, most overlap in size with the American marsupials, and many are likely to consume smaller animals as part of their diet. The bottom line is that specialist carnivores in the Americas are likely to face competition from other, unrelated specialists, as well as from an extraordinarily diverse range of opportunistic carnivores that eat some plant food too. Such competition could be reduced in two ways. The first is to become even more specialised and exploit prey to which other carnivores have little access. Specialisation on aquatic prey by the Water Opossum and lutrine opossums represents the most obvious example of this strategy, but it is possible that the snake-eating habits shown by some of the larger didelphids represent a further specialisation in the making. The second way to reduce competition is to adopt a degree of dietary generalism. In this case, carnivory can be pursued at times or in places where other carnivores are rare, with dietary switching taking place in situations where competition ramps up. This may be the strategy that most of the small American carnivorous marsupials have adopted. If this scenario is correct, we might expect to see complementary diet shifts taking place in diverse communities of carnivorous marsupials and their potential competitors, as each of the would-be combatants seeks to exploit food resources to which they have some surety of access. These ideas have yet to be tested, but offer some fascinating research possibilities!

Foraging behaviour: ambush versus search and destroy

As shown earlier, the carnivorous marsupials collectively hunt and eat a very wide range of prey, and use appropriately varied tactics, or behaviours, to do so. For the most part, however, they hunt alone rather than in groups. Because most prey types are small and usually dispersed, solitary carnivorous marsupials experience little risk in hunting them and reap all the energetic and nutritional benefits that these prey provide. A single beetle or cockroach, no matter how large, would be unlikely to satisfy a pack of ravenous dunnarts, shrew-opossums or any other self-respecting carnivorous marsupial! As always, there are exceptions. There is a brief window between emergence from the nest and independence in all marsupials, carnivorous or not, when young animals learn to forage in loose rag-tag groups with their mothers. In captivity, small groups of shrew-opossums, some didelphids, some dunnarts, mulgaras and Kowari will shred and share large food items. In the field, large groups of more than a dozen Tasmanian Devils have been seen gorging themselves at the carcasses of dead sheep or cattle. In all these situations, however, the animals have not cooperated with each other to kill their quarry, but have merely converged as individuals upon the food source after it became available. Alone among the carnivorous marsupials, the Thylacine (*Thylacinus cynocephalus*) is reputed to have hunted cooperatively – in pairs – especially when pursuing larger prey such as wallabies. The perceived (but seldom confirmed) attacks of this magnificent beast on livestock led to its deliberate and tragic extermination in the early 20th century, as discussed in Chapter 5.

In the specialist carnivorous marsupials, foraging behaviours are associated with obvious morphological modifications that allow efficient detection and processing of particular prey. The Numbat has the most specialised diet among the carnivorous marsupials. It requires many thousands of termites each day – up to 10% of an adult's average body weight – to meet its needs. Direct observations of foraging individuals show that they move slowly with their snout close to the ground. They are probably detecting buried prey by scent but possibly also by the faint rustlings termites make as they move inside their thin-walled feeding galleries. Once detected, Numbats break into these galleries and use their long (5 cm) tongue to collect their prey. The tongue is kept sticky by copious secretions of saliva to ensure that prey capture is maximised. Numbats have more teeth than other marsupials (48–52), indeed more than any other land mammals, but these are small and delicate; massive teeth are not needed when all prey is relatively soft-bodied.

Another specialist, the Water Opossum, shows distinct behaviours and morphological adaptations that are exquisitely attuned to exploiting aquatic prey. Because it swims with the top of its head above water while foraging, it is unique among the carnivorous marsupials in not using visual, auditory or (probably) olfactory cues to detect its prey. Instead, fish, frogs and freshwater invertebrates below the surface of the water are detected by touch. Prey capture is facilitated by the Water Opossum's long fingers and rough, sandpaper-like palms that help to prevent the escape of its slippery victims.

Dietary differences among the other, more generalist, carnivorous marsupials are often thought to result because animals are opportunistic hunters. That is, carnivorous marsupials eat whichever prey they encounter. Dietary variation arises because different habitats contain different suites of prey. However, this view is almost certainly too simplistic. In the first instance, several types of potential prey such as slugs and slaters are routinely avoided by most species. Slugs and other 'slimy' invertebrates such as planarians (flatworms) can be eaten. When this occurs, animals prepare them for consumption by scraping the invertebrates against the substrate or scratching away the surface mucus to increase their palatability. This behaviour has been observed in the Dusky Shrew-opossum (*Caenolestes fuliginosus*) and in at least two species of *Antechinus*, but probably occurs only if animals are hungry. Animals also avoid different prey types, depending on their age and experience. For example, older or larger hunters are more likely to tackle large or potentially dangerous prey than are juveniles, and have often learnt from prior experience to avoid prey types that are toxic or distasteful. This last response, termed 'conditioned taste aversion', is a valuable survival mechanism. It is also proving useful in an applied conservation context. The relentless westward march of the Cane Toad (*Rhinella marina*) across northern Australia over the last couple of decades has laid waste to populations of many predator species. Carnivorous marsupials such as the Northern Quoll (*Dasyurus hallucatus*) die quickly if they bite into the highly poisonous glands that toads carry on their shoulders. However, if quolls are fed minced toad flesh containing chemicals that cause nausea rather than death before their first encounter with a Cane Toad, they learn quickly that eating live toads is likely to cause discomfort and avoid them thereafter. This ground-breaking research, led by Rick Shine at the University of Sydney and Jonathan Webb

Fig. 9.9. The size of prey taken by carnivorous marsupials tends to relate to the size of the hunter. This dunnart (*Sminthopsis*) is tackling a common prey item – a juicy insect. Source: Steve Murphy.

at the University of Technology, Sydney, is providing the first hope that we can neutralise the impact of Cane Toads in areas where this invasive species has become established.

Differences in the types of prey that animals eat and what is available for them to eat can clearly arise by learned or innate avoidance. However, more importantly, or more frequently, such differences arise from active prey selection. That is, small species tend to prefer small types of prey and experience higher rates of energy return from hunting them. In contrast, larger species benefit more from targeting big prey. Different species also use varied foraging strategies to increase their chance of encountering particular prey, and often search only in those parts of the habitat where such prey occur. Thus, ningauis forage preferentially near or within hummocks of spinifex where their favoured small insects occur. Dunnarts are more active on open sand between hummocks where larger prey can be found. Some species also shift their diets in response to changing environmental conditions. For example, the desert-dwelling Lesser Hairy-footed Dunnart (*Sminthopsis youngsoni*) includes disproportionately more spiders in its diet when conditions are dry than after rain. Spider bodies have a relatively high water content, so selective consumption of spiders presumably assists the dunnarts in maintaining water balance. South-eastern Four-eyed Opossums similarly show preferences for juicy fruits when conditions are dry. In further, intriguing research, CSIRO ecologist Chris Pavey and colleagues uncovered strongly negative correlations between the proportion of vertebrates in the diet of the Brush-tailed Mulgara and both the frequency of use of torpor and the maximum duration of torpor bouts. Because torpor allows animals to reduce energy expenditure, an implication of Pavey's research is that this mulgara hunts vertebrates to reduce energy shortages. This is especially true in winter when energetically profitable invertebrate prey is scarce. Taken together, these studies indicate that at least some of the dietary plasticity we have seen in the carnivorous marsupials arises from their ability to track their preferred prey and to vary their intake according to changing environmental and resource conditions.

While foraging in different parts of the habitat should increase the chance that carnivorous marsupials will encounter preferred prey, recognising, capturing and efficiently killing these prey are the next critical steps in the hunting process. Numbats and a small number of other species that are active by day (e.g. dusky antechinuses, Orange-sided Opossum and Touan Short-tailed Opossum, *Monodelphis touan*) could be expected to use vision to help locate prey. However, direct observations suggest instead that this sense is deployed more to assist animals in navigating their way through the habitat, or to detect predators. If vision is used, it appears to be after a prey item has been detected; it is then used to track prey movements at close range. However, it must be admitted that we lack detailed understanding of the visual systems of carnivorous marsupials. In 2002, for example, the astounding discovery was made that the Fat-tailed Dunnart (*Sminthopsis crassicaudata*) possesses three classes of cone cells in the retina that are sensitive to different wavelengths of visible light. Along with some Primates (including ourselves), but unlike all other mammals, this finding appeared to mean that the Fat-tailed Dunnart sees the world in colour! Subsequent behavioural testing not only confirmed colour vision, but showed that this dunnart is also sensitive to UV light. Other marsupials, including the Elegant Fat-tailed Opossum (*Thylamys elegans*), have since been shown to have some capacity to distinguish colours. These species are largely nocturnal, so the benefits that might otherwise accrue from colour vision remain uncertain. It is also an open question as to whether colour vision plays any role in prey detection or recognition.

Because most carnivorous marsupials are active at night, prey detection must take place usually under conditions of low or zero illumination. Direct observations indicate that many species 'scan' the environment while slowly climbing or walking along search paths, pausing to sniff or listen for potential prey. These deliberate movements have been termed 'foraging walks', and probably serve to maximise the rate of encounter of predators with prey. Prey detection occurs via olfactory or auditory cues in many species, but in captivity can also include vision and combinations of sensory cues. These cues can be subtle and difficult to detect by the human nose or ear, but sometimes are quite obvious. We have watched Brush-tailed Mulgaras visiting ground-level nests of the Crimson Chat (*Epthianura tricolor*), locating them by the soft chirruping sounds made by newly hatched nestlings. The marsupial then moves in a straight line to the next nest to dispatch the occupants

Fig. 9.10. Just like us, the Fat-tailed Dunnart (*Sminthopsis crassicaudata*) sees the world in vivid colour. How many other carnivorous marsupials might have the same ability? Source: Steve Wilson.

there. The Gray Four-eyed Opossum is known to locate southern frogs (Family Leptodactylidae) similarly by zeroing in on their calls. These active foraging behaviours can be considered 'search and destroy' tactics, and are used most commonly by carnivorous marsupials. At the other extreme are ambush tactics. Here, the predator simply sits and waits until unwary prey are close enough to effect an ambush, and then springs upon them. This behaviour has been described in wild Spotted-tailed Quolls, as well as in the Tasmanian Devil. Ambush tactics are probably used rarely by smaller species under natural conditions in the field, but have been reported in captive Southern Ningaui (*Ningaui yvonneae*), Narrow-nosed Planigale (*Planigale tenuirostris*), Giles' Planigale (*P. gilesi*) and apparently in Dusky Shrew-opossums (*Caenolestes fuliginosus*).

Once a potential prey item is detected, carnivorous marsupials usually approach it closely and check its suitability by smell, and perhaps by taste or touch. Contact and prey capture can then occur in several ways. Terrestrial species will dig for buried prey items, and chase and pounce on free-moving ones. Arboreal and scansorial species will remove bark to expose or flush hidden prey or even make acrobatic leaps to pluck flying insects from the air. Larger species, such as Spotted-tailed Quolls, will also enter tree hollows to ensure their quarry has no escape. If a large or unfamiliar prey item is encountered, many species adopt a wait-and-see approach, often with the ears erect and the body flattened, before making a final approach. The next step is to subdue the prey and minimise any risk that might be entailed. For mobile or fast-moving prey, most carnivorous

Fig. 9.11. Young carnivorous marsupials, such as this juvenile short-tailed opossum (*Monodelphis*), learn about the world by degrees. Predatory efficiency, and knowledge of how to avoid predators, will increase with age – a steep learning curve for those animals that survive. Source: Thiago Semedo.

Fig. 9.12. Carnivorous marsupials sometimes dig for buried prey. This Red-cheeked Dunnart (*Sminthopsis virginiae*) has discovered and dug up a witchetty grub (the larva, or caterpillar, of large cossid moths) that it is greedily devouring. Source: Jiri Lochman.

marsupials pin the victim down with their forepaws or immobilise it in the jaws. Depending on the orientation of the prey, it is usually then manipulated so that it can be killed quickly by rapid bites to the head and neck. In shrew-opossums, which have very well-developed and forward-pointing lower incisors, prey items are held down while the victim is repeatedly stabbed. In most other species the canine teeth are used to puncture the prey. If a prey item is slow moving or sedentary, such as a caterpillar or worm, it may be picked up and placed in the side of the jaw where it is chewed with the cheek teeth, much like Bugs Bunny eating a carrot. Alternatively, bites may be inflicted at any points on the body. More complex sequences of behaviour are exhibited if prey items are poisonous (e.g. centipedes, scorpions) or large relative to the hunter (e.g. rats, *Rattus* spp.), and can involve rushes towards and retreats from prey to inflict multiple bites, removal of limbs or appendages. 'Death shakes' are used by some of the mid-sized and larger species such as mulgaras, Kowari and South-eastern Four-eyed Opossum. This behaviour involves a vigorous side-to-side shaking of the head, and helps to reduce the chance that a dangerous prey item such as a scorpion or a snake can turn the tables on the hunter. Chris Dickman has experienced death shakes on his index finger from mis-handling both the Dibbler (*Parantechinus apicalis*) and Brush-tailed Mulgara. These leave 'vvvv'-shaped lacerations of variable depth in the skin and underlying tissue; the author escaped, but smaller prey usually do not.

The predatory behaviour of carnivorous marsupials has sometimes been considered less flexible than that of comparable eutherian (placental) mammals. This is true in that no extant marsupials hunt cooperatively in packs, as do some members of the cat and dog families. However, we have shown nonetheless that 'predators with pouches' use a wide range of hunting and killing tactics to exploit their prey. There is also compelling evidence that they hone their skills with age and readily learn to exploit novel prey. For example, observations of captive Common Planigales (*Planigale maculata*), several species of *Antechinus* and some species of *Monodelphis* (e.g. the Orange-sided Opossum) show that predatory efficiency increases as animals learn with age. Juveniles that attempt to eat a caterpillar covered with irritating hairs, or a distasteful millipede, do not repeat the mistake if they encounter similar prey in future. Common Planigales also learn to avoid the invasive and highly toxic Cane Toad after a single encounter. They even learn to use chemical cues to distinguish between these prey and morphologically similar, but non-toxic, native frogs. In the field, young Mainland Dusky Antechinus (*Antechinus mimetes*) in summer and autumn hone their ability to find and catch particular prey that will be available during winter. This increases their chances of maintaining body mass at a time when food resources are scarce. Many species show great flexibility in learning to hunt small arboreal mammals and birds in artificial nest boxes, buried meat-based baits that have been set for larger pest species such as wild dogs, as well as a very wide variety of artificial baits that researchers use to tempt these often-recalcitrant species into their traps. Sometimes, even the traps themselves are cues for predatory attacks. We have seen larger carnivorous marsupials walking between pitfall traps and peering into them to check whether they contain potential prey. We have also observed Spotted-tailed Quolls extracting the occupants of folding metal traps, having first worked out how to open the trap doors. If this happens once, these fast-learning predators will check every trap in a trap line to free and then eat whatever unfortunate smaller mammals have been captured. Undoubtedly, further study of carnivorous marsupials will uncover further behaviours in their predatory repertoire.

Processing the prey: teeth and other morphological adaptations

If stalking and ambushing are tactics of the hunt, the teeth are the weapons of dispatch. They are needed to seize, immobilise and kill prey, and then cut it up into smaller chunks that can be swallowed. All carnivorous marsupials are dentally well endowed and, like other marsupials, show a pattern of tooth eruption and replacement that is unique among mammals: just the third upper and lower premolars are replaced by secondary teeth. As we noted earlier, the Numbat holds the record for sheer numbers of teeth (48–52). Like those of the marsupial moles (*Notoryctes* spp., 40–44 teeth), the teeth are small and separated from each other. The teeth also are not deeply rooted, and sometimes different numbers of teeth are present on each side of the jaw. These unusual features reflect the fact that teeth play a relatively unimportant role in subduing and processing the usually small prey on which these specialists depend.

The teeth of the generalist carnivorous marsupials are more robust and deeply rooted than those of the Numbat and the moles. The basic dental formula for didelphid marsupials is I 5/4 (five upper and four lower incisors in each half of the jaw), C 1/1 (one upper and one lower canine), PM 3/3 (three upper and three lower premolars) and M 4/4 (four upper and four lower molars). The teeth are symmetrical on each side of the jaw, with a total of 50 being present. The dentition of shrew-opossums is similar, but with four upper incisors and three or four lower incisors, giving a total of 46–48 teeth. Dasyurids also have four upper and three lower incisors, but sometimes only two upper and lower premolars, yielding just 42–46 teeth. In all these species the incisors are small, pointed or blade-like, and allow prey to be seized. The canines are well developed and large with a sharp cutting

Fig. 9.13. Many carnivorous marsupials eat invertebrates that have hard, crunchy exosekeletons. The teeth of the predator need to have plenty of sharp ridges and cusps to cut up and crunch such prey efficiently. These weapons are on display in this fat-tailed opossum (*Thylamys*) – ecologists handling such animals beware! Source: Thiago Semedo.

edge. In species such as the quolls and Tasmanian Devil, the canines are massive and highly developed and can deliver powerful stabbing wounds to prey. The canines are also well developed in smaller species such as the Patagonian Opossum, suggesting that they can readily subdue large prey and perhaps defend themselves quite vigorously if attacked. The cheek teeth in all species have sharp cusps, allowing prey to be sliced into chunks that can be easily swallowed. The carnivorous marsupials in general have higher bite forces for their size than other mammals, requiring not just sharp teeth but also robust jaws and powerful muscles to close the jaws and support the head and neck. Indeed, fascinating comparative studies by Steve Wroe at the University of Sydney (now University of New England) show that the Tasmanian Devil has the most powerful bite for its size of any mammal. However, the most formidable members of the 'Bite Club' were now-extinct Marsupial Lions, which were probably able to kill prey much larger than themselves and had serious bone-crushing capabilities (see Chapter 5).

The diets of carnivorous marsupials are generally rich in water, protein, vitamins and minerals, low in carbohydrates and variable in fat content. In captivity, small species such as dunnarts often show a preference for foods that have a high lipid content, whereas larger species such as the Spotted-Tailed Quoll and Tasmanian Devil prefer protein-rich diets that can be obtained from vertebrate flesh. The diets of all carnivorous marsupials tend to be readily digestible, and their stomachs and guts as a consequence are short and simple. The small intestine is the main site of digestion and food absorption, but food passes through quickly. Food remains can pass through the gut of small species such as planigales in 30–60 minutes, but may take 6 hours or more in larger carnivores such as the Eastern Quoll. The only gastric modification that has been described in carnivorous marsupials is a curious gland that opens just above the stomach in the Dusky Shrew-opossum (*Caenolestes fuliginosus*). This appears to secrete acids and enzymes that break down the large quantities of protein this species consumes. Similar glands have not been reported in other species, although all have diets that are protein-rich. The fast passage times of food through the gut and high rates of metabolism when active mean that most carnivorous marsupials need to eat a lot, and often. In captivity, dunnarts need to eat 10–15% of their body mass on a dry weight basis each day, and as much as 40% of their body mass if the food is fresh. They are likely to be even more ravenous in the wild, so that most active time is spent on the hunt for prey.

Population- and community-level effects of carnivorous marsupials

In this and previous chapters, we have touched upon some of the effects that carnivorous marsupials have on other animals with which they co-occur. Here we consider some of the effects that predation has on prey populations, as well as some of the more subtle and indirect effects that different species have on each other.

Direct effects of predation

Because carnivorous marsupials often show selectivity for different types or sizes of prey, we might expect that populations of heavily hunted prey taxa would be depressed or show well-developed behaviours that reduce predation impacts. In the short term, such effects can best be quantified by the use of controlled removal experiments of the predator and monitoring the responses of prey. In the longer term, selection pressure exerted by predators could be expected to cause shifts in the morphology, behaviour or ecology of prey. Although manipulative experiments have been carried out with a broad array of placental carnivores as the test subjects, only two have been carried out using carnivorous marsupials. Both studied species of *Antechinus*.

The first experimental study was carried out by Honours student Martin Predavec. Martin used 3 m × 3 m fences to exclude Brown Antechinus (*Antechinus stuartii*) from communities of leaf litter invertebrates in tall open forest habitat in Sydney. The invertebrates were sampled six times over 17 weeks using pitfall traps, and their biomass, numbers and composition were compared between the fenced exclusion plots and two sets of controls where the antechinuses had access. Although few consistent patterns were found, there was dietary evidence of selective predation on spiders and beetles. In the second study, Chris Dickman removed Mainland Dusky Antechinus from valley floor habitat in dense forest near Canberra, and sampled surface-active invertebrates using pitfall traps, as well as litter invertebrates using soil cores. Invertebrates were sampled at the time of antechinus removal and three times thereafter over 10 weeks, and invertebrate biomass compared between replicated removal and control areas. There was no consistent pattern of abundance for pitfall-trapped invertebrates, but soil invertebrates sampled in the soil cores showed a tendency to recover slowly in the areas from which mainland Dusky Antechinus had been removed. Dietary analysis showed that soil and litter invertebrates form the dominant component of the diet of this antechinus, thus helping explain these results. Neither experiment continued for long enough to identify whether the antechinuses would have affected invertebrate populations or communities over multiple seasons. This remains an open, intriguing question for future research.

Indirect effects of predation

Experimental manipulations involving large, non-marsupial predators have shown that populations of prey species usually increase when predator numbers are reduced, and decrease if predator numbers rise. Of course, this is not surprising: what else would we expect? However, some of these experiments have provided much deeper insights into the mechanisms that underlie predator–prey interactions. Prey numbers go down or up not just because more or fewer animals are eaten, but because prey individuals are often stressed and scared by the mere **presence** of predators. These situations have been called 'landscapes of fear', and sometimes they can be terrifying indeed. If an encounter with a predator is an ever-present possibility, prey are often less active and restrict themselves to parts of the habitat where they are most likely to be safe. Their stress levels are high, and they spend much time being vigilant and looking out for the predator over their shoulder. Little research has considered how carnivorous marsupials engender and respond to fear, but the topic has been the focus of recent investigations by the Desert Ecology Research Group at the University of Sydney. This work was stimulated because it appeared that fear could lead, unexpectedly, to dramatic increases in the numbers of species that co-occur in any given location. The findings of the research suggest that one species – the Brush-tailed Mulgara – spreads fear among smaller carnivorous marsupials, and in so doing, allows more species to coexist. This observation is clearly counter-intuitive: how does such a situation come about?

Initial observations of the Brush-tailed Mulgara in central Australia showed that, where this species occurs, other species of carnivorous marsupials are

Fig. 9.14. Higher order predators, such as the Brush-tailed Mulgara (*Dasycercus blythi*), are important in maintaining diversity in lower trophic levels that are present in the ecosystem. They do this by maintaining 'landscapes of fear' that prevent other species from becoming dominant and thus provide opportunities for more small species to co-occur. Source: David Nelson.

Fig. 9.15. Lesser Hairy-footed Dunnarts (*Sminthopsis youngsoni*) are capable of hunting and eating large prey, such as the snake-mimicking flap-footed lizard shown here. They fear mulgaras (*Dasycercus* spp.), but are in turn feared by numerous smaller carnivorous marsupials with which they co-occur. Source: Marie Lochman.

usually present too. In the sand dune habitats of the Simpson Desert, for example, this mulgara co-occurs on occasion with up to seven other species. On average, more than five species of other small carnivores co-occur at sites where this mulgara is present, compared with only three species at sites where it is absent. To examine whether mulgaras somehow elevate the numbers of other species, experiments were called for. Thus, if mulgaras do indeed have positive effects, the numbers of other species should decline if mulgaras are experimentally removed from an area. To achieve removal, three 1.5-ha fenced plots were constructed in the Simpson Desert to keep mulgaras from entering. Small holes in the fences allowed all other smaller species to come and go through the fences as they pleased. Exclusion of the Brush-tailed Mulgara from these plots led to a rapid decline in the local diversity of other carnivorous marsupials, from four species to just one. In contrast, the diversity of these species remained high in six other plots where mulgara numbers were not affected by fences. Following mulgara exclusion, only the Lesser Hairy-footed Dunnart could be found. Although hardly much larger than the other co-occurring species, this small (9 g) dunnart increased dramatically in numbers and then excluded all other carnivorous marsupials. Detailed dietary analyses showed that Lesser Hairy-footed Dunnarts comprised only ~1% of the diet of mulgaras, so they were not a major prey item, but further experiments confirmed nevertheless that dunnarts reduced their activity in the presence of mulgaras because they were afraid of bumping into them. The Brush-tailed Mulgara eats very few of the other smaller carnivorous marsupials in the Simpson Desert and, as a consequence, they are not afraid of encountering mulgaras on a dark night, but they are fearful of meeting the small and feisty Lesser Hairy-footed Dunnart.

What can we make of all this? In essence, the interactions between this cast of desert-dwelling carnivorous marsupials provide an example of the old adage that 'my enemy's enemy is my friend'. The small carnivorous marsupials are afraid of encountering the Lesser Hairy-footed Dunnart and avoid areas where this dunnart occurs. This dunnart, in turn, is afraid of meeting the Brush-tailed Mulgara and makes itself scarce wherever this larger predator occurs. As the small carnivorous marsupials have little to fear from mulgaras, they are then freer to come out and play if mulgaras are present, safe in the knowledge that they are unlikely to encounter their nemesis, the Lesser Hairy-footed Dunnart. It is surprising to think that the fear of encountering other species could lead to increases in species diversity, but such effects are being uncovered more and more frequently in studies of other mammalian predators and their prey. The recent discovery of these interactions is probably because they are subtle and indirect. They are harder to detect that the obvious 'blood and guts' results of a direct meeting between a predator and its prey. Carnivorous marsupials may live in landscapes of fear more often than we think, but relevant experiments to disentangle the webs of interactions are yet to be carried out. As with so many aspects of the ecology of these scrappy and beguiling beasts, exciting questions beckon the intrepid researcher!

VIGNETTE: GHOSTS IN THE DARKNESS: STUDYING THE ECOLOGY OF PALLID FAT-TAILED OPOSSUMS

By M. Soledad Albanese

Instituto Argentino de Investigaciones de las Zonas Aridas (IADIZA-CONICET Mendoza), Consejo Nacional de Investigaciones Científicas y Técnicas (CONICET), Argentina

When I moved to Mendoza Province (Argentina), more than 10 years ago, I never thought I would be fascinated by such a little creature. Actually, I didn´t even know this mouse-sized marsupial inhabited the desert landscapes of Argentina. However, with their large ears and big and expressive eyes, once I was introduced to the species, I was quickly hooked. I started my PhD with the main objective of studying several aspects of *Thylamys pallidior* ecology: where they live (habitat use); what they eat (foraging strategies); and how their population is structured throughout the year. But first I had to understand why they had been so challenging to study for other scientists. I soon learned it was not due to a lack of attention from researchers, but rather because they were quite difficult to catch. This marsupial coexists with many species of rodents, and although rodents were easily caught during trapping efforts, *T. pallidior* was so elusive to be almost mythical! I accepted the challenge and got on board for an often demanding, occasionally overwhelming, but always exciting project. Ten years later, I can say that it was totally worth it! The more I learned, the more I became fascinated with these intriguing little animals.

This mouse opossum, commonly called the White-bellied or Pallid Fat-tailed Opossum, or as I like to call it in Spanish 'Marmosa pálida' meaning pallid mouse opossum, has the largest distribution among all *Thylamys* species. It inhabits mostly arid or semiarid regions and is found in a great variety of landscapes, from low and flat Patagonian steppe to high elevations in the Andes of South America. Within these landscapes, the species inhabits diverse habitats such as Mesquite woodlands, shrublands, sand dunes, grasslands and can also be found near rocky habitats.

They have amazing adaptations that enable them to exist in harsh desert environments. The Monte Desert in Argentina, where this little marsupial occurs, is no exception. The climate is strongly seasonal, with freezing cold and dry winters where temperatures readily drop below 0°C. This contrasts the extremely hot and wet summers where temperatures commonly reach and sometimes exceed 40°C. I remember in summer afternoons during my PhD fieldwork in Ñacuñan Reserve (Mendoza), the feeling of opening the door of a giant oven every time we left the Field Station! Unlike us, to cope with these temperature extremes, mouse opossums are active only at night, when temperatures often drop 20°C or more. This is the time when many of their preferred prey are also active. About 70% of the diet of Pallid Fat-tailed Opossums comprises insects and spiders. The agility of these little carnivorous marsupials enables them to easily capture moths, crickets, grasshoppers, beetles and spiders, but perhaps their most impressive feat is the extraordinary accuracy and speed in catching and handling threatening

Fig. 9.16. The enigmatic Pallid Fat-tailed Opossum, *Thylamys pallidior*. Source: Dario Podesta.

opponents such as scorpions! During foraging trials performed in Ñacuñan Reserve, we wanted to evaluate which prey Pallid Fat-tailed Opossums consume and which they don´t. We offered the animals different kinds of insects and spiders, and I recall being a little bit concerned when it was time to test them with scorpions. However, that feeling only lasted a couple of seconds! In the blink of an eye this *Thylamys* immobilised its prey by grabbing the scorpion´s tail with one paw and the rest of its body with the other. It was all over in just a few frantic moments – dismembered scorpion! I was impressed to watch how this, in appearance shy and inoffensive animal, turned into a lethal and efficient predator at the instant a potential prey item appeared.

During the day, when Pallid Fat-tailed Opossums are not hunting for food, they remain in refuges. They use holes in trees, under the ground, or sometimes even occupy bird nests. In some cases the nests are abandoned, but in others the marsupials will prey on eggs they find inside. Even though this desert mouse opossum is mostly solitary, two or more individuals sometimes share the same nest. The composition of these groups is variable regarding sex and age. They can be composed of only males, only females or both sexes. On one memorable occasion during nest monitoring, we were stunned to find 19 animals of different sexes and ages in the same nest! It was a big, living ball of fur and tails! We don't yet know if these groups are formed by relatives sharing the same territory or if unrelated individuals just get together to take advantage of the benefits of group thermoregulation (keeping warm). In fact, this is one aspect of *Thylamys pallidior* biology we are investigating at the moment.

In deserts, winters are definitely an energetically demanding time of the year, when not only temperatures are low, but also resources are scarce. In this sense, Pallid Fat-tailed Opossums show interesting adaptations to deal with this challenging season. For example, they use daily torpor (i.e. decreased physiological activity) as a way to reduce energy expenditure by decreasing body temperature and metabolism. I will never forget my first experience with a torpid animal – the poor beast looked like it was frozen! Not a movement, not a sound and its little furry body was extremely cold to the touch. You would not even know it was breathing! I carefully wrapped it with my glove and put it in my pocket. Less than an hour later, the little opossum was completely awake and restless, ready to go back where I had found it.

Another interesting feature of this species is its tail. As their common name indicates, fat-tailed opossums are able to store fat in their tails to reabsorb as an energy source in lean times. The tail reaches maximum thickness by the end of autumn. By this time, tail width may have changed from 3–4 mm to 11 mm in only a couple of months. By the end of the winter, all animals show thin and wrinkled tails as a consequence of consuming their fat reserves. At this time of the year, individuals that survive winter are ready to reproduce. In Ñacuñan's population, reproduction occurs only once in a lifetime. Females are pregnant in spring and give birth in the most favourable time of the year, when resources are abundant. Most males disappear from the population by the beginning of summer and females just a couple of months later, after weaning their young. At this time, the population is totally composed of young animals from the same generation, born in the same season. This reproductive pattern has been found in other small opossums from South America, and is defined as a semelparous strategy. Even though this is well studied for Australian marsupials, more research is needed to understand how the process compares in their Neotropical counterparts.

Working with the pallid mouse opossum in the wild was certainly a great challenge for me. However, it was a very gratifying and enriching experience in many ways. When so little is known there is much to learn, and on this journey I have certainly learned a great deal about this secretive carnivorous marsupial.

10

REPRODUCTION AND LIFE HISTORIES

If they can acquire sufficient resources and survive the challenges thrown at them by the environment and other animals, carnivorous marsupials mature at 4–24 months of age. As adults, their continued persistence depends on finding a mate, or mates, and breeding successfully. This means not just mating, but ensuring that the young are well cared for and likely in their turn to grow and produce successful future generations of offspring. Here, we briefly sketch the basics of reproduction in carnivorous marsupials and then consider the diverse strategies, or 'life histories', they employ to maximise reproductive success.

Reproduction: the basics

We saw in Chapter 1 that marsupials produce tiny, immature young. This contrasts both with placental mammals that give birth to relatively well-developed young and with monotremes (the echidnas and Platypus) that lay eggs. Although there are many other features of the skull, skeleton and soft tissues that separate the two great groups of live-bearing mammals, the stand out difference lies in their reproductive anatomy and mode of reproduction. In all female marsupials, there are two lateral vaginal canals that serve to transport sperm after copulation, as well as a single central canal that opens during birth. The uteri are also separate, rather than fused into a single uterus as in placental mammals. In male marsupials, the testes are always housed externally in a scrotal sac, and are positioned above the penis. In placental mammals, the testes lie below the penis and are sometimes held within the body. In both sexes of marsupials, urine and faeces are expelled through a common opening called the cloaca. This is also the birth opening, and males insert the penis into the cloaca during copulation. Possession of this common opening distinguishes marsupials from placental mammals, but not from the monotremes, which share a similar anatomical arrangement.

Fig. 10.1. Carrying young is a serious burden for carnivorous marsupial mothers. When the joeys grow large, they increase their demands for milk, making meals for the mother a priority. This mulgara (*Dasycercus*) shows how multi-tasking is done with a supreme juggling act. Source: Jiri Lochman.

Marsupials also have short gestation periods. In most of the American carnivorous species, birth follows just 13–14 days after ovulation; in all the Australasian species studied so far gestation lasts 10–31 days. With such a short time for development, at birth the young – or 'joeys' – are poorly developed. They weigh individually as little as 5–10 mg and collectively less than 1% of the mass of the mother. Unlike their larger marsupial relatives that typically carry a single joey in the pouch, most carnivorous marsupials produce litters of two or more young. Marsupial moles (*Notoryctes* spp.) have only two nipples, and thus can carry only two young at most. Other species in the Australasian region have four to 12, or occasionally even 13, nipples, and often support a joey on each nipple. There is a strong inverse relationship between body size and nipple number: larger species such as the Numbat (*Myrmecobius fasciatus*) and Tasmanian Devil (*Sarcophilus harrisii*) have just four nipples, whereas the tiny Common Planigale (*Planigale maculata*) and Yellow-footed Antechinus (*Antechinus flavipes*) may sport a dozen nipples and carry litters to match. In the Americas, nipple numbers range from four in the monito del montes (*Dromiciops* spp.) and four to five in all the shrew-opossums to more than 20 in the Orange-sided Opossum (*Monodelphis dimidiata*) and Gray Short-tailed Opossum (*M. domestica*). Despite the impressive fecundity that these high nipple numbers imply, actual litter sizes seldom exceed 13 or 14. There is no clear relationship in the American carnivorous marsupials between body size and nipple number.

We often marvel that newborn deer or giraffe can stand and walk within minutes of their birth. In comparison with young placental carnivores or primates, this ability is indeed impressive. But now consider the feat performed by newborn marsupials. No larger than a grain of sticky rice and weighing as little as 5 mg – **six to seven orders of magnitude** less than newborn deer and giraffe – these animated pinkies have to get unaided from the mother's cloaca and find a nipple on which to attach. The hind limbs of these minuscule creatures are just undifferentiated buds, the lungs lack air sacs and the ventricles of the heart have not yet separated. The eyes are black pin pricks, the ears are closed and overall development of the organs may not be complete until at least 30 days of age. However, most crucially, newborn marsupials have relatively large heads and differentiated forelimbs. They move towards the mother's pouch area using swimming movements of the arms.

In some of the larger marsupials, such as kangaroos, the mother may lick the fur between the cloaca and the pouch to help smooth a path for the newborns to follow. This does not appear to happen in the carnivorous marsupials. In some, such as antechinuses, females stand on all four legs with their hips raised at birth, and the young climb rapidly down to the nipples. In others, such as the large American opossums (*Didelphis* spp.), the mother sits in a hunched position with the tail drawn forward. This reduces the distance from the cloaca to the pouch. There is some evidence that newborns use smell to guide them towards the nipples, especially if the nipples are enclosed within a pouch or obscured by hair. It can take newborns as little as 12–15 seconds to crawl to the pouch, and usually after 10 minutes all available nipples have been found. Once they reach their objective, the mouth of the young opens to envelop a nipple and effectively fuses with it for the next several weeks. Most of the carnivorous marsupials studied so far produce two to three times more young at birth than can be accommodated on the available nipples. The surplus newborns die, often eaten by the mother. As tiny neonates are not costly to produce, mothers probably benefit from producing a few 'spares' if this ensures that all their nipples are occupied.

Fig. 10.2. Even in the early stages of suckling, a cluster of joeys quickly grows short on room, jostling for space in their mother's 'pouch', as shown by these young Dibblers (*Parantechinus apicalis*). Source: Jiri Lochman.

As a consequence of being so small and under-developed at birth, young marsupials are entirely dependent on milk from their mother to grow to independence. The age at weaning – when young are no longer dependent on mother's milk – can be as short as 50 days, as in some of the short-tailed opossums. More commonly, weaning takes place after 70–90 days, with larger species such as the Eastern Quoll (*Dasyurus viverrinus*) continuing to lactate for 135–140 days and the Tasmanian Devil for as long as 240 days. In all marsupials, the composition of the milk shows great changes as the joeys grow. In the Eastern Quoll, for example, the percentage of solids in milk increases from ~12% in the week after birth to

Fig. 10.3. A tight squeeze. Brush-tailed Phascogale (*Phascogale tapoatafa*) joeys compete for space while still attached to the nipples on their mother's abdomen. Source: Jiri Lochman.

>30% twelve weeks later. Lipids form the greatest fraction of the milk overall and, together with protein, increase in concentration as the joeys grow. By contrast, carbohydrates decline in importance as lactation progresses. Very similar patterns are seen in the Gray Short-tailed Opossum and Virginia Opossum (*Didelphis virginiana*), except that the shift from dilute to gluggy, fat-laden milk occurs more rapidly to accommodate earlier weaning in these species. These astonishing changes in milk quality coincide with shifts in the suckling intensity of the young. For the first 49–56 days after birth, young Eastern Quolls are firmly attached to their mother's nipples. They suckle frequently and on demand, but as the joeys grow they are able to relinquish the nipple and are left for increasing amounts of time in a nest. Suckling then becomes less frequent, and the increase in concentration of milk at this time helps to keep the youngsters on a rapid growth trajectory.

As young carnivorous marsupials grow, they place increasing demands on their mothers. While still firmly attached to the nipples, growing joeys get to the point when they are too large to nestle comfortably against the mother's belly, whether or not they are ensconced in a pouch. Rather than lying side-by-side with each other, these young begin to flex their bodies and jostle for elbow room. At this stage, the joeys protrude from the mother's belly when she is active and, in most species that carry large litters, they get bumped along the ground as the mother attempts to move about. To avoid scraping her young on the substrate, the mother often adopts a curious gait in which the hind legs and feet are extended, giving her a stilted, stiff-legged appearance. The awkwardness of moving around in this manner provides a cue for mothers to ensure that the maternal nest is completed.

When joeys are first left in the maternal nest, the mass of a single joey averages just 5–15% of the mass of the mother, but by the time the joeys are weaned, their mass will have increased 4-fold. Depending on the species, each joey then weighs 20–60% as much as its mother. Even if carnivorous marsupials nurtured one young at a time, this would represent an impressive investment, but with litter sizes of a dozen

Fig. 10.4. A bumpy ride. Mothers must continue with their busy lives and constantly search for food, despite the exposure of ever-growing young to obstacles on the forest floor, as seen in this antechinus. Source: Jiri Lochman.

or more, maternal investment in lactation rises to a level that is truly astonishing. In smaller species, such as ningauis, dunnarts and antechinuses, females may spend a third of their lives lactating, with the mass of the litters they wean averaging three to four times the mass of the mother herself. It is not surprising that females lose condition and look ragged at the end of lactation, and also suffer high rates of mortality at this time. Not only have they been producing high-quality milk and depleting their own body reserves, they have shouldered the burden of carrying young when needed from nest to nest, and have led the young on their first risky forays into the outside world. Males do not escape the costs of reproduction scot-free, but contribute in other ways that are in some cases even more extraordinary. We turn to these contributions later in the chapter.

The basics of marsupial reproduction are now well known. Yet this knowledge has been acquired only after decades of painstaking and detailed inquiry by legions of dedicated researchers. Before this, marsupials were recognised as being very different to placental mammals, but their unique mode of reproduction long remained shrouded in mystery. Following the description by Vicente Pinzón of the female opossum that he acquired in Brazil in 1499 (see Chapter 6), the belief arose that an animal's young took refuge in the pouch but emerged to suckle from

Fig. 10.5. It won't be long before these youngsters are left in the nest. When this happens, the mother – a Mainland Dusky Antechinus (*Antechinus mimetes*) – will continue to forage, returning periodically during the day and night to suckle them. Source: Marie Lochman.

teats that were located elsewhere on the body. Another explorer, Francois Pelsaert, found the manner of reproduction of marsupials 'exceedingly strange'. In a posthumous report in 1648, he concluded that the young must grow directly out of the nipples. An account of the Brazilian fauna in the same year by Willem Piso compounded this error. Piso thought the opossum pouch was the animal's uterus and also the site where copulation must take place. Other naturalists propounded the correct idea that young were born via the birth canal and thence expelled from the cloaca, but got it wrong in believing that the joeys thereafter became glued in the pouch area by sticky secretions in the milk. Still others believed that the young were simply sneezed into the pouch following vigorous bouts of nasal copulation! Of course, all these notions are known now to be fanciful, but, even without them, there is little doubt that the independent and self-directed movement of young from cloaca to pouch is anything other than a fantastic journey, or the marsupial mother's investment in her demanding brood anything less than heroic.

It is satisfying to note that carnivorous marsupials featured in some of the most important detective work clarifying the marsupial mode of reproduction. James (J. P.) Hill carried out pioneering studies on the embryology, development and function of reproductive structures in the Eastern Quoll in the early 20th century. He was among the first to appreciate the hormonal basis for regulation of the reproductive process. Contemporary studies by Carl Hartman elucidated the hormonal role of the ovary and uterus in reproduction in the 'nearly carnivorous' Virginia Opossum, and also emphasised the novel insights and opportunities that marsupials bring to comparative studies of reproduction in mammals.

Life histories

What is a life history?

A 'life history' is, in essence, a catalogue of the major events that take place in the life of an organism and in the population or group to which it belongs. Consider a person's life history. We grow from babyhood through childhood to adulthood and, if we are fortunate, survive to see old age. Our major life history stages include childhood, puberty, employment, marriage, children, retirement and perhaps grandchildren later in life. The life histories of carnivorous marsupials can be thought of in similar terms as the thresholds that individuals must pass to become mature and reproduce. Life histories have several components, the most important being when individuals first reproduce, the number of times they reproduce, the number of young they produce on each occasion, and death. Other components might include the age at which individuals are weaned and the sex ratio of offspring they produce as adults. These components usually interact and constrain each other, and so cannot be considered in isolation. Reproduction, for example, is a costly process. Because of this, in most animals there is an inverse relationship between the effort expended in reproduction and longevity. Thus, in carnivorous marsupials, heavy investment in a single large litter

Fig. 10.6. Surviving to the reproductive phase of the life history is key for adults and for the population. This Monito del Monte (*Dromiciops gliroides*) has babies on board – a forearm and tail of the young are poking out of her pouch. Source: Marcela Franco.

early in life is likely to drain the energy reserves of the mother (fathers seldom contribute any resources to the care of young) and reduce her chance of surviving to produce more offspring in future. In contrast, females bearing small litters expend less energy in nurturing these young and are more likely to survive to breed on successive occasions. Schedules of life history events are often compiled for cohorts of individuals to produce a 'life table' for the population. Like the actuarial tables used by insurance companies, life tables can be used to specify the probability of death for individuals at a given age; with additional information on reproduction, they can be used further to describe whether a population is growing, shrinking or remaining static in size.

We saw earlier that carnivorous marsupials vary in the times they take to gestate, wean their young and become sexually mature, and they rear litters of different size. As we will see, they vary also in the numbers of litters they produce within and between breeding seasons, the sex ratios of the offspring they produce and in how long they live. What drives and maintains this life history variation? In some species, social factors are important: adults stop younger rivals from breeding and thus delay the actual time of first reproduction beyond when it might be otherwise achieved. In others, interactions with other species are involved. For example, Agile Antechinuses (*Antechinus agilis*) delay their breeding season by about a month in areas where they co-occur with the Mainland Dusky Antechinus (*Antechinus mimetes*), reducing competition with the larger species by so doing. Although such interactions explain some of the variation in life histories, one factor plays an overwhelmingly dominant role: the environment. Before we examine the life histories of carnivorous marsupials, we need to briefly examine how the environment exerts its effects.

Fundamental progress in thinking about the environment–life history nexus was made in the 1970s by American biologist Stephen Stearns and, in particular, by British ecologist Richard (T. R. E.) Southwood. These scientists emphasised the long time scales over which life histories evolve and the costs and benefits of adopting different tactics, such as fast or slow reproduction. In Southwood's view, three aspects of the environment have most influence in shaping these tactics. The first is the **durational stability of the habitat**, or the length of time that a given habitat can be exploited. Habitats with low durational stability include fire-prone heathlands or savannas that are prone to regular fires, riparian areas that flood regularly or hot deserts where extreme climatic conditions recur every few years. These habitats could be expected to support fast-breeding carnivorous marsupials that produce several generations of young in the short time that their living conditions are conducive to growth. Habitats with high durational stability include mature temperate or tropical forest, rock-outcrops or even the deep sandy soils exploited by marsupial moles (*Notoryctes* spp.). These habitats should remain stable and habitable for long periods, allowing their exploitation by slow-breeding species. Marsupial moles have the smallest number of nipples (2) and litter sizes (one to two) of any carnivorous marsupial; their habitat is highly stable, so there is little advantage in breeding very quickly.

The second key aspect of the environment identified by Southwood is its **temporal variability**. This describes the extent to which the resources and living conditions of a habitat vary during the time it is habitable. For carnivorous marsupials, this has a critically important influence on life histories. Except in equatorial habitats or those that experience some kind of environmental buffering, such as deep soils, invertebrates and other resources will vary at different times in the year. Winter is the season when food and shelter resources are likely to be most needed but in shortest supply, and summer the season of plenty. Southwood also nominated **spatial heterogeneity** (or habitat patchiness) as an influence on animal life histories. This aspect may affect the local densities or population structure of carnivorous marsupials, but it probably plays the least important role in shaping their

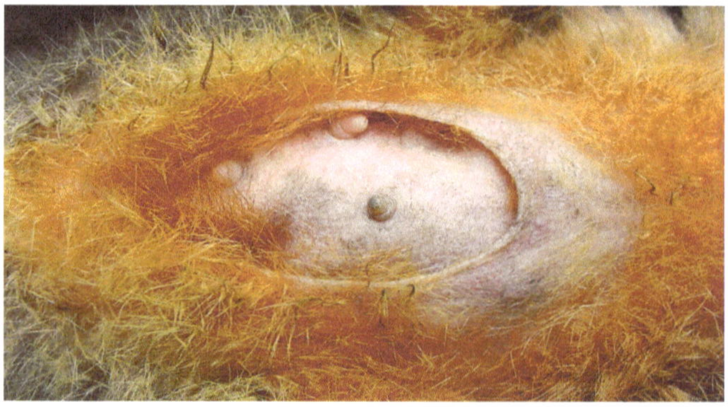

Fig. 10.7. The abdomen of a reproductively active female Large Lutrine Opossum (*Lutreolina crassicaudata*) – the nipples and pouch are clearly visible. Source: Gustavo Scrocchi.

life histories. Nonetheless, these three aspects of the environment together influence when and how often carnivorous marsupials breed, at what age they become sexually mature and how long they are likely to live. We use the term 'life history strategy' to refer to the combinations of life history components that allow carnivorous marsupials to exploit the many and varied environments where they occur.

How do life histories differ?

In comparison with other marsupials, the carnivores generally produce the largest litters and have the shortest lives. With the exception of the ever-anomalous marsupial moles, most species attempt to rear litters of three or more young and produce one to three litters per year. If circumstances conspire to reduce a litter to just one or two joeys, the mother usually kills and removes them, conserving her energy for a subsequent occasion when a full litter can be nurtured. Very few carnivorous marsupials live more than 3 years, and most rear litters of four to 12 at each bout of reproduction. This strategy has been characterised as 'living fast and dying young'. With some exceptions, such as the Water Opossum (*Chironectes minimus*) in which longevity appears to be less than 3 years, large carnivorous marsupials tend to live longer than their smaller counterparts and nurture smaller litters. The Numbat, for example, rears just two to four joeys each year and may do so until 4–5 years of age. Still larger species such as the Tasmanian Devil also have relatively sedate life histories, bearing two to four young each year and surviving up to 6 years in the wild. Fragmentary evidence suggests that the now-extinct Thylacine (*Thylacinus cynocephalus*) also reared two to three young each year. Its lifespan in the wild was never recorded, but, on the basis of its size, we might expect it to have lived somewhat longer than any of its smaller relatives. Captive Thylacines are recorded to have lived up to 8 years.

There are some 136 species of carnivorous marsupials, and the life histories of perhaps a third of these have been evaluated in greater or lesser detail. Rather than describe these one-by-one, it is more interesting and informative to ask if there are any broad patterns that we can define. In an attempt to find such generality, three researchers – Tony Lee at Monash University, Pat Woolley at La Trobe University, and Dick Braithwaite at CSIRO – proposed a novel arrangement of key life history components that allowed them to identify six distinct strategies. These components are: the frequency with which females enter oestrus (or become sexually receptive) within a single breeding season; age at sexual maturity; seasonality of breeding; and the duration and timing of male reproductive effort (Table 10.1). Although based originally on dasyurids, the categorisation proposed by Lee and colleagues has great utility and can be extended readily to encompass all carnivorous marsupials.

TABLE 10.1. A FRAMEWORK FOR CLASSIFYING THE LIFE HISTORIES OF CARNIVOROUS MARSUPIALS.

Key factors in this scheme are whether females enter oestrus once (monoestrous) or more than once (polyoestrous) per breeding season, whether males survive beyond their first breeding season, the length and timing of the breeding season, and the age of animals at sexual maturity. This classification scheme was first devised by Tony Lee and colleagues in 1982, and the table here is redrawn and modified from that work.

Oestrous pattern	No. of seasons per male	Duration of breeding season	Seasonality of breeding	Age of sexual maturity	Type of strategy
Monoestrous	Annual	Restricted	Seasonal	11 months	1
Monoestrous	Perennial	Restricted	Seasonal	11 months	2
Monoestrous or polyoestrous	Perennial	Restricted	Seasonal	11 months	3
Polyoestrous	Perennial	Extended	Seasonal	8–11 months	4
Polyoestrous	Perennial	Extended	Seasonal	4–7 months	5
Polyoestrous	Perennial	Extended	Aseasonal	?	6

Strategy-1

In this strategy, both sexes reproduce at 11 months of age. Males achieve sexual maturity several weeks before mating, at 8–10 months, but copulations are delayed until females enter their only period of oestrus for the year. Curiously, sperm production in some species, most notably antechinuses, shuts down ~2 months before the mating period, with sperm being stored until needed. The benefits that might accrue from this pre-mating disruption to sperm production remain uncertain; indeed, this odd phenomenon is not known to occur in any other mammals. In most dasyurids, the mating period is highly synchronised within populations and takes place over the same 2–3 week period each year, usually in winter. In didelphids, the mating period is seasonal, but appears to be less tightly orchestrated than in their Australian counterparts. After mating, all males die. Maximum male life expectancy is therefore slightly less than 12 months. Males expend a lot of effort in finding and copulating with females during the annual breeding period, and it is the effort of this hyper-activity that contributes to their death. All or almost all females breed in their first year, and a few that survive the rigours imposed by nursing their litters may reproduce a second time at the age of 23 months. Typically, second-year females comprise as little as 10% of the breeding female population in small species that exhibit strategy-1, but the percentage may increase to 50% or more in larger species. Litter size in species exhibiting this first life history strategy varies from six to 12 (and very occasionally 13) among its Australian exemplars and from eight to 14 in its South American exponents.

Fig. 10.8. Some of the American opossums exhibit 'big-bang' reproduction. The short-tailed opossums (*Monodelphis* spp.) are one such genus, but only some species within the genus adopt this life-history strategy. It is uncertain why it is not exhibited by all members of the genus. Source: Thiago Semedo.

Let us pause for a moment. We have just described, in brief, the most extreme life history strategy known in the Class Mammalia. Not only is the timing of the breeding period of strategy-1 species synchronised like clockwork every year, but males enter this period with relatively old, stockpiled sperm, and then die as a consequence of their intense efforts to inseminate as many females as possible. Can we be sure these bizarre events really take place? And, if they do, what drives them? The first clue that breeding occurs seasonally and is concluded by the death of all males came from the pioneering studies of Pat Woolley. Before pursuing Brush-tailed Mulgaras (*Dasycercus blythi*) in central Australia (Chapter 8) and contributing to development of the life history framework we are using here, Pat studied the reproductive cycles and characteristics of the Agile Antechinus, Yellow-footed Antechinus and, to a smaller extent, the Mainland Dusky Antechinus. Working then at the Australian National University for her doctoral thesis, Pat documented clear and predictable seasonal changes in the gonads, sexual accessory glands, body mass and behaviour of her study animals. Crucially, she also noticed that males vanished from her study sites in the field after the breeding season. Until publication of her revelatory results in 1966, the syndrome of post-mating male death had been thought to be the preserve of annual plants, some fish such as salmon, and a few unfortunate invertebrates such as black widow spiders and praying mantises where males are eaten by their mates. If Nobel Prizes were awarded for biological discoveries (as they should be!), Pat Woolley's landmark research would have been in hot contention. Confirmation of the abruptness and totality of post-mating male death came shortly after, via the doctoral work of Don Wood (Chapter 7). Wood went to great lengths (and heights) to find male Subtropical Antechinus (*A. subtropicus*) after the breeding season, placing large numbers of traps on the forest floor and in the rainforest canopy. He found none: all males were absent by the time females gave birth. Subsequent studies by many researchers in the Australian region have shown that this life history strategy – sometimes dubbed 'big-bang reproduction', or more technically with respect to the male die-off, ***semelparity*** – extends to all species of *Antechinus*, *Phascogale*, the Kaluta (*Dasykaluta rosamondae*) and some populations of the Dibbler (*Parantechinus apicalis*) and Northern Quoll (*Dasyurus hallucatus*). In South America, big-bang reproduction has been documented in the Orange-sided Opossum, Northern Three-striped Opossum (*Monodelphis americana*) and some populations of Pallid Fat-tailed Opossum (*Thylamys pallidior*), as well as in several non-carnivorous opossums such as the Gray Slender Opossum (*Marmosops incanus*). In both the Orange-sided Opossum and the Brazilian Slender Opossum (*Marmosops paulensis*), complete semelparity (i.e. in both sexes) has been recorded: all males die soon after breeding, and all females follow after weaning their litters.

Most species of carnivorous marsupial that exhibit strategy-1 life histories occupy habitats with high durational stability and strongly seasonal food resources. Antechinuses and phascogales, for example, are restricted to forest and woodland habitats where peaks in the abundance of invertebrate and small vertebrate prey occur reliably in spring and summer. By mating in winter, females in these species ensure that their litters will be born in early spring and suckled and weaned during the period of the year when food resources are most abundant and reliable. As the growing young place high energetic demands on mothers, ensuring the correct timing of breeding reduces the risks that they will go short of food and suffer reproductive failure. If food peaks only in spring and summer, any females attempting to rear a second litter after the first would have to do so in autumn and winter after the first litter had been weaned. Scarcity of food resources in autumn and winter would make second breeding attempts very risky. Not surprisingly, in environments where invertebrates and other food resources peak during a restricted period each year, females therefore achieve sexual maturity at 11 months and have time to produce just one litter. In males, corticosteroid (stress) hormones ramp up to very high levels in the circulatory system before and during the mating period. These hormones permit the frenetic activity that allows males to find and spend long periods in copulation with females during the short window when females are receptive, but when the mating period is over, the males die. The loss of males after mating dramatically reduces adult population sizes in strategy-1 species, often by more than half. Populations remain small and comprised only of females with their dependent young for the next 2–3 months, increasing again only when juveniles become independent and leave their maternal nests.

The clockwork precision and finality of these reproductive phenomena in strategy-1 carnivorous marsupials have stimulated much research and debate among evolutionary biologists. Indeed, attempts to explain the origins of the unique events that characterise this life history strategy have taken up well over 100 person-years of effort in doctoral studies and other research. Some of this research is touched upon by the writers of several vignettes (Chapters 4, 6, 7, 9 and 10). After we consider the remaining five life history strategies exhibited by carnivorous marsupials, we explore further this most extreme strategy.

Strategy-2

Life history events exhibited by strategy-2 species are similar to those in strategy-1, the major difference being that males do not necessarily die after their first breeding season (Table 10.1). Consider the Sandstone Pseudantechinus (*Pseudantechinus bilarni*). Both sexes breed in winter at 11 months of age, when females enter their single period of oestrus for the year. Young are born a month or more later. In stark contrast to their strategy-1 counterparts, however, about a quarter of the males in their first breeding season survive for another year to potentially breed again. A few males even survive into a third year, living up to 32 months. In further contrast to strategy-1 animals, males with strategy-2 life histories can produce sperm in their second year. In the field, of course, all strategy-1 males die so the question of whether they could produce sperm again would seem academic. Well, yes and no. The testicular changes that cause the cessation in sperm production before males' first and only breeding season are complete and irreversible. How can we be sure? Males that have been captured 3–4 months before the breeding season and maintained in captivity sometimes survive beyond the usual time of mating, and may even live for 2–3 years, but they are completely sterile: the testes never recover their function in these older animals. Are there any

Fig. 10.9. The pseudantechinuses, such as this Woolley's Pseudantechinus (*Pseudantechinus woolleyae*), are strategy-2 breeders – males do not necessarily die after their first breeding season, and typically not all females are pregnant at its conclusion. Source: Marie Lochman.

exceptions? Intriguingly, perhaps there are. A few studies have reported occasional survival of strategy-1 males beyond 12 months in two species of *Antechinus*. In some, but not all, of these very rare cases, these longer living males have lost their testes due to natural hazards, or in fights or trapping accidents. For the survivors that retain their testes, it is interesting to speculate whether any residual testicular activity might remain.

Not all strategy-2 females breed each year. In the Sandstone Pseudantechinus, as few as two-thirds of females have been recorded breeding within a season, with some reproducing for the first time when they reach 23 months of age. A few females survive to live for more than 3 years, although they appear not to breed at this age. Females have six nipples and bear litters of four to five. Up to 15% of these young may be lost in the first month or two of life, reducing litter sizes even further. Other species have similar population and reproductive patterns to the Sandstone Pseudantechinus, and thus also exhibit strategy-2 life histories. These include other pseudantechinuses such as the Fat-tailed Pseudantechinus (*Pseudantechinus macdonnellensis*), some populations of the Dibbler, Northern Quoll, White-footed Dunnart (*Sminthopsis leucopus*) and probably monito del montes (*Dromiciops* spp.). The survival of some males and females to a second breeding season, and the relatively small litter sizes, suggest that reproductive effort in these species is reduced, compared with that of their strategy-1 counterparts. Because of the higher survival of both sexes after mating, population fluctuations are also much less marked in strategy-2 than in strategy-1 species.

Strategy-3

Species with this life history strategy achieve sexual maturity at 11 months, and mate for a short period in winter or spring. Females usually enter oestrus only once a year when matings occur, but differ from their counterparts in strategies 1 and 2 in having the ability to enter oestrus for a second time in the same season (Table 10.1). This facility is termed **polyoestry**. Although polyoestry is not common, the second oestrus appears to be stimulated if the first does not result in successful mating or if the pouch young fail to attach to the nipples, or die prematurely. This strategy has been interpreted as providing some insurance against reproductive failure. If correct, it is not clear why the need for insurance might be greater in strategy-3 species compared with those in strategies 1 and 2, or alternatively why species using these last two strategies do not have the ability to re-enter oestrus in the same season. In strategy-3, both sexes survive their first breeding season and may reproduce in a second, and occasionally even third, year. Strategy-3 species are represented only among larger sized dasyurids such as the mulgaras (*Dasycercus* spp.), Eastern Quoll and Spotted-tailed Quoll (*Dasyurus maculatus*). Litter sizes are usually five to six, but can be as high as eight in the Crest-tailed Mulgara (*Dasycercus cristicauda*). It is unknown whether any American carnivorous marsupials employ this strategy.

In the Simpson Desert, we have recorded female Brush-tailed Mulgaras showing signs of pouch development in August each year and carrying litters of five to six young in September or October. The joeys are then deposited in a nest at the end of a deep burrow and suckled until they are weaned in late November and December. Breeding is annual and usually synchronous from year to year. We have recorded just two exceptional breeding events that disrupted this pattern. On both occasions, females failed to mate or lost their young very soon after birth, but re-entered oestrus and bore full litters of six almost 2 months after other females in the population. Small numbers of both sexes survived to breed in a second year. In Tasmania, research by PhD student Janet Godsell showed similar patterns of reproduction in the Eastern Quoll. Females mated in May or June and produced litters of four to six in late June or July each year, and newly weaned juveniles first emerged from their maternal nests in November. About 33% of females survived until their second breeding season, as did 25% of males. In New South Wales, by contrast, historical records show that females were more likely to re-enter oestrus later in the same season, resulting in a secondary pulse of juveniles in the population after summer.

It is intriguing to note that species using strategies 1–3 often co-occur, yet only females employing strategy-3 can enter oestrus for a second time. If this ability is potentially beneficial for the strategy-3 quolls, why not for the antechinuses and White-footed Dunnarts that live alongside them? The durational stability of the habitat will be the same, or very similar, for females of all these species. However, these species may perceive both the temporal variability of food resources and the spatial heterogeneity of the environment in different ways. Unlike their smaller strategy-1 and 2 relatives, female quolls include more vertebrate flesh in their diets. Because this source of food is not as seasonally limited as invertebrates, quolls are less likely to be constrained to breed at precisely the same time each year. Quolls also forage in patchy forest and open habitats where they are likely to encounter a broader range of prey than is available to strategy-1 and 2 species. Access to more prey is again likely to reduce the risk of failure of a second litter born later in the same season, thus giving strategy-3 females a fighting chance of rearing some young successfully.

Strategy-4

In environments where food resources are abundant and available for longer periods each year, females are no longer confined to a short breeding season and do not need to enter oestrus on a second occasion only as an insurance measure. Instead, females in strategies 4–6 exhibit more regular polyoestry, and both sexes can breed twice in a season (Table 10.1). The breeding season of strategy-4 species often lasts 6 months, and very occasionally up to 8 months. In dasyurids, breeding usually extends between late winter and summer, but in didelphids reproduction coincides with the wet season. In Mexico, Central and northern South America, this occurs from May or June through to October, but at latitudes south of the equator the wet season is reversed and occurs from October or November through to March. Sexual maturity is achieved at 8–11 months of age in strategy-4 species and, because of this, neither sex achieves adulthood quickly enough to breed in their season of birth. Litter sizes vary from four to 12.

Giles' Planigale (*Planigale gilesi*) provides a good example of this strategy. Restricted largely to cracking clay soils in arid regions, breeding in this species takes place between mid-July and late January. Gestation lasts only 15–16 days and young are weaned at 65–75 days, so successive litters may be produced only 80–90 days apart. Although, theoretically, this could allow females to produce three litters within a season, there is no evidence that this maximum is ever attained; females produce two litters per season at most. Females have 12 nipples, but usually rear litters of just six to eight young. By producing fewer young per litter and reducing the potential number of litters per season, Giles' Planigale reduces the effort that it allocates to reproduction each season. This has a clear pay-off: although tiny (~10 g), both sexes of this planigale can survive to participate in a second and, exceptionally, third breeding season.

Like Giles' Planigale, several other species that exhibit strategy-4 life histories occur in arid environments. These include at least two other species of *Planigale*, all three species of *Ningaui*, the Kultarr (*Antechinomys laniger*), Kowari (*Dasyuroides byrnei*) and several species of dunnart such as the Sandhill Dunnart (*Sminthopsis psammophila*), Hairy-footed Dunnart (*S. hirtipes*) and Lesser Hairy-footed Dunnart (*S. youngsoni*). In some arid environments, such as hummock grasslands, the durational stability of large areas of habitat is likely to be low owing to the frequent onset of extreme climatic conditions, including wildfires. Insects and other prey consequently undergo 'booms' and 'busts' in their populations, in turn increasing the temporal variability of the food resource for carnivorous marsupials. Under such uncertain conditions, females often face the prospect of losing their litters, but can spread the risk of failure by breeding twice in the season. If conditions remain poor, many animals make long-distance movements to find food resources elsewhere in the patchy desert landscape (see vignette, Chapter 8). Conversely, if conditions remain benign and productive throughout the breeding season, females will often rear both an early and a later litter.

Fig. 10.10. Kultarrs (*Antechinomys laniger*) exhibit a strategy-4 life history – both sexes can breed twice in a season. Source: Jiri Lochman.

It is not clear how many American opossums exhibit strategy-4 life histories, because the age of sexual maturity is unknown for most species. Nonetheless, fat-tailed opossums, such as the Elegant Fat-tailed Opossum (*Thylamys elegans*) and some populations of Pallid Fat-tailed Opossum (*T. pallidior*), are likely candidates, producing one or two litters over breeding periods that last up to 6 months. The Big Lutrine Opossum (*Lutreolina crassicaudata*) is an even stronger candidate for inclusion in this life history. Although gestation length and developmental times are not known, it is unlikely that joeys of this large carnivore mature in the season of their birth. It is polyoestrous: field studies have shown animals to breed in spring and again later in the year after the young of the first litter have attained independence. Inclusion of the Big Lutrine Opossum within the fourth life history strategy expands the range of habitats where these species occur. The long breeding season of the Big Lutrine Opossum probably reflects the reliability of aquatic food resources that are available to it. Indeed, occasional records of reproduction outside the usual breeding season indicate that this species can exploit exceptionally favourable conditions at any time.

Strategy-5

This strategy is similar to strategy-4, with the only difference being that females have the potential to breed in the season of their birth. To allow such early breeding, females have to develop rapidly. Some carnivorous marsupials achieve this by being small, and thus reducing the amount of growth needed to reach adult size. These species include the Fat-tailed Dunnart (*Sminthopsis crassicaudata*), Common Dunnart (*S. murina*) and Stripe-faced Dunnart (*S. macroura*), all of which weigh 25 g or less. Other carnivorous marsupials achieve early sexual maturity by rapid growth. These include the 50–60 g Hooded Red-sided Opossum (*Monodelphis palliolata*) and the much larger (~500 g) Gray Four-eyed Opossum (*Philander opossum*) and South-eastern Four-eyed Opossum (*P. frenatus*). In all these species, females have short gestations of 2 weeks or less and wean their young at 65–75 days. Sexual maturity is achieved at just 4 months in female dunnarts and by 5–7 months in all the other species. Litter sizes are four to six in the larger species and six to 10 in the smaller ones. Some individuals exhibiting this strategy survive to participate in a second breeding season, with smaller representatives living up to 18 months and larger species up to two and a half years.

As in strategy-4 species, adults that breed early in the season could potentially produce second and third litters at intervals of 80–90 days. And, some do. However, even though the breeding season might extend up to 8 months, both the dunnarts and smaller opossums usually attempt to rear just two litters, spacing them out and taking some time to replenish depleted maternal resources. Relatively few individuals attempt to produce a third litter. However, the larger opossums are different. The Gray Four-eyed Opossum, for example, can pump out litters at intervals of 90 days and produce up to four litters in a season. Such fecundity is possible because this larger opossum produces fewer young per litter than its smaller counterparts, and thus reduces the energetic costs of lactation. In addition, production of four litters per year occurs only in mature tropical forests in Central America and Peru where resources last long enough to allow breeding to extend for 9–10 months of the year. In southerly parts of the species' range in Brazilian Amazonia, females with young have been recorded only in the late rainy season in February and March. In situations where the breeding season is extended and juveniles are recruited almost continually, populations in many strategy-5 species are relatively stable and do not usually show dramatic fluctuations in size within or between years.

Despite the potential for strategy-5 young that are born early in the season to mature and reproduce by the end of the season, field studies show that this potential is seldom realised. It is possible that early maturing young lose out in competition for mates with older and more experienced adults. If this is so, the early breeding potential in strategy-5 species might be realised only in populations at low density where competition is reduced. Unfortunately, this possibility remains untested: field studies on low density populations are difficult and unattractive for researchers because they yield few captures. If strategies 4 and 5 turn out to be functionally the same, what then might be the advantage of producing early-maturing young? For the desert-dwelling dunnarts, we speculate that the early maturation of young produces a 'pool' of animals that can rapidly track and exploit changes in food or habitat resources. Temporary population irruptions that have been observed in both the Fat-tailed Dunnart and Stripe-faced Dunnart (Chapter 7) might represent examples of this opportunistic strategy. The Common Dunnart also appears to exploit its environment in this manner. Fires frequently sweep its preferred heathland habitats, creating patches of fresh and abundant food resources in their wake. Young dunnarts are among the first responders, moving in from the surrounding 'pool' to exploit the bounty. Early maturation of young may bring similar benefits for the forest-dwelling representatives of strategy-5, although temporal changes in food and habitat resources should be less marked in forest habitats.

Strategy-6

If food resources are available reliably and consistently throughout the year, we might expect temporal constraints on reproduction to be relaxed and thus see year-round breeding. This is exactly what occurs in strategy-6 carnivorous marsupials. Females are polyoestrous, males stay in reproductive condition for long periods, and both sexes potentially contribute to the production of several litters over their lifetimes. Only one species is known to exhibit this life history strategy in Australia: the northern form of the Common Planigale. At least seven species of New Guinean carnivorous marsupials appear to breed throughout the year, as judged by captures of females that are pregnant, carrying pouch young or in later stages of lactation. These include the New Guinea Quoll (*Dasyurus albopunctatus*), Speckled Dasyure (*Neophascogale lorentzii*), Narrow-striped Dasyure (*Phascolosorex dorsalis*) and at least four of the five currently described dasyures in the genus *Murexia*. Similar numbers of American marsupials also breed year-round, at least in some parts of their geographical ranges. These include the Tlacuatzin (*Tlacuatzin canescens*), Gray Short-tailed Opossum, Long-nosed Shrew-opossum (*Rhyncholestes raphanurus*) and several larger species such as the Water Opossum, Brown Four-eyed Opossum (*Metachirus nudicaudatus*), Gray Four-eyed Opossum and McIlhenny's Four-eyed Opossum (*Philander mcilhennyi*).

Many strategy-6 species live in stable forest habitats where food resources show little seasonal variation and durational stability of the habitat is high. Under these constantly favourable conditions, females appear to maximise their reproductive output by producing small litters over extended periods. Thus, in the equatorial forests of New Guinea, most strategy-6 species have four nipples and rear litters of three to four young. The New Guinea Quoll is exceptional in having six nipples but usually rears no more than four young at a time. Litter sizes are also small in the Long-nosed Shrew-opossum (maximum 4 joeys) and in the larger Water Opossum (one to five) and four-eyed opossums (five to seven) of South America. This strategy allows females to reduce their investment in each litter, with the compensation being that more litters are reared over their lifetimes. In the remaining strategy-6 species, however, maximum litter sizes are much larger (12–14), so a different interpretation is required. For the northern Australian planigale, the durational stability of the habitat is low because its habitat floods each year during the summer monsoon. Here, the large litter size (12), frequent reproduction and seasonal movement of animals to higher ground allow this species to exploit an environment that is extraordinarily productive but only patchily habitable for part of the year. The high fecundity of the Tlacuatzin most likely reflects a similar response to its environment. This small and enigmatic opossum exploits varied open savanna, scrub, dry forest and even agricultural habitats that are highly productive, but also prone to disturbance. In the Gray Short-tailed Opossum, females vary their litter size and do not always rear the maximum number possible. They produce small litters of six to eight in buffered and stable rock outcrop habitats where their survival prospects are likely to be high, and larger litters in other habitats where longer term survival is less assured.

Other life history strategies – variations on a theme

These six strategies neatly encompass most of the known variation in life histories of carnivorous marsupials. There are of course many species whose life histories we do not know, but for the most part it is reasonable to expect that rare or unstudied species will not differ greatly from those of their better known counterparts. The standout exceptions are the marsupial moles (*Notoryctes*). The low nipple number and small litter sizes of both the cryptic Kakarratul (*N. caurinus*) and Itjaritjari (*N. typhlops*) suggest that these species' life histories are quite distinct from those of all other carnivorous marsupials. The shrew-opossums also retain a distinct aura of mystery about their reproductive habits. Despite the best efforts of biologists to breed them in captivity, none of the seven species has ever done so. Intriguingly also, animals have never been found in the field with attached pouch young. This might suggest that joeys are dropped in the maternal nest at an early stage but, frustratingly, no nests have ever been found to confirm this suspicion. Apart from the Long-nosed Shrew-opossum, fragmentary evidence suggests that the Incan Shrew-opossum (*Lestoros inca*) and perhaps two of the five species of *Caenolestes* breed seasonally. It would be premature to place these species into any of the six life history strategies at this stage.

The carnivorous marsupials are nothing if not contrary, and it would be a mistake to think that even the better known species can always be pigeon-holed. We noted earlier that the post-mating death of all males occurs in some – but not all – populations of the Dibbler and Northern Quoll, so they exhibit strategy-1 or strategy-2 life histories depending on the location. Not only do they show variation between populations, both species sometimes oscillate between these two life history strategies in the **same** population in different years. Several species of opossum also vacillate between life history categories in different parts of their distributions. As shown by some of the larger species, such as four-eyed opossums, this flip-flopping is usually between strategies 5 and 6 and relates to whether animals breed for most or all of the year. The South-eastern Four-eyed Opossum provides a good example of this variability, breeding seasonally in coastal moist broadleaf forest in south-eastern Brazil, but year-round in other forested habitats at similar latitudes inland.

Several other carnivorous marsupials 'bend the rules' with respect to their life history categories. Consider the Numbat. Most matings occur over a short period in early January each year, but if females fail to conceive or lose their litter at an early stage, they can enter oestrus and mate for a second or third time in the same season. This qualifies the Numbat for membership in the strategy-3 life history club. Yet, unlike all other members of this group, males take their time to develop and do not become sexually mature until their second year. Another strategy-3 member, the Tasmanian Devil, is even more tardy.

Fig. 10.11. The Northern Quoll (*Dasyurus hallucatus*) is not easily classified into a discrete life history category. This species oscillates between category 1 and 2 life histories. Post-mating death of males occurs in some, but not all, populations, and not in every population each year. Source: Alex Hartshorne.

Females achieve sexual maturity at 2 years or earlier, but males may not mate until they are 3–4 years of age even if they are sexually mature at 2 years. Weaning occurs at 150–240 days in the Tasmanian Devil, the longest for any carnivorous marsupial. These two species compensate for their slow development, once-annual reproduction and small litter sizes, by breeding in successive years. In contrast to other carnivorous marsupials, these life history characteristics qualify the Numbat and Tasmanian Devil for membership of their own sub-group within strategy-3: we might characterise it as 'living slow and dying old'.

How do we obtain life history information?

Before leaving our treatment of how the life histories of carnivorous marsupials differ, we need to ask how this ostensibly arcane information is obtained. The earliest knowledge was gained by native peoples in the Americas and Australasia – people for whom an intimate understanding of their living environment was essential for their persistence. This knowledge, although extensive, was communicated orally and has now been diminished and fragmented to a large degree following the arrival of new settlers from Europe (see Chapter 6). With the arrival of Europeans, other means of gathering knowledge began to be employed. Before the 20th century, most studies were thus built upon anecdotal observations of wild animals made by collectors and naturalists. These intrepid explorers returned specimens to museums that were then inspected by researchers to yield information on species' anatomy, development and reproduction (see Chapter 6). Later workers, such as James Hill and Carl Hartman, gained more detailed insight into reproductive cycles by maintaining animals in captivity and recording changes in their

Fig. 10.12. Look carefully and you will see the silver antenna of a tracking device extending horizontally from behind the back of this Monito del Monte (*Dromiciops gliroides*). Such sophisticated devices have allowed us to uncover crucial information about home range, daily movement patterns, habitat use and reproduction in carnivorous marsupials. Source: Francisco Fonturbel.

organ systems and breeding behaviour as they grew and matured. This approach continues to bear fruit today. Somewhat later, in the 1930s and 1940s, skilled naturalists such as David Fleay began to spend more time in the field recording the behaviour and habits of animals under natural conditions. Soon after, the advent of mass-produced live-capture traps such as the Sherman trap (commercially available from 1955) and Elliott trap (1965) greatly expanded opportunities for researchers to study species' life cycles. The post-war period saw a rapid uptake in the use of live-capture traps, and this in turn has allowed unprecedented insight into the secret lives of carnivorous marsupials and other previously mysterious creatures.

Live-capture traps are simple devices. Usually made of wire mesh or sheet aluminium, they are set in sites where target animals are expected to be active. When an animal passes through an open trap entrance, it triggers closure of the trap door and thus confines the animal within the body of the trap until it is released by the researcher. Animals are usually enticed into the trap by an appropriately juicy or smelly bait that the researcher has placed inside. Once through the trap door, animals can spring it shut by stepping on a trip wire or treadle, or by tugging at the bait and releasing a door stop. Bedding material is also often used. This gives captured animals a place to rest, and keeps them comfortable if cool or inclement conditions are anticipated. Another commonly used live-capture trap is the pitfall trap. As the name implies, these traps are pits that are dug into the ground to capture animals scurrying about unwarily on the ground surface. The pits are usually reinforced with plastic or PVC sides to stop soil falling in, and given a mesh floor to prevent captured animals from digging their way to freedom. These traps are usually also provisioned with a drift fence of wire mesh or plastic sheeting that is erected across the top of the open pitfall trap. The fence serves to intercept animals that are running about in the vicinity of the trap and guide them into the pit.

Different traps are needed to capture the various species of carnivorous marsupials. For example, most dunnarts are exceedingly suspicious of shiny metal traps and can be captured reliably only in pitfall traps. By contrast, fat-tailed mouse-opossums – probably the closest to dunnarts in terms of appearance and ecological similarity – enter metal Sherman traps readily. Field ecologists have come up with an ingeniously wide range of trap designs to improve the chance of capturing their particular target animals. These include pipe traps on trees for arboreal species, funnel traps that imprison animals after they have crawled through one-way doors, and horizontal pipe traps made of PVC to capture and contain such powerful animals as the Tasmanian Devil. Baits are also tweaked to improve capture success. Although the 'universal bait' for small carnivorous marsupials is a gooey mixture of oats and peanut butter, researchers often add honey, golden syrup, pieces of bacon or other dry meat, vanilla, aniseed, vegetable oil, and other ingredients in the knowledge (or hope) that they will catch more animals. Dry cat food is a popular alternative. For larger carnivorous marsupials that include more flesh in their diets, baits of sardines, chopped liver, chicken wings or chicken heads are *de rigueur.* These baits have the advantage of becoming smellier and more attractive to the target animals over time, even if they become somewhat less attractive to all but the most hard-core researchers!

The most important point about using live-capture traps is that they are set to catch and keep the target animals without harm until the researcher arrives. Ideally, trap checks are made at or towards the end of the main period of activity. This minimises the amount of time that animals spend in the traps, and allows them to be released in time to return to their nests or burrows to rest. We are sometimes asked by curious, if not sceptical, members of the public why researchers go to so much trouble to capture animals in the first place if our intention is just to release them again as quickly as possible. The answer, of course, is that we seek to obtain as much information as possible from captives **before** releasing them. Basic information that is usually recorded on each animal includes its specific identity, sex, body size, reproductive state and body condition: its vital statistics. Mass is the most commonly recorded measure of animal size in the field, but measurements of the length of the head, body, tail, ears or feet may be taken, especially if these help to distinguish species that are similar in appearance. The sex of carnivorous marsupials is unambiguous: the scrotal sac is obvious in males even when young are still attached to their mother's nipples. Reproductively active males can be distinguished if they have relatively large testes and extrude the penis during handling. In females, pregnancy is signalled by a loss of hair in the abdominal area around the nipples, the growth of longer, pale-coloured hairs around this depilated area, and a granular appearance of the skin as the underlying tissues gear up for lactation. After birth, the joeys can usually be counted and sometimes sexed while in the pouch. They cannot be removed without harm, but common assessments of their growth stage include their crown–rump length (distance from the top of the head to the rump), hairiness, whether the digits have separated, and the state of development of the ears and eyes. After the joeys have been deposited in a nest, lactating females

Fig. 10.13. Chris Dickman checking an Elliott trap, with an open pitfall trap at his foot. Lines of traps are usually set to maximise the chances of catching animals, and here a line of paired Elliott and pitfall traps can be seen running out from the shrub. In open areas such as this, early trap checks are essential to ensure that animals are released and traps closed before the sun gets too high. Source: Paul Zborowski.

can be identified by the bare, pink, puffy appearance of the pouch area and enlarged nipples. In some species, gentle palpation will even allow milk to be expressed. At the end of lactation, females that have reared a litter can be readily distinguished from those that have not by their larger size, orange-brown staining of fur in the abdominal area, and nipples that are clearly distinct and visible. Many researchers also make notes on the outward appearance of captured animals, such as whether they have missing fur, injuries, or external parasites such as fleas or ticks.

With a carnivorous marsupial in the hand, much more information can be obtained. We saw in Chapter 7 that radio-transmitters and other tracking devices can be attached to animals before they are released, enabling researchers to later track their fine-scale movements, habitat use, den sites and patterns of activity. Faecal samples can be collected to allow examination of the diet or the presence of any internal parasites. Samples of hair likewise might be taken to identify longer term shifts in diet from the changes that different foods leave as their chemical signature along the lengths of individual hairs. Small blood or tissue (ear clip) samples can be used in a wide range of physiological or genetic analyses. For life history research, once an animal's vital statistics have been recorded, the most critical next step is to mark the animal before release so it can be recognised if it is recaptured. Most researchers use passive integrated transponder (PIT) tags – a chip as small as a grain of rice inserted into the scruff of the neck, like the vet does for your pet dog or cat. Each chip has a unique number that can be electronically scanned even while the captive is still inside a catching bag. Alternately, ears are notched in various combinations to ensure individual identification (see vignette, Chapter 8). Provided that researchers are able to return to the study area and set traps on a regular basis over a period of a year or more, deep insight into species' life histories can be obtained. Young animals marked soon after independence can be tracked to reveal how quickly they grow, when they become sexually mature, when they reproduce, how many young they rear during each bout of reproduction, and how long they live. If many individuals are marked and tracked on each sampling occasion, all the key life history characteristics that underpin the six strategies can be elucidated.

Sounds easy, doesn't it? In some ways it is. Indeed, for many biologists there is nothing better than being in the field, capturing and handling their study animals and trying to figure out how they live. However, there are risks for the unwary or the impatient, as we will show with a couple of examples from earlier stages in the life history of one Chris Dickman.

Example 1: the unwary. In December 1985, small populations of one of the world's rarest carnivorous marsupials were discovered on two tiny islands north of Perth, Western Australia. That species, the Dibbler, was known at the time from a handful of localities in the south-west of Western Australia. The species had been presumed extinct until the chance discovery by wildlife photographer Michael Morcombe of a pair of Dibblers at Cheyne Beach, near Albany, in 1967. Soon after, Pat Woolley showed that the species breeds once a year and some males survive into their second year. Not only this, older males retain the ability to produce sperm and thus potentially participate in two breeding seasons. This is clearly a category-2 life history strategy. Discovery of the island Dibblers was incredibly exciting and presented an outstanding opportunity to uncover more aspects of the secret life of this elusive and almost legendary beast. Using both pitfall traps and Elliott sheet aluminium traps, Chris was able to capture and mark some 30 individual Dibblers on a first visit to the islands in March 1986. Qualified now as a Dibbler-wrangler *par excellence*, fame and fortune seemed assured! However, a trip just a month later produced a huge shock. The Dibblers had almost vanished, and the few animals that were captured were all females. For a young biologist hoping to get a few runs on the board, the thought that you might have single-handedly knocked off the largest known population of a severely threatened species was somewhat enervating; the path from Dibbler-wrangler to Dibbler-mangler, perilously short. In hindsight, the fact that all animals captured on that second trip were females and showing signs of pregnancy should have given pause for more positive reflection. Perhaps, just perhaps, male island Dibblers die after mating, in contrast to their counterparts on the mainland. The discovery of a single male later in that field trip confirmed this possibility: the (un)fortunate animal was shagged out and moribund, its tongue and penis drooping on the sand where it was lying.

Fig. 10.14. The Dibbler (*Parantechinus apicalis*) exhibits male die-off after mating in some years on Boullanger Island, off the coast of Western Australia, but does not show die-off on the mainland. Here, Dibbler joeys are being counted and their sex recorded by Chris Dickman and Phil Withers (of the University of Western Australia) during a 3-year study of the Dibbler on Boullanger Island. Source: Chris Dickman.

The animal's whole appearance was that of a classic strategy-1 male at the end of the mating period. Subsequent sampling confirmed that many females were in fact still around, but because they were now carrying young they had become shy about exploring the traps and thus were harder to catch. Juveniles of both sexes turned up in healthy numbers later in the year, further confirming the big-bang strategy of this island population. After an unwary and disconcerting beginning, this field program went on to add a new page to our understanding of the Dibbler's life history.

Example 2: the impatient. Relatively few studies of populations run for more than 10 years. For those that do manage to go the distance, you might think this is an ample length of time in which to discover the factors that increase or decrease the size of the population, and to accurately characterise the species' life history. Such expectations would be quite reasonable for most small mammals, but for our ever-contrary carnivorous marsupials, patience can be needed to maintain studies of even longer duration. Take the research program started by Chris Dickman and colleagues in the Simpson Desert in early 1990. By 1999, the program had run for 10 years and amassed nearly 1000 captures of three species of carnivorous marsupials in the course of live trapping each year. In July 1999, the Australian Mammal Society held a symposium on the biology of carnivorous marsupials, and proposed subsequently to publish papers from this event in a special theme issue of the journal *Wildlife Research*. This provided a wonderful opportunity to synthesise the decade-long results of the desert research and present them to a broad audience.

One of the three carnivorous marsupials in the study, the Wongai Ningaui (*Ningaui ridei*), had been shown from previous studies of captive animals and from careful analysis of museum specimens to be polyoestrous, breeding between August/September and February. With no evidence that breeding occurs in the season of birth, these observations seemed to define this ningaui clearly as a strategy-4 species. However, analysis of the Simpson Desert results told a very different story: here, breeding began in October, young were born in November and became

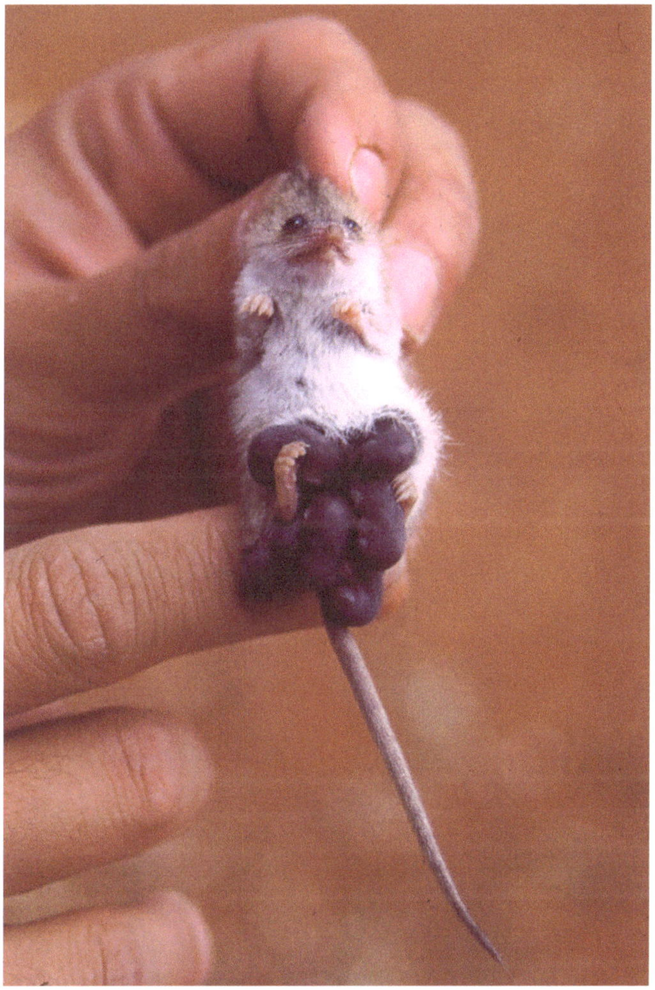

Fig. 10.15. A 10-year study of the Wongai Ningaui (*Ningaui ridei*) in the Simpson Desert suggested that this diminutive carnivorous marsupial produced just one litter per year, in late spring. In the 12th year of the study, two individual females were found to have produced a second litter. The animal shown here is one of those two fecund females, and has six joeys attached to her nipples. Source: Chris Dickman.

independent by the following February or March. Females bred once, all at the same time, and carried litters of five to eight joeys. Neither sex died off after breeding, but nor did they survive to a second breeding season. Eureka! We had discovered the first carnivorous marsupial with a strategy-2 life history living in an open desert environment. There is something perversely satisfying in thinking that you have gleaned new insights into how an animal pursues its life, even if it is a minuscule creature that nobody has ever heard of. The research describing these revelations was duly written up and accepted for publication in the special issue of the journal. The last step in the publication process is to check and return the proofs of the typeset paper to the journal editor, and we were able to do this just before leaving for the next field trip to the desert, in March 2001. Job done! Except, of course, it wasn't. The first two ningauis that we encountered on that field trip were females that had been captured earlier in the season with large pouch young. Both had new litters – their second in the same breeding season. By the time we returned from the field a couple of weeks later, our hot-off-the-press paper was already out of date. The Wongai Ningaui is a strategy-4 species because it has the **capacity** to rear two litters in a breeding season, even if in most years it raises only one. Perhaps we should have been more patient and waited until the research program was finalised. It is sobering to think that even 10 years may not be enough to reveal the 'true' life history of a carnivorous marsupial, but the ningaui example reminds us that life histories can sometimes be more flexible and less fixed than we appreciate.

Big-bang theories

Before closing this chapter, we return to two of the most puzzling but important aspects of strategy-1 carnivorous marsupials: their synchronous annual breeding cycles and the post-mating death of all males. The question of when to breed is particularly crucial for seasonal breeders because they have only a short window of opportunity to exploit seasonal peaks in food resources. Often, they must initiate mating when conditions are cold and miserable, many weeks before the energetic demands of lactation become acute. Breeding in some carnivorous marsupials seems influenced by changes in temperature or rainfall, or increases in the food supply, but these cues are not reliable from year to year and provide only a general guide to improving conditions later in the season. They cannot explain the clockwork year-by-year precision in timing that characterises big-bang breeders. Instead, most species in strategy-1 appear to use photoperiod, or day length, as their major trigger. Changes in day length are very consistent and predictable in any locality, irrespective of the prevailing weather conditions or how much food might be available at the time. In the Gray Short-tailed Opossum and several species of dunnarts, for example, reproductive activity begins in spring when day length is increasing and slows or ceases after the summer solstice. The day length effect has been confirmed in several dunnart species by exposing captive animals to increased artificial light. This prompts testicular enlargement in males and oestrus in females, even at times outside the usual breeding period.

Using increased day length (or decreased night length) to initiate breeding activity makes sense because longer days herald improving conditions in spring and summer. However, this cue is not sufficient to account for the precise year-to-year timing of breeding in carnivorous marsupials such as antechinuses and phascogales. In eastern Australia, all species in these genera breed earlier in spring at high latitudes and progressively later in spring at lower, more tropical latitudes. This is counter-intuitive. If increasing day length alone is important, why don't animals breed at any time between mid-winter and mid-summer when days are lengthening? And why is breeding delayed at low, tropical latitudes, where the spring flush of food resources needed by lactating females could usually be expected to arrive early?

Day length is still important in triggering synchronous spring breeding in antechinuses and phascogales, but it is used in a way that is not immediately obvious. Instead, the critical trigger for reproduction is the **rate** at which the day length increases during spring. At low latitudes, day length differs little between seasons, so changes in day length are small and take place at a leisurely pace. In contrast, at high latitudes the change from short winter days to long summer ones means that day lengths in spring change quickly, so a given rate of change in day length is attained earlier at more southerly latitudes. If the average time of ovulation of females is taken as the time of breeding in antechinuses along Australia's east coast, different species are seen to ovulate when a specific, but different, threshold in the rate of change is reached. For the Yellow-footed Antechinus, ovulation is triggered when day lengths increase at 77–97 seconds per day. Thresholds for other species are higher: 92–99 seconds per day for the Subtropical Antechinus, 99–111 seconds per day for the Brown Antechinus and 127–137 seconds per day for the Agile Antechinus. Responding to these different rates of change allows each species of antechinus to rear litters in spring or summer, and to do this earlier at southerly latitudes. Even the most northerly distributed species, the Fawn Antechinus (*Antechinus bellus*) and Cinnamon Antechinus (*A. leo*) appear to respond to rate of change of day length but do so at the slower rates of 32–47 seconds per day.

From a human perspective, we know that days in spring get longer, but few (if any) among us would be able to measure increments of just a minute or two each day. These differences are scarcely perceptible, but antechinuses and phascogales have an extraordinary, almost supernatural, ability to perceive and respond to these tiny increments in the vernal photoperiod. We still do not know precisely how they do it, but there is some suspicion that the pineal gland, which sits at the base of the brain, is involved in some way. Another possibility stems from the fact that male antechinuses take a couple of months before breeding to reach full sexual development, and females almost as long. Thus, both sexes are perhaps reading and responding to declining rates of change of day length before the winter solstice. Some experimental work on captive Agile Antechinuses indicates that females will enter oestrus early if exposed to shortening day lengths, thus supporting this idea. However, it may not apply universally, because populations at low, tropical latitudes breed 3 months or more after the winter solstice and must therefore initiate reproductive development when day lengths are increasing slowly in early spring. Although rate of change of day length clearly triggers reproduction, its mode of action may differ between species and locations. So far, the only mammals apart from strategy-1 carnivorous marsupials that are thought to respond to the rate of change of day length in spring are Rock Hyraxes (*Procavia capensis*) in southern Africa. However, the year-to-year synchrony of breeding in these miniaturised elephant-relatives is looser than that exhibited by antechinuses or phascogales, which seem to have no peer in their daylight-reading abilities.

Okay, now we know how most strategy-1 species achieve their remarkable annual synchrony in timing of breeding. We then need to ask why. This part of the puzzle was solved by some elegant detective work by Steve Van Dyck at the Queensland Museum. Steve argued that – unlike in most other parts of the world – there is a travelling peak, or wave, in insects and other invertebrates that starts early at cool, high latitudes and moves progressively northward towards the tropics as spring wears on. He proposed that invertebrate populations increase early at higher latitudes in southern New South Wales and Victoria because they respond to warmer temperatures after the cold winter months. At lower latitudes in Queensland, by contrast, where winter temperatures are mild, invertebrate flushes are driven by increased rainfall, which usually arrives in late spring or early summer. This simple, but ingenious, explanation has since been quantified by Diana Fisher at the University of Queensland. Diana and her colleagues showed not only that food resources occur at the right times to explain when strategy-1 carnivorous marsupials breed, but also that the year-to-year predictability of the food peaks applies equally to South American exemplars of this strategy (see vignette, present chapter).

Our last word here goes to the phenomenon of post-mating male death in strategy-1 carnivorous marsupials. Several ideas have been proposed to explain the cause, or causes, of this abrupt and unique event, and there is still some debate about it among researchers. How do males die? Most of the research on this question has focused on antechinuses, and it is not clear yet how conclusions based on these carnivorous marsupials might apply to other species with a type-1 life history. Antechinuses have featured prominently in big-bang research, in part, because this was the genus in which the male death syndrome was first discovered. Convenience also plays a role – several species of *Antechinus* are abundant and accessible close to population centres where most researchers live. Furthermore, antechinuses are incorrigibly curious animals: they cannot resist checking out shiny new objects such as metal live-traps, and thus many species can be captured and studied without too much effort.

In addition to the precision in timing of ovulation that is achieved when a certain threshold is reached in the rate of change of day length, both sexes are primed for reproduction in the lead up to the breeding season by an internal 'clock'. This annual rhythm prompts increases in body condition before breeding and, in males, stimulates an increase in circulating stress, or corticosteroid hormones. These hormones increase also in females, but the secretion in females of proteins termed corticosteroid binding globulins reduces the effects of the increased stress hormones. Castrating males before the breeding season similarly reduces the concentration of circulating corticosteroids. Cortisol is the most potent of these hormones, and has the twin effects of suppressing appetite and promoting conversion of protein into sugars so that animals can engage in sustained activity. As a consequence, males interact aggressively with each other, search for mates by day and night, and mate with many females (up to 16 have been recorded) during the mating period. Copulations can last 2–12 hours, occasionally even up to 14 hours. Despite increasing the activity of males and maximising their access to mates, elevated cortisol levels have some strongly negative side effects. These can be seen in the stunning changes in appearance of males over the mating period. Sleek, muscular, alert and agile at the start of breeding, males lose up to 40% of their body mass by the end of the mating period. They sustain cuts and gashes and lose fur and body condition. If they retain enough strength to enter a live-capture trap, upon extraction they will lie languidly on the researcher's hand and just gasp for breath as the end draws nigh. Cortisol suppresses the immune system and inflammatory responses, and autopsy of males after mating further reveals the dramatic and systemic nature of its effects. Males suffer massive haemorrhaging from ulcers in the stomach and intestines, organ failure, and elevated burdens of internal and external parasites that have taken advantage of the collapse of the immune system. Muscle wastage has occurred; the lungs contain fluid. By the end of the 2-week period of mating, death is not only swift and inevitable, but probably welcome too!

This physiological explanation of how antechinus males die is well and good, but we are still no closer to answering the more difficult question of **why** they die. What are the selective forces that have led to the phenomenon of abrupt and total male death? This question remains contentious and several answers have been proposed. The earliest hypothesis suggested that males altruistically remove themselves from the population to reduce competition for food with females and their young. The idea that males should act in such a selfless and noble manner is superficially appealing, but unfortunately not viable. It envisages a situation that is unstable and untenable from an evolutionary point of view: altruists would quickly lose out to 'cheats' that survived and fathered young in subsequent breeding seasons. Another idea proposes that, with the seasonality of food allowing just a single litter to be reared each year, males should invest all their effort in siring young at their first opportunity because it is likely to be the only chance they get. That is, being small, their chance of surviving to a second breeding season is very low. This idea suggests that males compromise their

Fig. 10.16. Going out with a bang! This male Endangered Black-tailed Dusky Antechinus (*Antechinus arktos*) was captured by Andrew Baker, measured and released just a couple of weeks before it was destined to die in agony (and ecstasy?) at the close of a 2-week breeding frenzy in mid-September. Source: Andrew Baker.

immediate survival prospects to increase their chance of paternity. By viewing male behaviour as selfish, rather than altruistic, this explanation is clearly more plausible than the 'altruistic suicide' idea. However, it too has a flaw: females in many big-bang species sometimes survive to a second breeding season and, given this, it is unclear why only males should expire so abruptly after mating.

Two further proposals also stem from the notion that male behaviour is fundamentally selfish, and assume that males gain such strong benefits from their intense reproductive efforts that these outweigh the costs of premature death. How do these proposals work? The first recognises that females in all big-bang species store sperm after mating and may not release it to fertilise their eggs for at least a week or more after copulation. Because females usually mate with several males during the mating period, the joeys in each litter often have several different fathers. In effect, the males are competing with each other to sire young, but are doing so via their sperm rather than directly among themselves. From the female perspective, fostering sperm competition is probably advantageous because mothers ensure a degree of genetic diversity in their offspring and increase the chances that at least some of their young will survive. From the male perspective, the mating period is too long to stay with a single female and guard her from the approaches of other males. The only way to ensure paternity is thus to inseminate as many females as possible during the mating period, even if the intense effort of doing this causes death. The second proposal recognises that females face the energetically costly prospect of raising large litters, and that some will not survive the task. A male that mates with a single female may thus have little certainty that he will father any young; if the female dies, so too do all his offspring. From the male perspective, mating with many females can thus be seen as 'insurance' that some of their offspring will be born to females that survive. At present, the weight of evidence favours this sperm competition hypothesis. The notion posits that males with large bodies and large testes will sire more young than smaller males, and this prediction has been borne out to date in two species of *Antechinus*.

The next steps in unravelling the drivers of the intriguing strategy-1 life history will be to study populations in which male deaths occur in some years but not others, such as the Dibbler and Northern Quoll, and to focus more attention on the South American taxa that show the big-bang strategy. It will be fascinating to know whether males in these species have sharply elevated levels of cortisol during the breeding season, and whether these stimulate the intense reproductive effort and death that we see in their Australian counterparts. Such comparative studies promise to be exciting indeed!

VIGNETTE: COMPETITIVE COITUS: EVOLUTION OF THE MALE DIE-OFF

By Diana Fisher
University of Queensland, Brisbane, Australia

If you have ever grown annual vegetables or grains you will have noticed that these plants die after they produce seeds: this is semelparity. Some other plants, insects, molluscs and fish such as salmon also die after the first time they reproduce. In some spiders, just the males exhibit suicidal reproduction. This male-only semelparity is very rare in nature, but it happens in a handful of Australian carnivorous marsupials (dasyurids). In other dasyurids, males can breed several times (iteroparity). The mechanism of male die-off is escalation of stress hormones causing immune system collapse. Males stop producing sperm a month before mating and their testes disintegrate – the next generation relies on stored sperm in their epididymis, and males have one chance to contribute.

How this bizarre reproductive strategy could have evolved in these species (but no other mammals) has intrigued researchers since it was discovered in the 1960s. A popular reason given in nature documentaries is altruism. This argument that males sacrifice themselves for the good of the species so there is more food to go around is incorrect, because the promiscuous, mobile males can't advantage just their own young this way. It

Fig. 10.17. Does sperm competition drive the evolution of the spectacular male die-off that features in the life history of some carnivorous marsupials, such as this Tasmanian Dusky Antechinus (*Antechinus swainsonii*)? Source: Gary Cranitch.

doesn't pay to increase the evolutionary fitness of your competitors by sacrificing yourself! We suspected that a clue to the evolution of male suicidal reproduction is in the extreme sexual behaviour of these species. For example in *Antechinus*, matings are up to 14 hours and both sexes are promiscuous. Perhaps competition between males has become so fierce that they need lethal effort to succeed. Yet this would not be the whole story – why has male competition escalated so much?

Two ecologists first proposed a promising theory to explain it in the 1970s. They suggested that climate is the basis of die-off, because seasonality of food drives mating time. Their reasoning was that species with die-off live in temperate forests in southern Australia, where insects have a predictable peak of abundance at the same time each year. Females must synchronise mating so that young are raised and weaned when food is plentiful. Marsupials are born at a very early stage, so lactation lasts for months. Unlike non-marsupial small mammals, they can't squeeze in a second litter before winter. Males may as well compete to death, because they have a better chance of fathering more young that way than saving their effort and trying to survive for another year.

By analysing 52 species of insect-eating marsupials, and long-term insect monitoring at sites where they live, we found that male post-mating survival was lower in places where prey abundance peaked strongly at a predictable time of year. This supports part of the climate hypothesis. Seasonality of insect abundance is associated with short, synchronised single mating seasons. This intense competition equals low male survival.

In mammals generally, large testes with lots of sperm-producing tissue and long mating times reflect sperm competition, where males don't fight before mating but instead allow their sperm to compete after the female mates with multiple partners. Our previous experiments have shown that female *Antechinus* benefit from sperm competition. In our species comparison, we found that semelparous male dasyurids have larger testes for their body size and longer matings than other species, indicating sperm competition.

We concluded that shutting down sperm production before mating, foregoing future reproduction to use all possible energy for mating gives these males an advantage in sperm competition. Die-off is advantageous because males gain reproductive success at the expense of survival.

11

CONSERVATION

In the preceding chapters, we have provided a glimpse of the extraordinary diversity of the world's carnivorous marsupials and uncovered some of the secrets that underlie their success throughout Australasia and the Americas. We hope that our journey has engendered a sense of appreciation for these small, but charismatic, beasts, and perhaps conveyed some of the sense of astonishment and awe that even we – as seasoned professionals – still feel about our study subjects. If we are to maintain our appreciation and unlock some of the deeper secrets that the carnivorous marsupials have yet to divulge, we need to ensure that the existing species and their habitats will endure. What is the status of the 136 species that we have described and what does the future hold for them? We conclude our narrative by considering the conservation of carnivorous marsupials.

Fig. 11.1. Conservation is challenging for many species of carnivorous marsupial, such as this threatened Silver-headed Antechinus (*Antechinus argentus*). How can we manage the multiple threats of climate change, changed burning practices and feral predators that might mean extinction of such species? Source: Gary Cranitch.

Conservation in the Anthropocene

At the time of writing, the global human population has surged to well over 7.5 billion, and is expected under some projections to exceed 10 billion by 2050. Such numbers place immense pressure on the world's natural systems to provide enough food, water, shelter and other material resources to maintain even basic standards of living for people in many parts of the world. The burgeoning human population also means that there is less and less room for creatures other than the livestock and pets we keep to meet our immediate needs for food and companionship. The consequences of human overpopulation are severe. Detrimental environmental changes are estimated to be taking place in the modern world at rates 10–100 times faster than during the last great period of global change, some 66 mya. Natural habitats are being lost and land, air and water resources polluted. Climate change alone is predicted to wipe out one species in six as the Earth continues to warm. We are collectively doing very little to tackle these problems, apart from hoping that science and technology will somehow give us more elbow room. These dire changes and prognoses for an oppressive future have ushered in a new epoch that is known informally as the Anthropocene. New epochs are usually marked by climatic and geological changes. Climate change is upon us, of course, but what about geological change? Tongue-in-cheek assessments suggest that in future we will look back and recognise the Anthropocene as a thin smear of plastic and concrete in the geological record!

With the spectre of dystopia upon us, what is the future for carnivorous marsupials? Will there be room on the planet for these creatures, and in any case, why should we care? As biologists, it could be argued that the authors of this book simply have vested interests in seeing our study animals thrive. Of course, we do, but more broadly, we see carnivorous marsupials, other mammals, other living things, ourselves, as part of the natural world. Progressive losses of links and nodes in the biological web reduce the complexity and resilience of natural systems and reduce their capacity to sustain all life – including people. Still, conservation remains a relatively new concept that dates back little more than a century, and it is by no means universally accepted. Much has been written about why we should act as conservators of the natural world, and we visit some of the key arguments here before we return to consider the carnivorous marsupials specifically.

In the first instance, there is the simple argument that other species have a right to exist independently of human beings. That is, their existence should not have to depend on whether they have values or uses for people, or whether they live in habitats for which we have no use. This is a compassionate, perhaps even a morals-based perspective. It is philosophically compelling but, unfortunately, not usually persuasive in our often self-centred modern society. Another perspective stems from that very self-centredness. Nationalism is often frowned upon now, because it is perceived to slow the free movement of people and products across the globe, but for carnivorous marsupials – and most other living things – free movement is not an option. The species that occur in the various parts of Australasia and the Americas are endemic to those regions: that is, they occur nowhere else. Their endemism places unique responsibility for their care and stewardship on the people who share their habitats. National pride and a sense of ownership become involved. If Australians cannot conserve the Tasmanian Devil (*Sarcophilus harrisii*) or Chileans the Long-nosed Shrew-opossum (*Rhyncholestes raphanurus*), who can?

Other arguments for conservation place values of different kinds on living creatures and their habitats. These values range from the aesthetic to the mercenary. Aesthetic arguments are based on the joy that people feel when they see living things they perceive as interesting or cute, or beautiful landscapes that appeal to and stimulate the senses. Such feelings are not usually conjured by the sight of modern 'habitats' such as graffiti walls, car parks or freeways. Aesthetic pleasure derived from the living world can also provide inspiration, cultural attachment and even health benefits. For example, proximity to natural environments, and even artificial green spaces that provide wildlife habitat, are well known to improve both physical and psychological wellbeing. Regular visits to green spaces by urban dwellers multiply these benefits. These feelings of connection to the environment perhaps tap into the deep wellspring sourced from our evolutionary past, when we lived **in** the world rather than merely on it.

For the more hard-nosed among us, conservation imperatives can be illustrated by estimating the practical or monetary values we derive from the natural world. Such values are sometimes termed 'ecosystem services'. They include, for example, the benefits that flow from earthworms aerating and returning nutrients to soils, honey bees that provide critically important pollination services for agriculture, aquatic plants that filter fresh water, or the microorganisms that decompose our wastes. The most recent global estimates of the value of these ecosystem services range from US$125–145 trillion (yes, trillion!) per year. Mind-boggling as these figures are, they do not even include the extra values that accrue from the use of plant and animal species in foods, drugs, clothing, medical research and a plethora of other direct uses, or in such endeavours as wildlife tourism. The monetary values associated with carnivorous marsupials have not been costed, but could be expected to be substantial. In varied circumstances, they reduce the impacts of insect pests, act as vectors for pollen and seeds, provide models that are used in biomedical and evolutionary research, and act as magnets for ecotourism. The Tasmanian Devil is the most obvious attraction for many visitors to Tasmania, for example, but local tourist enterprises are also built around the chance of seeing smaller and equally charismatic quolls (*Dasyurus* spp.). If we continue to eat into the world's natural capital, the services we receive will diminish, and the accelerating loss of species means that restoration will not be possible.

In summary, we should be concerned for the future of carnivorous marsupials for all the reasons we have cited. Indeed, we argue that the question for human society should not be **why** to conserve our disappearing fauna, but how best to do so?

Assessing conservation status

All the countries that sustain carnivorous marsupials have laws that are intended to protect them, as well as the many other species and habitats for which these nations are custodians. In some of the larger countries, such as Australia and Brazil, there may be additional legislation that seeks to check untrammelled development and exploitation at local, state or regional levels. These multiple layers of environmental 'green tape' could be expected to provide effective protection for native species and their habitats, and in some cases they do. In the Australian state of New South Wales, for example, species, populations and ecological communities (distinct groups of species) that are faring poorly can be identified and listed for protection under the *Biodiversity Conservation Act 2016*. At the national level, similar protections are provided by the *Environment Protection and Biodiversity Conservation Act 1999*. Listings of species can become complicated if they happen to cross state borders. Thus, the Crest-tailed Mulgara (*Dasycercus cristicauda*) and Western Quoll (*Dasyurus geoffroii*) are both listed as presumed extinct in New South Wales, but are considered to be only 'near threatened' elsewhere in Australia. Other marsupials enjoy an even more schizophrenic status. Take the Black-striped Wallaby (*Notamacropus dorsalis*). This small and ostensibly innocuous wallaby is listed as endangered in New South Wales and receives full protection there, but it is shot as an over-abundant pest just north of the border in Queensland! This strange situation arises because only a small part of the range of this wallaby occurs in New South Wales; its stronghold is Queensland. State and regional boundaries often have a degree of arbitrariness about them. Indeed, this will occur every time you see straight line boundaries between different places on a map. To surmount jurisdictional difficulties that can arise when species cross lines on maps, it is most effective to gauge how particular species are faring by considering their status across their entire geographical ranges.

How should assessments of conservation status be carried out? Although we focus here on carnivorous marsupials, an ideal scheme for assessing the status of different species should be one that has broad applicability. It should, ideally, provide criteria that allow us to judge reliably whether a caenolestid species is of greater or lesser conservation concern than an earthworm that forages in the same soil, or whether the spiders that fall prey to a dasyurid are at greater or lesser risk of extinction than the hunter. Enter the IUCN (see Chapter 2). Founded in 1948, the IUCN is the world's largest and most diverse conservation and environmental network. After much debate in its early years, in 1964 the IUCN created the first inventory of selected threatened species. Since then, the organisation has sought to expand the range of species whose conservation status it has examined and, most importantly, it has refined the criteria that are used to make decisions about conservation status and ensured that these are applicable, as far as is possible, to all living things. The IUCN criteria and categories are now seen as the gold standard for how assessments of conservation status should be completed. The list of threatened species generated under the IUCN scheme – the Red List – is used as a

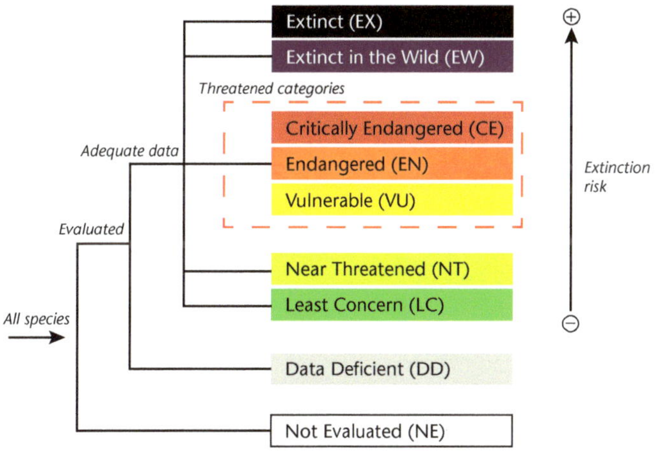

Fig. 11.2. The International Union for Conservation of Nature (IUCN) has devised a classification scheme that evaluates species against objective criteria, such as population size and distribution and their trends, thus allowing them to be placed into one of the categories shown here. Species for which there is adequate information can be evaluated as at greater or lesser risk of extinction. At highest risk are species that can be presumed Extinct or are thought to be Extinct in the Wild. At lowest risk are species evaluated as being of Least Concern. Species that are not well enough known to assign to a risk category are considered Data Deficient. Species in all of these categories can be found on the IUCN Red List (http://www.iucnredlist.org/). Other species may be known but have not yet been evaluated, for example, because they have been recently described. Source: Reproduced with permission from IUCN (2012).

TABLE 11.1. CONSERVATION STATUS OF CARNIVOROUS MARSUPIALS.

Status categories are those defined by the International Union for Conservation of Nature (IUCN), and species assignations are based on IUCN criteria (species listed as Least Concern are not shown).

Part (a) Australia and New Guinea	
Near Threatened	
Atherton Antechinus	*Antechinus godmani*
Crest-tailed Mulgara	*Dasycercus cristicauda*
New Guinea Quoll	*Dasyurus albopunctatus*
Western Quoll	*Dasyurus geoffroii*
Spotted-tailed Quoll	*Dasyurus maculatus*
Bronze Quoll	*Dasyurus spartacus*
Broad-striped Dasyure	*Murexia rothschildi*
Red-tailed Phascogale	*Phascogale calura*
Common Brush-tailed Phascogale	*Phascogale tapoatafa*
Carpentarian Pseudantechinus	*Pseudantechinus mimulus*
Kakadu Dunnart	*Sminthopsis bindi*
Julia Creek Dunnart	*Sminthopsis douglasi*
Vulnerable	
Fawn Antechinus	*Antechinus bellus*
Kowari	*Dasyuroides byrnei*
Northern Brush-tailed Phascogale	*Phascogale pirata*
Butler's Dunnart	*Sminthopsis butleri*
Sandhill Dunnart	*Sminthopsis psammophila*
Endangered	
Northern Quoll	*Dasyurus hallucatus*
Eastern Quoll	*Dasyurus viverrinus*
Numbat	*Myrmecobius fasciatus*
Dibbler	*Parantechinus apicalis*
Tasmanian Devil	*Sarcophilus harrisii*
Critically Endangered	
Kangaroo Island Dunnart	*Sminthopsis aitkeni**
Not Evaluated	
Silver-headed Antechinus	*Antechinus argentus*
Black-tailed Dusky Antechinus	*Antechinus arktos*
Buff-footed Antechinus	*Antechinus mysticus*
Tasman Peninsula Dusky Antechinus	*Antechinus vandycki*
Data Deficient	
Woolley's Three-striped Dasyure	*Myoictis leucura*
Tate's Three-striped Dasyure	*Myoictis wavicus*
Chestnut Dunnart	*Sminthopsis archeri*
Grey-bellied Dunnart	*Sminthopsis fuliginosus*
Extinct	
Tasmanian Tiger, Thylacine	*Thylacinus cynocephalus*
Part (b) Central, South and North America	
Near Threatened	
Gray-bellied Shrew-opossum	*Caenolestes caniventer*
Pygmy Opossum	*Chacodelphys formosa*
Monito del Monte	*Dromiciops gliroides*
Long-nosed Shrew-opossum	*Rhyncholestes raphanurus*
Paraguayan Fat-tailed Opossum	*Thylamys macrurus*
Dwarf Fat-tailed Opossum	*Thylamys velutinus*
Vulnerable	
Condor Shrew-opossum	*Caenolestes condorensis*
Blackish Shrew-opossum	*Caenolestes convelatus*
Sangay Shrew-opossum	*Caenolestes sangay*
Reig's Opossum	*Monodelphis reigi*
Karimi's Fat-tailed Opossum	*Thylamys karimii*
Endangered	
None	
Critically Endangered	
Single-striped Opossum	*Monodelphis unistriata*
Not Evaluated	
Panchos's Monito del Monte	*Dromiciops bozinovici*

Mondaca's Monito del Monte	*Dromiciops mondaca*
Arlindo's Short-tailed Opossum	*Monodelphis arlindoi*
Gardner's Short-tailed Opossum	*Monodelphis gardneri*
Santa Rosa Short-tailed Opossum	*Monodelphis sanctaerosae*
Touan Short-tailed Opossum	*Monodelphis touan*
Data Deficient	
Handley's Short-tailed Opossum	*Monodelphis handleyi*
Ihering's Three-striped Opossum	*Monodelphis iheringi*
Olrog's Four-eyed Opossum	*Philander olrogi*
Tate's Fat-tailed Opossum	*Thylamys tatei*
Buff-bellied Fat-tailed Opossum	*Thylamys venustus*

**Although considered a full species by the IUCN, this dunnart is treated in this book as a subspecies of the Grey-bellied Dunnart (*Sminthopsis fuliginosus aitkeni*).*

primary source of information for conservation and management. Indeed, although there is no requirement to do so, many countries are co-opting IUCN assessment guidelines and enshrining them with little or no modification into their own national regulations or legislation.

The IUCN identifies nine categories into which species can be placed (Fig. 11.2), and prescribes quantitative criteria to identify which pigeon-hole is most appropriate for them. Let us now examine each of the categories to see where the carnivorous marsupials fit.

The first category, **Extinct**, is applied when there is no reasonable doubt that the last representative of a species has died. To remove any reasonable doubt, surveys need to be carried out at appropriate times and in habitats that the target species used or preferred throughout its historically known range. If it fails to show up after exhaustive survey effort, it is deemed to have expired. Fortunately, only one carnivorous marsupial presently falls into this dismal category, the Thylacine (*Thylacinus cynocephalus*) (Table 11.1 and see Chapter 5). This magnificent beast was put to death by the Tasmanian Government and its supporters at the behest of the island's sheep farmers. Sheep are still chased and killed in the paddocks in Tasmania, but the main culprits are wild dogs – as they were all along. The last Thylacine died a miserable and lonely death in a small cage in the Hobart Zoo in 1936. Despite its unnecessary and untimely demise, the Thylacine still generates income and goodwill for the Tasmanian state economy. It is used as a symbol of the wild places in Tasmania that tempt intrepid tourists and explorers to visit. Images of the Thylacine abound on products from T-shirts to bottles of beer. The species even stars from time to time in films. In *The Hunter*, for example, Willem Dafoe is dispatched by a biotech company to collect tissue samples from a Thylacine. His mission in this film was undertaken in the belief that the Thylacine secreted an immobilising anaesthetic that might have had military application. Although such nefarious usage for Thylacine tissue is exceedingly unlikely, we will never know what other opportunities have been missed because of our thoughtless extermination of this striking carnivorous marsupial. However, it is interesting to speculate how much more income might have been generated for the sad Tasmanian economy if the species had been allowed to persist. No other carnivorous marsupial is listed as Extinct to date, but the Single-striped Opossum (*Monodelphis unistriata*) may be perilously close, if indeed it still persists. We discuss this species in more detail later. Unfortunately, many other carnivorous marsupials are probably on pathways leading to extinction.

If you carry out dedicated and exhaustive searches for a species and fail to find it in what had been its usual haunts, it has very likely become extinct. Yet, we sometimes hear stories of species coming back from the dead. We saw in Chapter 10 that the Dibbler (*Parantechinus apicalis*) could be viewed as a Lazarus-like marsupial. Thought extinct for 83 years, this species turned up unexpectedly in 1967, and then was found at a few more localities within its historical range. The Eastern Quoll (*Dasyurus viverrinus*) may represent another example. Although this species hangs on in Tasmania, it once had an extensive distribution on the eastern Australian mainland from South Australia to northern New South Wales. The last known mainland population hung on in the eastern suburbs of Sydney until the mid-1960s. With the demise of this population, the mainland Eastern Quoll was considered extinct. Then, in 2016, a specimen was handed in to the Australian Museum. It was an Eastern Quoll. It had been collected some 200 km north of Sydney in the Barrington Tops National Park in 1989, and stuffed soon after collection by an amateur taxidermist. The trophy had then sat in the collector's home before making its way to the museum. Could this species persist in the region today? Bill Dowling, a long-term resident of the Barrington region and a naturalist with an encyclopaedic and unparalleled knowledge of the local fauna and flora, claims to have seen Eastern Quolls in the depths of the Barrington wilderness. Such 'back from the dead' examples are heartening, but sadly rare. The great majority of species on the extinct list have been given a one-way ticket to oblivion.

The second IUCN category is **Extinct in the Wild**. Before expiring in the last remnants of their natural range, it is sometimes possible to capture a few of the surviving individuals of a species and put them on life support in captivity. The idea here is to keep the species going, grow its numbers under the sheltered conditions provided by a zoo or other facility, and then restore the original population at some time in the future when its prospects in the wild have been secured. No carnivorous marsupials are listed in this category. This is not to say that threatened carnivorous marsupials are not currently housed in captivity; in fact, many species are maintained in breeding colonies in parks and zoos. The difference is that these colonies are used to bolster small or declining populations that still occur in the wild. The Tasmanian Devil provides a good example. In the 1990s, wild populations in Tasmania began to succumb to debilitating, and ultimately fatal, tumours that grew on the heads and inside the mouths of affected animals (see vignette, present chapter). In addition to searching for the cause, and cure, of this disease in the wild population, researchers set up captive breeding colonies of uninfected Devils on offshore islands and in safe enclosures in Tasmania and on the Australian mainland. If the Tasmanian Devil had shuffled off its mortal coil in its natural habitats, it would have been eligible for listing in this second IUCN category. Fortunately, Devils appear to be staging their own recovery, but the species remains under close watch to ensure that its conservation status does not decline further.

The next three IUCN categories together define species that are considered under the umbrella term, 'threatened' (Fig. 11.2). The first of these threatened categories is **Critically Endangered**. As this label suggests, species that are so listed are in dire straits and at high risk of imminent extinction. As defined by the IUCN, this category includes species with one or more of the following:

- Populations that have declined in size by 90% or more over the last 10 years or three generations, and where the causes are known and have now ceased.
- Populations that have declined in size by 80% or more over the last 10 years or three generations, and where the causes are not known and may continue to operate.
- Geographical distributions that cover less than 100 km^2, or areas within these distributions where the species occupies less than 10 km^2, and where further declines are likely.
- Small populations containing fewer than 250 mature individuals and where further declines are likely.
- Tiny populations of 50 or fewer mature individuals.
- A 50% or greater probability of extinction in the wild within 10 years or three generations, as predicted in quantitative analyses.

Each of these criteria comes with further specifications about how they should be measured and interpreted, thus ensuring consistency in the listing of species under this category. For example, population sizes can be measured directly, from signs of species' presence such as scats or images taken on remotely set cameras, or from knowledge of the habitat that is used. It also helps to know whether the species fluctuates in numbers or distribution, and whether it faces known threats. In quantitative analyses, reliable prediction of a 50% or greater likelihood of extinction requires information on a population's size, the age at which individuals reach sexual maturity, their reproductive output and survival rates at different ages, and the degree of variation that can be expected in these parameters between years. Using this information, population models can then be constructed to predict a species' viability for 10 years or more into the future. To be listed as Critically Endangered, a species has to meet just one of the several criteria, even if it does not meet any others. Thus, a species that is down to its last 50 mature, reproductively capable individuals would be considered Critically Endangered even if these were spread over an area of more than 100 km^2 and had shown no recent population declines.

Two species of carnivorous marsupial are listed by the IUCN as Critically Endangered. The Kangaroo Island Dunnart is the first. Although listed as *Sminthopsis aitkeni* by the IUCN, taxonomic revision suggests that this insular dunnart is actually a subspecies of the Grey-bellied Dunnart (*Sminthopsis fuliginosus*). Its correct designation should then be *Sminthopsis fuliginosus aitkeni*. Irrespective, this dunnart is a distinct form that occurs only on Kangaroo Island where it status is parlous. It has been captured only 32 times, and since 1991 has been recorded from an area of less than 100 km^2 at the far western end of Kangaroo Island, off the coast of South Australia. The population size is not known, but is likely to be very small. Although all recent records come from Flinders Chase National Park, where overt habitat destruction from grazing or agriculture is prohibited, this dunnart remains at risk. Wildfires, predation by feral House Cats (*Felis catus*) and loss of the species' preferred heathland habitat to a fungal pathogen (*Phytophthora cinnamomi*) constitute potential threats to the species, because they all cross National Park boundaries unchecked. A post-doctoral researcher, Rosemary Hohnen, is currently setting both live-capture traps and remotely triggered cameras to improve our knowledge of the Kangaroo Island Dunnart, but until we know more its conservation status will remain dire.

The Single-striped Opossum is the second species classified as Critically Endangered. This cryptic species is known from just two specimens: the first taken in Sao Paulo state, Brazil, in 1821, and the second in Misones, Argentina, in 1899. Despite their age, the two specimens suggest that this opossum has distinctive fur markings that would make it easy to identify in the field. Yet, field surveys at the two collection localities and across much of the intervening 400 km of country between them have failed to turn up any further Single-striped Opossums. Except for uncertainty about its true distribution, this species would seem to be an equally good candidate for listing as Extinct. Its IUCN entry in fact lists it as 'Critically Endangered (possibly extinct)'. Whatever its true category might be, this attractive opossum is clearly in bad shape.

Sometimes species can be discovered only just before they were going to disappear. This appears to be the case for some antechinuses recently described by a research team led by Andrew Baker. The Silver-headed Antechinus (*Antechinus argentus*) and Black-tailed Dusky Antechinus (*A. arktos*) are both found in south-eastern Queensland, Australia and are currently being considered for federal listing as endangered (see vignette, Chapter 4). These beasts may be among the rarest living carnivorous marsupials, barely hanging on in the face of multiple threats – climate change, inappropriate fire regimes and feral pests. The challenge here is knowing enough about their distribution, rarity and ecological

Fig. 11.3. Can we intervene to save the recently discovered Silver-headed Antechinus (*Antechinus argentus*) before it is sent to extinction? Source: Gary Cranitch.

needs to designate an appropriate threatened category, because both species have only been discovered in the last 5 years. Researchers are desperately trialling a range of detection strategies to better understand just how rare these species are, so we can work out how to save them before they spiral to extinction. These strategies include training detection dogs to sniff out the species in the wild places where they may persist. This method proved successful in August of 2017 when Black-tailed Dusky Antechinus were rediscovered at a historic site during the first deployment of a detection dog in Border Ranges National Park (New South Wales). This antechinus species had not been seen in New South Wales since the 1980s, despite thousands of cumulative research-hours expended trying to trap these animals in the remote cloud forests they call home. Another method being investigated is to collect some of the natural scent lure that females produce during the mating season and mass produce it in a laboratory to use as a species-specific attractant in traps. Such detection strategies tap into the remarkable capacity of mammals to use their noses – this may represent a brave new world for detecting rare mammals to help conserve them. Hopefully, such strategies will help researchers fast-track the science to intervene before the populations of each antechinus species reach their tipping point.

Species that are faring poorly, but not so much so that they are at Death's door, might meet the criteria for listing as **Endangered**. The criteria here are slightly more relaxed and, as defined by the IUCN, this category includes species with one or more of the following:

- Populations that have declined in size by 70% or more over the last 10 years or three generations, and where the causes are known and have now ceased.
- Populations that have declined in size by 50% or more over the last 10 years or three generations, and where the causes are not known and may continue to operate.
- Geographical distributions that cover less than 5000 km^2, or areas within these distributions where the species occupies less than 500 km^2, and where further declines are likely.
- Populations containing fewer than 2500 mature individuals and where further declines are likely.
- Small populations of 250 or fewer mature individuals.
- A 20% or greater probability of extinction in the wild within 20 years or five generations, as predicted in quantitative analyses.

Five species of carnivorous marsupial are listed currently as Endangered, all from Australia. Except for the Dibbler, these species include some of the larger carnivores as well as the Tasmanian Devil (Table 11.1). Two species of quoll, the Eastern Quoll and Northern Quoll (*Dasyurus hallucatus*), are good exemplars of what populations of endangered species look like. Both have suffered large declines in their geographical ranges: the Eastern Quoll from the Australian mainland and Northern Quoll from lower rainfall areas across the Top End of the continent. These declines occurred during the 20th century and

Fig. 11.4. A road-killed Eastern Quoll (*Dasyurus viverrinus*). The effects of road mortality on populations of this species are not known, but are unlikely to improve in any way its Endangered status in its last stronghold in Tasmania. Source: Brett Dennis/ Lochman LT.

would be of less concern if the species' declines had slowed and numbers stabilised. Unfortunately, populations of both species are continuing to crash, with numbers plummeting by at least 50% over the last decade and those of the Northern Quoll predicted to decline further in future. The causes differ, but are no less aggressive. As Tasmanian Devil populations have crashed in recent years due to the facial tumour disease we noted earlier (leading to the listing of the Devil itself as endangered), populations of other predators have moved in to fill the niche of this large predator. Domestic dogs and, especially, feral cats have benefitted from the decline of Devils, and in consequence appear now to be suppressing Eastern Quolls. Many animals also die as a direct consequence of development, such as during urban expansion, forestry operations or on roads. Northern Quolls, by contrast, have been decimated in the eastern parts of their range by the arrival of the Cane Toad (*Rhinella marina*). Despite desperate efforts to slow the toad's continuing westward advance, the last major strongholds of the Northern Quoll in the north Kimberley and the Pilbara in Western Australia look set to succumb over the next few years. This quoll does occur still on several large islands off the northern Australian coast. Provided that these can be kept toad-free, there is hope that the Northern Quoll will survive, albeit in much reduced populations across a fraction of its former range.

Species listed as **Vulnerable** under the IUCN scheme are not as far down the slippery slope to extinction as those in the other two threatened categories, but they are still of concern. Ten carnivorous marsupials fall into this category, five each from Australia and South America (Table 11.1). Species meeting the criteria for this category are those with one or more of the following:

- Populations that have declined in size by 50% or more over the last 10 years or three generations, and where the causes are known and have now ceased.
- Populations that have declined in size by 30% or more over the last 10 years or three generations, and where the causes are not known and may continue to operate.
- Geographical distributions that cover less than 20 000 km^2, or areas within these distributions where the species occupies less than 2000 km^2, and where further declines are likely.
- Populations containing fewer than 10 000 mature individuals and where further declines are likely.
- Small populations that comprise 1000 or fewer mature individuals, or that occur in small and restricted areas that are prone to the effects of human activities and other disturbances.
- A 10% or greater probability of extinction in the wild within 100 years, as predicted in quantitative analyses.

The carnivorous marsupials listed currently as Vulnerable have achieved their status by meeting variations of these criteria. Two species that occur in the Top End of Australia – the Fawn Antechinus (*Antechinus bellus*) and Northern Brush-tailed Phascogale (*Phascogale pirata*) – have populations that are estimated to have declined by 30% or more over the last 10 years. Both species have been affected by an increase in the frequency and intensity of fires in their forest and woodland habitats, loss and fragmentation of these habitats, increased predation from feral cats in the wake of fires, and the arrival of the Cane Toad. In South America, populations of the Blackish Shrew-opossum (*Caenolestes convelatus*) and Karimi's Fat-tailed Opossum (*Thylamys karimii*) are also suspected to have declined by at least 30% over the last 10 years due to loss and severe fragmentation of their habitats. The preferred wet forest habitats of

Fig. 11.5. Karimi's Fat-tailed Opossum (*Thylamys karimii*) is threatened by deforestation and agriculture. Source: Thiago Semedo.

the Blackish Shrew-opossum are being destroyed because of increasing livestock production, agriculture and road building. The species probably has a distribution that now covers less than 20 000 km^2. Similarly, the preferred open grassland, shrubland and dry forest habitats of Karimi's Fat-tailed Opossum are undergoing rapid conversion to monocultures centred largely on soybean crops. These crops are inimical to the persistence of many native species. Although Karimi's Fat-tailed Opossum has a broad geographical distribution, the large-scale and intensive nature of the soybean industry does not bode well for its future – or for the future of other species that have the misfortune to live in the path of the soybean bulldozers.

In contrast to species that have become vulnerable to extinction because their populations have declined, Australia's Kowari (*Dasyuroides byrnei*) is listed as Vulnerable due to its already-small population size – between 1000 and 10 000 individuals – and what appears to be a downward trend in its numbers. This tenacious predator is confined to the remote stony deserts of western Queensland and northern South Australia. Here, changes in land use due to livestock grazing and the establishment of feral cats in Australia's arid regions have seen its distribution and numbers decline. Another threat comes from the Great Artesian Basin Sustainability Initiative. First established by several state governments and the Australian Government in 1999–2000, this scheme seeks to improve the efficiency of water-use by the pastoral industry after water has been extracted from the Great Artesian Basin. It provides funds for piping water into dry areas that are currently the preserve of the Kowari. As cattle move into the newly watered areas, they are accompanied inevitably by feral cats and European Red Foxes (*Vulpes vulpes*). These threats together reduce Kowari numbers and the quality of its habitat, and are likely to worsen its conservation status in future.

Other species are listed as Vulnerable because they meet some of the distributional criteria for this category. Thus, Butler's Dunnart (*Sminthopsis butleri*) and the Sangay Shrew-opossum (*Caenolestes sangay*) have geographical distributions that cover less than 20 000 km^2, as well as declining and severely fragmented populations that are scattered among a small handful of localities. Butler's Dunnart seems to have vanished from one of its former haunts, at

Fig. 11.6. The Sangay Shrew-opossum (*Caenolestes sangay*) is listed as Vulnerable because it has a limited distribution (less than 20 000 km^2), as well as declining and fragmented populations. Source: Reed Ojala-Barbour.

Kalumburu in the north Kimberley, but persists on the Tiwi Islands (Bathurst and Melville) off the coast of the Northern Territory. Threats on these islands include changes to the fire regimes, habitat change due to invasion by weeds, habitat loss due to the establishment of plantations, and probably predation from feral cats. The Sangay Shrew-opossum is known from three to five locations on the eastern flank of the Andes in southern Ecuador, and has a geographical range of perhaps just 6000 km^2. As with Butler's Dunnart, the already-small range of this species is being further compromised by the conversion of forest for agricultural usage and, to a lesser extent, urban settlement. The Sandhill Dunnart (*Sminthopsis psammophila*) meets a slightly different distributional criterion to achieve its status as Vulnerable. Its geographical range extends from two localities in the Great Victoria Desert in Western Australia to a single locality on Eyre Peninsula in South Australia. The area that it occupies within each locality is small, and probably amounts in total to considerably less than 2000 km^2. Recent doctoral research by Amanda McLean at the University of Adelaide confirms that the species occurs in low numbers, but also that populations increase following high rainfall. Amanda's work suggested that large fire events and predation from feral cats and Red Foxes will keep populations from growing in future.

The two remaining species categorised as Vulnerable are very poorly known and, we think, their status is somewhat debateable. Reig's Opossum (*Monodelphis reigi*) has been recorded in just two localities, both in the Guiana shield region on the border between eastern Venezuela and western Guyana. Its listing as Vulnerable reflects the very restricted area over which it occurs – less than 20 km^2 – and the fact that it is at potential risk from fires, mining, tourist activities and selective logging. As extensive surveys have been carried out in ostensibly suitable habitat and Reig's Opossum remains known from a small handful of specimens, it may be more at risk than its status category suggests. Similar arguments can be made for the Condor Shrew-opossum (*Caenolestes condorensis*). This enigmatic beast is known from a single location at 2080 m on the border between Ecuador and Peru. It has been assessed as having a population smaller than 1000 mature individuals and as occupying an area less than 20 km^2. Just three specimens have been captured, all in a restricted type of transitional habitat between dense forest and low heath. Further surveys are needed to clarify the range limits of this shrew-opossum, especially to know whether it occurs in areas that are protected from development and other disturbances. Whatever the case, it clearly has a very small range and could slip quickly towards critical endangerment or even extinction if its habitat were to be modified or fragmented.

The next IUCN category, **Near Threatened**, allows for placement of species that do not quite meet the minimal requirements for listing as Vulnerable but nonetheless exhibit concerning trends in population size, distribution or other indicators of status. Nine species of Australian carnivorous marsupial fall into this category, as do three species from New Guinea and six from South America (Table 11.1). As in our previous examples, these species can be categorised similarly, but for varied reasons. Thus, populations of the Brush-tailed Phascogale (*Phascogale tapoatafa*) and New Guinea Quoll (*Dasyurus albopunctatus*) are suspected to be declining at rates of at least 20–25% due to losses of habitat and the impacts of introduced predators. These declines, both over the last 10 years and projected to continue over the next decade, are

Fig. 11.7. This brood of young Brush-tailed Phascogales (*Phascogale tapoatafa*) will grow into independent adults that rely on trees and hollows in old growth forest to hunt prey and use as nesting sites. Unfortunately, Australia continues to have the worst land clearing rate of any developed nation and species such as this suffer as a direct result. Source: Jiri Lochman.

Fig. 11.8. Western Quolls (*Dasyurus geoffroii*) are listed as Near Threatened by the IUCN. They still can be found in low density throughout Jarrah (*Eucalyptus marginata*) forest in south-western Western Australia. They suffer from habitat clearing, predation by feral predators such as cats and foxes, illegal shooting and being hit by cars. Source: Jiri Lochman.

worrying, but not quite sufficient to qualify for categorisation as Vulnerable. The Crest-tailed Mulgara and Bronze Quoll (*Dasyurus spartacus*) are also suspected to be undergoing moderate numerical declines, but their categorisation as Near Threatened arises because their total population size is probably just above the threshold (10 000 mature individuals) that would qualify them as Vulnerable. By contrast, the Atherton Antechinus (*Antechinus godmani*) and Broad-striped Dasyure (*Murexia rothschildi*) are considered Near Threatened because they have restricted and fragmented distributions that are continuing to decline. Intriguingly, all Near Threatened carnivorous marsupials in South America are so listed because their populations are thought to have declined at rates of 20–25% over the last decade due to the continuing destruction of their habitats for agriculture, soybean plantations or logging. Their collective distributions encompass parts of Brazil, Paraguay, Argentina, Chile and Ecuador, suggesting that vast swathes of natural habitat in South America are subject to disturbance for production or exploitative enterprises.

Now for some good news: the most positive IUCN category, that of **Least Concern**, contains more species of carnivorous marsupial than any other. Australia has 37 species in this category, New Guinea has 10, and the Americas 33 species. Overall, 59% of carnivorous marsupials thus appear to be secure, and yet this security is more apparent for some species than others. For example, several species of forest-dwelling antechinuses have large populations that would number in the hundreds of thousands or millions; the Brown Four-eyed Opossum (*Metachirus nudicaudatus*), Water Opossum (*Chironectes minimus*) some dunnarts and short-tailed opossums have geographical ranges that span millions of square kilometres. On the other hand, several species of poorly known dasyures are categorised as Least Concern on the grounds that they occur in widely scattered localities in New Guinea. Although little researched and seldom seen, these cryptic carnivores occupy protected habitats that secure their populations from any obvious external threats. This is presumably also the rationale for the categorisation of Pine's Opossum (*Monodelphis ronaldi*) as Least Concern. Only one specimen of this species has been captured, in 1994, in mixed forest habitat in south-eastern Peru. However, this region has not been subject to much biological survey work, and the species is thought likely to occur more broadly. Its single known capture locality is in a protected area with no threats, thus strengthening the rationale for considering it to be secure.

The final IUCN category, at least for species that have been subject to review, is that of **Data Deficient** (Table 11.1). This is reserved for species that are considered too poorly known to assign status with any reliability, even after all known information about them has been gathered. Nine species fall into this category, one in Australia, two in New Guinea plus a third that is shared with Australia, and the remainder in South America. You might think that species as elusive as Pine's Opossum would fit well into this category, so let us consider what it takes to be regarded as Data Deficient. The Chestnut Dunnart

Fig. 11.9. A species currently rated as Least Concern by the IUCN is Rory's Pseudantechinus (*Pseudantechinus roryi*). This is not to say the species does not suffer from the usual threats of habitat disturbance and feral predators, but rather that its overall status is deemed secure at present. The genus *Pseudantechinus* is undergoing taxonomic revision and Rory's Pseudantechinus may ultimately be considered a western variant and synonymised within the species, Fat-tailed Pseudantechinus (*P. macdonnellensis*). Such taxonomic changes mean that the conservation status of various species need to be reassessed. In the case of these pseudantechinus species it results in an increased distributional range, which is likely to make the species more secure. However, the opposite can also occur. Consider the antechinuses (*Antechinus*) and monito del montes (*Dromiciops*), both of which have recently been revised and contain new species with more limited distributions that require listing. Source: Jiri Lochman.

(*Sminthopsis archeri*) is a good example. This small, innocuous species is known from ~30 specimens collected from the lowland savannas of southern New Guinea and from widely separated woodland localities in north-eastern Australia. Despite extensive surveys in potentially suitable habitats on Cape York and in wooded areas further south, no more captures have been recorded. The species is believed to occur sparsely and patchily in tropical savannas, and may be at risk from predation by feral cats and habitat degradation due to mining and overgrazing by livestock. Woolley's Three-striped Dasyure (*Myoictis leucura*) and Tate's Three-striped Dasyure (*M. wavicus*) are each known from small numbers of specimens collected from the rugged Central Cordillera in New Guinea. The former dasyure has been recorded at two main regions in undisturbed lowland and montane forest, the latter from a single location. Populations are probably secure given limited levels of disturbance in the vicinity of the capture localities, but actual threats – if any – remain unknown. In South America, small numbers of captures, lack of knowledge about population size and ecological requirements, as well as uncertainty about the existence of credible threats characterise the five Data Deficient species. Olrog's Four-eyed Opossum (*Philander olrogi*) provides a case in point. A few specimens were collected from widely scattered localities in north-western Peru and eastern Bolivia 30–40 years ago. No further specimens of this moderately large and distinctive opossum have turned up since. The species may be at risk of losing forest habitat due to agricultural expansion in Bolivia, but this threat is speculative rather than proven.

Table 11.1 shows there are still a few species of carnivorous marsupials that have yet to be assessed. These have all been recognised recently and are in the queue to be reviewed. When they are, the assessment process will often trigger further review of the status of currently listed species. This is because many of the **Not Evaluated** species were incorporated formerly within previously recognised taxa, and when new species are split off the known distributions of the original species may decline. For example, the Mainland Dusky Antechinus (*Antechinus mimetes*) was thought formerly to extend as far north as south-eastern Queensland and south to Tasmania. Following taxonomic revisions, the northern-most populations are now known to represent the newly described Black-tailed Dusky Antechinus (*A. arktos*), which, as we have seen, is currently slated for federal Endangered listing. An IUCN listing will follow the Australian federal listing for this species. The Tasmanian populations represent two species: the Tasmanian Dusky Antechinus (*A. swainsonii*) and the Tasman Peninsula Dusky Antechinus (*A. vandycki*). In this example, the Tasman Peninsula Dusky Antechinus apparently has a very restricted distribution that suggests it will be considered threatened when assessed. The Mainland Dusky Antechinus also now has a dramatically reduced distribution, although it is unlikely this species meets any of the criteria for listing. The monito del montes provide a similar example from the Americas. Panchos's Monito del Monte (*Dromiciops bozinovici*) and Mondaca's Monito del Monte (*D. mondaca*) were included within *Dromiciops gliroides* until 2016. Then, the *mondaca* form was found to be distinct and to occupy a small area in coastal Chile, while the *bozinovici* form was found to replace *gliroides* throughout the northern half of the distribution of the original species in Chile and western Argentina. The Monito del Monte (*D. gliroides*) is currently categorised as Near Threatened. However, with the original distribution now split three ways, each new species will likely move into the Vulnerable (or higher) category when the next formal assessment is completed.

How often are new species recognised? The 10 Not Evaluated species in Table 11.1 provide a clue: field surveys and taxonomic revisions are ongoing, and new species are being recognised all the time. We are aware of active investigations of species identities within the genera *Antechinus, Murexia, Myoictis, Planigale, Pseudantechinus* and *Sminthopsis* in the Australasian region, and within *Monodelphis* and *Thylamys* in the Americas. Several new species, at least, are likely to be described over the next few years. These are exciting times – watch this space!

When new species crop up, or indeed when any new information comes to hand about the population size, distribution or threats to carnivorous marsupials, how long do we have to wait for conservation assessments to be done? And who does them? The IUCN has an expansive international membership of professional scientists, fauna and flora experts and other specialists. The criteria and information that it uses to place species within the various status categories are set using input from the international scientific community, and represent the product of much research, debate and revision. To make status assessments, the IUCN first appoints experts who have personal experience with particular species or groups of species. There may be just two or three people in a specialist group, or as many as several dozen, and it is the task of these groups to keep the status of 'their' species under review. The Australasian Marsupial and Monotreme Specialist Group has comprised 50–65 members at different times, with other specialists being co-opted to the group as needed. The New World Marsupial Specialist Group is smaller (~19–20 people), but again consults with further specialists as required. Formal meetings are sometimes arranged to review the status of all the marsupials within the purview of each group, but applications can be made at any time to change the status of species when new information is obtained. Of course, people always have their biases for or against the listing of particular species, but the arguments that are mounted must adhere to the IUCN rules. All assessments are checked by IUCN Red List authorities to ensure compliance, and assessments are then posted to the IUCN website (http://www.iucnredlist.org/). The clear structure and evidence base needed for evaluating species' status provides great confidence that the listing process for carnivorous marsupials, and other species, is as objective as it can be.

Improving conservation status

Despite the good news that many carnivorous marsupials appear to be of little concern from a conservation perspective, the picture is not quite so

Fig. 11.10. The Stripe-faced Dunnart (*Sminthopsis macroura*) is listed as Least Concern by the IUCN but, like many of the carnivorous marsupial groups, it likely harbours several cryptic species. Understanding this hidden biodiversity and then preserving it will ensure better conservation outcomes and ecosystem health. Source: Jiri Lochman.

rosy for those that have made their way onto the lists of threatened species. For these, there is real concern that they may slide further towards extinction if appropriate steps are not taken. What are the problems that these threatened species face, and how might they be alleviated? In 1994, wildlife ecologist Graeme Caughley proposed that conservation biology was heading in two quite different directions. In the first place, Caughley noted that some species are at risk of extinction because they have small populations. Their small population size means that they have little resilience to bounce back after a disturbance such as a wildfire or flooding event that destroys the habitat they need. Small populations are also at risk of becoming inbred; that is, the only potential mates for most animals are close relatives. Inbreeding tends to bring 'bad genes' to the fore, causing congenital defects or developmental problems for the young, and limited resilience to challenges from disease organisms or changing environmental conditions. Conversely, Caughley argued that other species face extinction because their populations are being degraded by external threats such as habitat loss or direct over-exploitation by people. The populations and geographical distributions of these declining species may be quite large, but if their trend is inexorably downwards, they will sooner or later face oblivion unless the trend can be reversed. These two directions in conservation biology, which in some threatened species are both important factors, have been incorporated into the IUCN criteria, and also provide a helpful framework for deciding how best to bring these species back from the brink.

How can threats to small or declining species be identified? The researchers and other specialists who carry out IUCN assessments usually have a good idea of the threats that affect the species they study, and these threats are listed during the assessment process. For threatened species, in particular, it is crucial that the key threats are correctly identified. If an unimportant threat is identified by mistake, and then neutralised, it will do nothing to assist the threatened species of interest. Indeed, the species is likely to be worse off if the resources needed to help improve its status have been wasted. This is no different from getting the correct diagnosis for an ailment from your doctor. If you wish to get better, you need to know exactly what is wrong with you so that the appropriate course of treatment can be prescribed. In general, ecologists and conservation biologists prefer experimental confirmation of the importance of a suspected threat. If, for example, you suspect that a predator is reducing the numbers of a threatened species, this can be confirmed by removing the predator from parts of that species' range. If the threatened species increases in places where the predator has been removed but continues to decline in places where it remains, the evidence for predator-impact is compelling. Other kinds of evidence that help clarify the importance of different threats include historical or anecdotal records, and occasionally even direct observations of the effect of a threat on a population. Much has been written about how threats can be, and should be, identified. We do not dwell on this body of work here, and assume instead that the IUCN-listed threats are those on which we need to focus.

Conserving small populations

We saw earlier that several carnivorous marsupials have small populations with a few individuals that are either scattered among several distant locations or confined mostly to just one or two small areas. Such species may be naturally rare and confined to patchily distributed habitats, islands, or landforms such as rock outcrops. In the absence of obvious external threats, surveillance monitoring of these species may be sufficient to ensure they persist. If surveys are carried out often enough, they should be able to identify any unexpected drops in population numbers or new threats that arise. Several species that occur in the Top End of Australia are subject to regular monitoring, such as the Northern Quoll and Kakadu Dunnart (*Sminthopsis bindi*), as are the Tasmanian Devil and the two species of quoll that co-occur with Devils in Tasmania. Surveillance monitoring is sometimes seen as mundane and costly for the returns that it provides, but, without it, the widespread decline of mammals across the Top End would not have been noticed, nor would the rise of facial tumour disease in Tasmanian Devils. The work of ecologists Menna Jones in Tasmania (see vignette, present chapter) and Dick Braithwaite and John Woinarski among others in the Top End has been of incalculable value in this regard, alerting us to these problems and showing that swift management interventions were needed.

If monitoring detects a novel threat to an already-small population, what can be done? If the threat is an external one, action can often be taken to remove it, as we'll see later. But what are our options if the threat arises from within, such as from inbreeding, or a failure of animals to find mates, annual male die-off or even a chance dip in numbers due to poor weather conditions? On the one hand, provided the ecological conditions are right, new sites could be found to translocate animals and thus establish new populations. The most successful examples of this approach are from Western Australia. Here, small and remnant populations of the Dibbler, Western Quoll, Numbat (*Myrmecobius fasciatus*), Red-tailed Phascogale (*Phascogale calura*) and many non-carnivorous marsupials have been moved to new locations where immigrant populations have then established. The Dibbler is particularly vulnerable to predation from feral cats, and disappeared from some its previously known haunts on the south-western mainland after its rediscovery there in 1967. Small numbers of founder individuals were moved to the cat-free and appropriately named Escape Island from 1998 to 2000 and have successfully established there. Additional Dibbler colonies have been set up in three localities on the mainland, including one in a large (380 ha) fenced enclosure designed to keep out both feral cats and Red Foxes. In most cases, animals are translocated to sites where they once occurred. However, if suitable sites are not available, animals can be moved to new sites where the habitat, climate and other aspects of the environment are similar to those supporting the original source population. Thus, Escape Island represents a new site for the Dibbler because there is no evidence that the species ever occurred there. The advantage of the translocation approach is that it increases the overall number of individuals of a given threatened species and the number of sites they occupy. Translocations provide insurance: if a species occurs at a single site it may face a high risk of failing, but if it occurs in multiple sites across a large regional area it should prevail even if populations at one or two sites fail for any reason.

On the other hand, animals may have such specific habitat requirements or face such pervasive threats that translocation sites cannot be found. In this situation, animals may need to be captured and brought into captivity, and returned to the wild at some time in future when the threats they face have passed. We know of no carnivorous marsupials that have required such drastic management attention, at least not yet, although this has been the fate of several other vertebrates. In 1987, for example, the last surviving wild individuals of both the American Black-footed Ferret (*Mustela nigripes*) and California Condor (*Gymnogyps californianus*) were captured and brought into captivity. Grave fears had been held that their wild populations were so small, and faced such threats, that they would soon be extinguished without swift and effective management intervention. The captive breeding programs were successful, as were attempts to find safe areas in which to release animals back into the wild. Both species now have small, but steadily increasing, populations that provide hopeful prospects for the future. An expansive captive breeding program is in place to provide insurance populations of the Tasmanian Devil should wild populations continue to decline. Fortunately, however, the Devil appears to be staging its own recovery (see vignette, present chapter), and there are no plans to entirely remove the last Devils from the wild to assure the conservation of this iconic species.

Except in cases such as the Black-footed Ferret and California Condor, captive breeding should not be seen as a last-gasp conservation effort, but rather as part of a broader proactive approach to ensuring species' survival. Many carnivorous marsupials are maintained in captivity with the aim of breeding them and increasing their numbers. Surplus animals then can be used to supplement small wild populations if and when needed, or to act as founders for new populations in translocation programs. This approach has been used with the Western Australian species noted earlier, as well as for the Tasmanian Devil. Of course, unless carefully managed, captivity can reduce the range of natural behaviours that animals show. However, it also has one key advantage: captive animals can be selectively paired and pedigrees and family trees maintained to reduce any effects that might otherwise accrue from inbreeding. Careful placement of genetically typed animals into wild populations can increase genetic diversity and enhance prospects for individual wellbeing and population persistence.

Conserving declining populations

Perusal of the IUCN Red List indicates that the loss and fragmentation of habitat, impacts from invasive species and, to a lesser extent, direct persecution and over-exploitation by humans are the key threats driving declines of carnivorous marsupials. Each of these threats arises directly or indirectly from human activity, and consequently there is often some tension in balancing immediate human interests and conservation outcomes. Climate change represents a potent threat that will affect many carnivorous marsupials – not to mention people and all other living things – in future.

We saw in Chapter 7 that carnivorous marsupials exhibit clear preferences for particular habitats, so it is no surprise they fare poorly if these habitats are destroyed or degraded. Habitat loss can arise from the conversion of native vegetation to crops, especially if these are farmed intensively as monocultures over large areas. Palm oil, soybean, cocoa, coffee and cotton are among the most rapidly expanding agro-industries in Australasia and the Americas, and have resulted in the clearing of many millions of hectares of native vegetation over the last decade. Trees and shrubs are also removed for livestock production in many regions. The Australian state of Queensland holds the unfortunate regional record for officially sanctioned land clearing, giving the green light for the bulldozing of over 750 000 ha of land in one 12-month period in 1999–2000. Between 1997 and 1999, the immense loss of habitat was estimated to kill ~100 million native vertebrates each year, with 1.25 million of these being antechinuses, dunnarts and other carnivorous marsupials. This disgraceful state of affairs is only marginally better now, with almost 300 000 ha cleared in the same state during 2014–2015. Habitat is also destroyed by urbanisation, mining and other development activities that remove native vegetation. The areas affected can be very large. For example, Queensland (again!) is currently seeking to open a coal mine that covers an area of 447 km^2, despite intense opposition. Brazil has recently opened up an astoundingly large area of over **46 000 km^2** for mineral extraction in the northern states of Pará and Amapá. Beyond the areas obliterated by these

Fig. 11.11. Although species such as the Yellow-footed Antechinus (*Antechinus flavipes*) are classified as Least Concern by the IUCN, they need substantial blocks of forested habitat to survive. As habitat clearing continues at a rapid rate in many countries, even common species such as this one have declined at both local and regional scales in recent decades. Source: C. A. Henley.

enterprises, additional settlements plus roads, rail, port facilities and other transport infrastructure lead to the fragmentation of habitat over still larger areas.

In the face of such onslaughts, carnivorous marsupials are likely to persist only if there are designated parts of the landscape that are set aside for habitat protection, and if other non-protected areas are subject to low-intensity land use. About 13% of the land area of Australia, or just over a million square kilometres, has protected status. Wilderness and national parks provide the highest level of protection from exploitative and extractive uses, but Indigenous protected areas, state and regional government reserves also add importantly to the reserved estate. Most areas of protected land are publicly owned, but an increasing amount of land has been purchased in recent years by private conservation organisations. In New Guinea, with a land area of 785 750 km^2, just over 3% is formally reserved. However, both Papua in the east and Irian Jaya in the west have wildlife management and conservation areas that are maintained and protected by local indigenous people. The first of these areas – the YUS Conservation Area on the Huon Peninsula – was set up only in 2009. Although it is too early to evaluate how well these land-protection initiatives are working, the inclusion of local people will undoubtedly be crucial to their success. Protected area coverage in the Americas shows great variation, from less than 3% of the land area in Uruguay to more than 22% in Venezuela. Protected areas there vary also in the degree of protection they offer, from prohibiting development of any kind to allowing almost any disturbance to occur. For example, the immense area recently opened for mining in Pará and Amapá states in Brazil had been a reserve from 1984 until this status was unilaterally abolished in 2017. Nonetheless, with the exception of the Single-striped Opossum, some of the ranges of all currently threatened carnivorous marsupials coincide with reserved areas or places that are otherwise protected by virtue of their remote location. Over half of the Near Threatened, Data Deficient and Not Evaluated species also occur in protected areas, thus providing considerable hope that the world's carnivorous marsupials have some secure areas to which they can retreat.

Despite the advantages that protected areas confer for individual species and ecosystems, vigilance is still needed to ensure that these places maintain their functions. If reserved areas are too small and surrounded by land that has been developed, many species will become 'trapped' and unable to disperse. The problems associated with small population sizes will then follow – shallow gene pools and higher susceptibility to detrimental chance events. In addition, protected area boundaries may be recognised by people, but they are unlikely to act as barriers for outside forces that encroach into the protected areas. Habitats on the edge of nature reserves and national parks are often invaded by weeds and are subject to disturbance from traffic, rubbish dumping, severe winds and even localised heat effects if they are close to urban areas. Perhaps most importantly, protected areas are often subject to invasion by generalist predators. These include feral cats and Red Foxes specifically in the Australian region, and domestic dogs everywhere. Except for Tasmanian Devils, all carnivorous marsupials are potentially susceptible to these larger mammalian predators. Some, such as the Eastern Quoll and Numbat, have been shown to be at considerable risk. What can be done to reduce the impacts of these predators?

In Australia, where the problems posed by feral cats and foxes are most acute, protected areas are often managed with a view to reducing the numbers of these predators. Trapping and shooting are sometimes carried out, but require much effort over long periods. Poison baiting is also used, and can be very effective. Following experimental confirmation that Red Foxes were reducing populations of rock-wallabies in Western Australia, in 1996 the government set up a large-scale baiting program to reduce the impact of foxes on rock-wallabies and other mammals. This extraordinary initiative, dubbed 'Western Shield', has proven to be one of the most successful conservation programs in Australian history, and is certainly the most effective with

Fig. 11.12. The Gray Four-eyed Opossum (*Philander opossum*) is broadly distributed and occurs in a range of habitat types. However, the species needs taxonomic reassessment – several subspecies probably need to be recognised as full species and cryptic taxa may await discovery. These will each need independent re-assessment under threatened species listing criteria. Source: Hugh Tyndale-Biscoe.

respect to the recovery of threatened carnivorous marsupials. At least a dozen species have shown dramatic population increases, and the ecosystems they inhabit have regained much of the health that they had lost due to the impacts of the marauding fox. Western Shield works for several reasons.

First, the key threat to native animals was diagnosed correctly and at an early stage. Jack Kinnear and his team at the Western Australian Department of Conservation and Land Management (now Parks and Wildlife) showed with little doubt that the Red Fox was cutting a swathe through rock-wallaby populations, and was most likely responsible for declines of the Numbat, Western Quoll, Red-tailed Phascogale and a host of other threatened native species. A second key step was identifying a toxin that could be used to selectively kill foxes, and to do so quickly and relatively humanely. This was the compound known as '1080' or, more technically, sodium monofluoroacetate. 1080-poison occurs naturally in many species of native plants, but importantly is found at high concentrations in Western Australian heartleaf plants in the genus *Gastrolobium*. Native animal species have co-evolved with these plants and have high levels of tolerance to 1080. Thus, 1080-laced baits could be

Fig. 11.13. Can broader use of fenced areas to exclude feral predators benefit smaller species such as this Hairy-footed Dunnart (*Sminthopsis hirtipes*)? Source: David Nelson.

distributed to kill foxes but have little or no effect on native species. A third step was to scatter baits regularly over a sufficiently large area – 3.9 million hectares, all in protected conservation estate – to ensure that Red Fox populations were reduced to such low levels that native species could recover. The program has been modified in recent years by scattering baits for feral cats to ensure that populations of this species do not increase as a consequence of fox declines. A final reason why Western Shield has worked is that it has been carried out by large teams of dedicated people with strong government support. Personnel who work for state conservation agencies are almost always deeply committed to the conservation tasks with which they are charged, but government support for them sometimes vacillates. The Western Australian experience provides a powerful example of what can be achieved in conservation if the necessary support is maintained. Similar large-scale baiting programs have been initiated in South Australia and Victoria following the successes in Western Australia, and are also showing very promising recoveries of native animals.

Another approach to blunting the effects of predation from the Red Fox and feral cat has been to build fences to keep them out of important conservation sites. Such fences are used currently to protect populations of Dibblers, Numbats and Red-tailed Phascogales, but surveillance monitoring has shown that other species of carnivorous marsupials, such as dunnarts, derive benefit too. Fences that can keep cats and foxes out are expensive to build and maintain, and are sometimes criticised as being little more than open-area zoos. Nonetheless, for some species, fenced enclosures provide a degree of safety they would not experience beyond the fences, and for this reason alone they provide an important component of the conservation toolkit. Fenced compounds are likely to play a key role in sustaining some species until a time comes that the threats beyond the fences can be managed and mitigated.

New ideas – the brave and the bold

Moving threatened species around, breeding them in captivity and finding safe, protected habitats are standard and currently very important ways of achieving conservation goals. However, on their own, they may not be sustainable because of the increasing costs they incur. Are there other means by which we could cost-effectively stave off the threat of more species' extinctions in future? Two broad approaches have been suggested, but would need considerable discussion and debate before any 'social licence' was granted for them to proceed. Both approaches would provide potential solutions to the difficult problem of how to manage the impacts of invasive species on threatened native animals.

The first approach would use genetic engineering. In broad terms, this means tampering with the genetic makeup of target species to make them behave as we humans wish. This sounds scary, and there is no doubt that it can be. Indeed, any mention of the acronym 'GMO' (genetically modified organism) is enough to send media outlets into a frenzy and get concerned citizens marching in the streets. But let's provide some perspective. We all eat genetically modified foods, wear clothes made from materials such as cotton that we have artificially selected, walk our pets that we have modified to look nothing like their original wild counterparts, and are kept alive by varied drugs that are helpfully produced by microorganisms that we have designed for just that purpose. GMOs are everywhere. We just get worked up about certain applications of them.

From a conservation perspective, two kinds of genetic engineering have been proposed. The first would take a microorganism such as a bacterium or virus that is known to move quickly and readily between individuals of the invasive species we wish to manage. The cat 'flu' virus is often mentioned in popular discourse as a means to keep populations of the feral House Cat under control. Feral cat populations in most parts of the world already carry this virus and have high levels of immunity to it, so it doesn't actually kill many individuals directly. This makes it useless for proponents who advocate using it to kill cats, but it could be useful from another perspective. Because it is highly transmissible, the virus could act as a potential vehicle for carrying modified genes between cats. Although the cat flu virus has not been tested for this purpose, experimental trials have investigated this potential use for viruses that move between other invasive species such as the European Rabbit (*Oryctolagus cuniculus*) and the House Mouse (*Mus musculus*). How would this approach work? Consider the House Mouse. The mouse is commonly infected by murine cytomegalovirus, and animals with the virus seldom show obvious clinical effects. The genes that encode for proteins that form part of mouse eggs or sperm are known, and can be sliced from the mouse genome and inserted into the virus. When the genetically modified virus makes its way to a new mouse, the host then begins to manufacture antibodies against the viral invader. That is, the virus provokes the immune system of the mouse, and is the normal response of any healthy individual. But here is the clever part: if the virus is carrying genes that construct the mouse's own eggs or sperm, the mouse manufactures antibodies against these too. Put simply, it sterilises itself. This process, termed 'immuno-contraception', has been shown to prevent breeding in mice held under laboratory conditions. A modified form of the European Rabbit virus, myxoma-virus, has also been engineered to contain the rabbit's own reproductive genes. There would be risks inherent in liberating genetically modified viruses into wild populations, but there are great costs also in allowing invasive species to continue to run riot over their new landscapes. At some point we will need to be brave enough, collectively, to evaluate the costs and benefits of such GMOs in mature debate.

The other kind of genetic engineering is more direct and involves gene 'editing'. Here, genes that we like are inserted into the genome of a target species, and those that we do not are excised. This sounds extreme, but is in fact beginning to be used commonly to modify crop strains and even human cell lines as a means of inactivating various genetic diseases. The approach goes by the name of **clustered regularly interspaced short palindromic repeats,** or CRISPR, for convenience. It has not yet been used as a means to control invasive species, but there has been discussion of using CRISPR technology on several target species. It is possible, for example, that gene editing could be used to remove those parts of the Cane Toad (*Rhinella marina*) genome that allow toads to produce their very toxic poison. Another emerging technology, termed the 'gene drive', could then be applied to ensure that the newly modified genes spread rapidly through the population. If the poison genes were removed from the toad population, native species would be able to use toads as safe, if rather unappetising, prey items. How likely is it that such a high-tech approach would ever get approval for use on invasive species? In the USA, the National Park Service is evaluating the potential for using CRISPR technology to reduce the effects of many invasive species that are now laying waste to native species and the natural values of its protected areas. Again, with public consultation and mature discussion about costs and benefits, this novel approach to management could at some time get the green light to proceed.

Meanwhile, a simpler ecological solution would appear to be at hand, at least in the Australian context. We saw in this chapter, and in Chapter 5, that feral cats and foxes are responsible for the extinction of many Australian native mammals. These ferals remain a potent threat to other species – especially small- and mid-sized mammals – that are still managing to hang on. The proposed solution would be to allow a top predator, such as the Dingo (*Canis dingo*) or Tasmanian Devil, back into the broader landscapes where these small- and mid-sized threatened species occur. Much research on the Dingo has shown that it can suppress the activity of cats, and especially foxes, and by so doing alleviate the predation pressure that these smaller carnivores place on native species. Of course, the Dingo eats native species too, but it prefers larger prey such as kangaroos that are not under threat of extinction, and generally eats far fewer small- and mid-sized native species than either the fox or cat. Like the Brush-tailed Mulgara (see Chapter 9), the Dingo therefore has an indirectly positive effect on small native mammals by suppressing the negative effects of the invasive predator species. Bringing back such relatively large species as the Dingo and Tasmanian Devil to achieve positive conservation goals has been termed 're-wilding'. Research by Mike Letnic and colleagues at the University of New South

Fig. 11.14. One of many small mammals at risk each night of predation by cats and foxes: the Common Planigale (*Planigale maculata*). The estimated size of the Australian feral cat population is several million. If each cat is consuming individuals of several native species per night then many million native animals are being lost each week to cats alone. Source: Steve Wilson.

Wales suggested that returning the Dingo alone to regional areas throughout Australia (where it is currently persecuted) would provide a net benefit to threatened native mammals over an area of ~2.4 million square kilometres.

So why do we not explore the re-wilding solution? In brief, Dingoes attack sheep. Even though every other country that maintains sheep flocks manages to live to some degree with its native predators – using tried and trusted techniques such as guardian dogs to protect their livestock – Australia, to a large extent, does not. Despite vociferous opposition from the powerful agro-industry in Australia about any possible role for Dingoes in the landscape, we feel again that it is time for reasoned debate about this issue. Business-as-usual is not good enough when we continue to lose native species and see others slide relentlessly onto the IUCN Red List. Now is the time to stand up for our beleaguered native fauna and show some boldness and bravery.

What can you do?

Our aim in writing this book was to showcase the marvellous carnivorous marsupials and uncover their secret lives. We are enthralled by the strange and beguiling ways of these creatures and we hope, by now, that we have piqued your curiosity as well. If we are collectively concerned about the plight of carnivorous marsupials and the rest of our native fauna, what can we do to assure their persistence in future? We conclude by offering a series of actions you can take in this most important enterprise.

- Get up close and personal with carnivorous marsupials to appreciate them at first-hand. Check the nocturnal house at your local zoological park or, if you are fortunate enough to live in an area where these species occur, take time to watch out for them and learn more of their habits.
- Take photographs using your phone. If you have access to remotely triggered cameras, set these in

places where it is safe (and legal!) to do so, and see if you can identify the species whose images you capture. Share them with friends and colleagues to inform them about the cryptic fauna that shares their neighbourhood.

- Contact your local or state parks and wildlife service if you take images of species that you do not recognise. Not only can you solicit their assistance to identify what you have photographed, but you can provide help to these authorities by providing them with information about the distribution of native animals. In many cities and rural areas, there are 'citizen science' programs that provide huge amounts of useful information.
- If you have time, volunteer to help in any surveys of carnivorous marsupials and other fauna that are being undertaken by local authorities, educational institutions, university students or others. Help in the offices of these organisations is often appreciated, and can be very rewarding. We have volunteers in our laboratories who spend many enjoyable hours looking at photographs that have been taken by our remote cameras. There is great mutual benefit to this.
- Habitat loss and fragmentation destroy the homes of many native species and kill large numbers of animals and plants. Lobby your friends, neighbours, colleagues and local political representatives to make them aware of the consequences of rampant and unnecessary development. Handwritten letters, while old-fashioned, actually carry more weight with politicians than do 'form' letters.
- Irresponsible pet ownership results in domestic cats and dogs roaming at will in areas that support carnivorous marsupials and other native fauna. Again, lobby your network to ensure that irresponsible pet owners know the consequences of their actions. A lot of people simply don't realise that their cute little couch potato may cause mayhem for local fauna if left to roam unsupervised.
- Join a local conservation group, if one exists, or think about establishing one yourself. Many successful groups start at grass roots and maintain large and active memberships via social media.
- If you are thinking of further study, consider studying biology if you are not doing so already. And if you are thinking of further study at the tertiary level, we hope that we have provided enough teasers throughout our text to entice you to look no further than research on carnivorous marsupials.

As we battle to ensure that life on Earth will continue and see out the Anthropocene, the stage is set for this to be the 'biologist's century'. Carnivorous marsupials are a small but vital component of the extraordinary web of life that has evolved on Planet Earth over 3.7 billion years. Your engagement, and everyone's, is now critical to ensure that life continues.

We leave you with a stirring call to arms from Menna Jones, our final vignette contributor, to conserve the iconic carnivorous marsupials that roam the Tasmanian wilderness.

VIGNETTE: CONSERVING TASMANIA'S CARNIVOROUS MARSUPIALS: THE DEVIL IS IN THE DETAIL

By Menna Jones
University of Tasmania, Hobart, Australia

The global stronghold for the flesh-eating carnivorous marsupials is the island of Tasmania, which supports three species living together: the Tasmanian Devil (*Sarcophilus harrisii*) and the Spotted-tailed (*Dasyurus maculatus*) and Eastern Quolls (*Dasyurus viverrinus*). Elsewhere, there are three further species of quolls distributed across mainland Australia, two quolls in New Guinea and three largely flesh-eating carnivorous marsupial species in South America. All living species are mid-sized; the large marsupial carnivores, the Australian Thylacine and the borhyaenids and sabretooths of South America, are lost to extinction. The

Fig. 11.15. Is the Tasmanian Devil (*Sarcophilus harrisii*) a 're-wilding' salvation species that could help control feral cat populations on southern mainland Australia? Source: Jiri Lochman.

marsupial and placental carnivores that evolved in southern and northern continents are a classic case of convergence in body form and ecological niche. Both radiations produced carnivores that were dog-like (Australian thylacinids) and cat-like (the marsupial lions: thylacoleonids in Australia and the sabretooth *Thylacosmilus* in South America), hyaena-like bone-eaters (the Tasmanian Devil), small, elongated carnivores (the quolls in Australia) and bear-like carnivores (giant rat kangaroos, *Propleopus*) (see Chapter 5).

Sadly, much of this unique diversity has been lost to extinction and the remaining species are in peril. Synergistic effects of habitat fragmentation and degradation for agriculture, changing fire regimes and invasive pests such as foxes, cats and Cane Toads have led to the decline and widespread local extinction of all quoll species on the Australian mainland. Islands play a crucial role in the survival of many Australian mammals because threatening factors are often absent. Tasmania is the largest of Australia's islands, lacks the Dingo and Red Fox, and has suffered less habitat destruction than the mainland. The Tasmanian Devil and Eastern Quoll and, until the last century, the Thylacine, survived only in Tasmania.

Tragedy struck in the 1990s with the emergence in Tasmanian Devils of an unusual transmissible cancer: devil facial tumour disease (DFTD). Cancers usually arise and die within their individual host, but transmissible cancers are immortal cell lines in which live tumour cells are the infectious agent. They spread from animal to animal through injurious contact, biting in the case of Devils. The tumours occur mostly around the head and inside the mouth and become horrendous, ulcerated, open lesions up to 12 cm across that can, although rarely do, obstruct eyes and teeth. Nearly all Devils become infected once they reach sexual maturity and fight and bite, leading to death within 6 months of tumours appearing. The disease has spread across most of the Devil's range in Tasmania, causing more than 85% total population decline and in excess of 90% local declines. Diseased populations comprise mostly young animals.

The loss of even one species is a calamity, but when that species has a pivotal role in the food web, its decline can cause cascading changes in the abundance of species at other trophic levels. Predators are masters of persecution, eliminating or harassing their smaller competitors. Devils are not large (a squat 7–14 kg) but they are the apex or largest predator in Tasmania's ecosystems since the demise of the Thylacine. When the Devil is away, cats may play – the competitive release of cats is linked with declines of smaller native mammals and the rise of invasive rodents (Black Rats, *Rattus rattus*), that can sustain high levels of predation by cats because they have much higher reproductive rates than native species. These ecological changes in Tasmania mirror the mammal declines that occurred on the Australian mainland 100 years ago. Quolls are also at risk. The Eastern Quoll suffered major decline in the early 2000s, probably due to a change in the weather: a series of warm, wet winters when they like it cold and dry. Their populations have not yet recovered and cats may be playing a role by shifting their daily activity time to hunt the few quoll babies now produced each year. Cats may also be outcompeting the similar sized Spotted-tailed Quolls, which are now much less abundant in eastern Tasmania where there are few Devils and many cats.

So what can we do about this whole devilish catastrophe? First, we have to recognise that cats (and foxes on the Australian mainland) are here to stay; we will never eradicate them. Second, traditional methods of control of disease or invasive predators are not effective at broad landscape scale and require costly, intensive management implemented in perpetuity. What we need to do is harness the power of natural ecological interactions in food webs and the capacity for animals to evolve rapidly in response to the threats. Restoring apex predators, such as Tasmanian Devils and Dingoes, to suppress cats might achieve cat control with little ongoing intervention or cost. The reintroduction of Devils to the Australian mainland is part of a big vision for restoring fauna and functional ecosystems across Australia. We know more-or-less why Devils became extinct on the continent: a combination of extreme climate, a rise in human impacts and the introduction of the Dingo. So even though this happened 3000 years ago, there is no reason why this element of the fauna, the Devil, which could play a crucial functional role, cannot be restored. Restoration of top-down influence of the Devil in Tasmanian ecosystems is more problematic. The Devil will likely solve the problem itself though. New and emerging infectious diseases are usually highly lethal, but evolve with their hosts to eventually become endemic, leading to coexistence and host recovery. There are encouraging signs that Devils are evolving rapidly in response to their disease epidemic. Changes in genes associated with cancer and immune function have occurred in as little as four to six generations (8–12 years) following local disease outbreak. These rapid evolutionary responses correspond with cessation of population decline, reduced force of infection, healthy Devils that never become infected and recoveries of individual Devils, whose tumours regress in wild populations.

Other natural ecological and evolutionary processes can be manipulated also to help quolls and other wildlife live with cats. Cats are very difficult to control using traditional methods, but high cat abundance is often supported by high densities of rabbits. Rabbits can be controlled effectively by shooting, poisoning and warren ripping, as well as by biocontrol methods such as rabbit haemorrhagic disease virus (RHDV). Restoring the fine-scale structural elements of habitats (low dense tussocks and shrubs, logs and rocks) can increase survival by providing wildlife with places to hide to avoid or escape predation. Wildlife can be taught to recognise invasive pests; quolls to avoid eating Cane Toads and other wildlife to avoid cats. Some species are evolving to recognise foxes, of which there is no equivalent predator in the native Australian fauna.

Australia's marsupial carnivores are exquisitely beautiful creatures. Their conservation status is of great concern, but we are on the brink of realising new approaches to their conservation that are grounded in ecological science. In combination with investment in creating islands and fenced reserves that are free of invasive predators, these new approaches may help to achieve a vision of restoring native marsupial carnivores, and also their beautiful prey species, at broad landscape scales outside fences.

Appendix: List of carnivorous marsupial species

EXTANT CARNIVOROUS MARSUPIALS (6 families; 32 genera; 136 species)

Families: Dasyuridae: 75 spp.; Didelphidae: 48 spp.; Caenolestidae: 7 spp.; Notoryctidae: 2 spp.; Myrmecobiidae: 1 sp.; Microbiotheriidae: 3 spp.

Collectively, Didelphidae and Caenolestidae comprise the American marsupials, or ameridelphians (cohort Ameridelphia), while the remaining families comprise the carnivorous Australian (and New Guinean) marsupials, or australidelphians (cohort Australidelphia).

All species included in this account are listed alphabetically by genus and species within subfamily/family.

1. Australian endemic carnivorous marsupials (3 families; 15 genera; 61 species)

Carnivorous (predominantly insectivorous) species (61 species)

Order Dasyuromorphia

Family Dasyuridae (so-called 'carnivorous marsupials') (58 species)

Subfamily Dasyurinae

Dasycercus (mulgaras) – 2 species: Brush-tailed Mulgara, *D. blythi*; Crest-tailed Mulgara, *D. cristicauda*

Dasykaluta – 1 species: Kaluta, *D. rosamondae*

Dasyuroides – 1 species: Kowari, *D. byrnei*

Dasyurus (quolls) – 4 species: Western Quoll, *D. geoffroii*; Northern Quoll, *D. hallucatus*; Spotted-tailed Quoll, *D. maculatus*; Eastern Quoll, *D. viverrinus*

Parantechinus – 1 species: Dibbler, *P. apicalis*

Pseudantechinus (false antechinuses) – 6 species: Sandstone Pseudantechinus, *P. bilarni;* Fat-tailed Pseudantechinus, *P. macdonnellensis;* Carpentarian Pseudantechinus, *P. mimulus*; Ningbing Pseudantechinus, *P. ningbing*; Rory's Pseudantechinus, *P. roryi*; Woolley's Pseudantechinus, *P. woolleyae*

Sarcophilus – 1 species: Tasmanian Devil, *S. harrisii*

Subfamily Phascogalinae

Antechinus (antechinuses) – 15 species: Rusty Antechinus, *A. adustus*; Agile Antechinus, *A. agilis*; Silver-headed Antechinus, *A. argentus*; Black-tailed Dusky Antechinus, *A. arktos*; Fawn Antechinus, *A. bellus*; Yellow-footed Antechinus, *A. flavipes*; Atherton Antechinus, *A. godmani*; Cinnamon Antechinus, *A. leo*; Mainland Dusky Antechinus, *A. mimetes*; Swamp Antechinus, *A. minimus*; Buff-footed Antechinus, *A. mysticus*; Brown Antechinus, *A. stuartii*; Subtropical Antechinus, *A. subtropicus*; Tasmanian Dusky Antechinus, *A. swainsonii*; Tasman Peninsula Dusky Antechinus, *A. vandycki*

Phascogale (phascogales) – 3 species: Red-tailed Phascogale, *P. calura*; Northern Brush-tailed Phascogale, *P. pirata*; Common Brush-tailed Phascogale, *P. tapoatafa*

Subfamily: Sminthopsinae

Antechinomys – 1 species: Kultarr, *A. laniger*

Ningaui (ningauis) – 3 species: Wongai Ningaui, *N. ridei*; Pilbara Ningaui, *N. timealeyi*; Southern Ningaui, *N. yvonneae*

Planigale (planigales) – 4 species: Giles' Planigale, *P. gilesi*; Long-tailed Planigale, *P. ingrami*; Common Planigale, *P. maculata*; Narrow-nosed Planigale, *P. tenuirostris*

Sminthopsis (dunnarts) – 16 species: Kakadu Dunnart, *S. bindi*; Butler's Dunnart, *S. butleri*; Fat-tailed Dunnart, *S. crassicaudata*; Little Long-tailed Dunnart, *S. dolichura*; Julia Creek Dunnart, *S. douglasi*; Grey-bellied Dunnart, *S. fuliginosus*; Gilbert's Dunnart, *S. gilberti*; White-tailed Dunnart, *S. granulipes*; Hairy-footed Dunnart, *S. hirtipes*; White-footed Dunnart, *S. leucopus*; Long-tailed Dunnart, *S. longicaudata*; Stripe-faced Dunnart, *S. macroura*; Common Dunnart, *S. murina*; Ooldea Dunnart, *S. ooldea*; Sandhill Dunnart, *S. psammophila*; Lesser Hairy-footed Dunnart, *S. youngsoni*

Family Myrmecobiidae (1 genus; 1 species)

Myrmecobius – 1 species: Numbat, *M. fasciatus*

Family Notoryctidae (marsupial moles) (1 genus; 2 species)

Notoryctes (marsupial moles) – 2 species: Kakarratul, *N. caurinus*; Itjaritjari, *N. typhlops*

2. New Guinean endemic carnivorous marsupials (1 family (shared with Australia); 6 genera (2 shared with Australia); 15 species)

Order Dasyuromorphia

Family Dasyuridae [continued] (carnivorous marsupials)

Carnivorous (carnivorous, insectivorous) species (15 species)

Subfamily Dasyurinae [continued]

Dasyurus (quolls) – 2 species: New Guinea Quoll, *D. albopunctatus*; Bronze Quoll, *D. spartacus*

Myoictis (three-striped dasyures) – 4 species: Woolley's Three-striped Dasyure, *M. leucura*; Müller's Three-striped Dasyure, *M. melas*; Wallace's Three-striped Dasyure, *M. wallacii*; Tate's Three-striped Dasyure, *M. wavicus*

Neophascogale – 1 species: Speckled Dasyure, *N. lorentzii*

Phascolosorex – 2 species: Red-bellied Phascogale, *P. doriae*; Narrow-striped Dasyure, *P. dorsalis*

Subfamily Phascogalinae [continued]

Murexia (murexias, dasyures) – 5 species: Habbema Dasyure, *M. habbema*; Short-furred Dasyure, *M. longicaudata*; Black-tailed Dasyure, *M. melanurus*; Long-nosed Dasyure, *M. naso*; Broad-striped Dasyure, *M. rothschildi*

Subfamily Sminthopsinae [continued]

Planigale – 1 species: Papuan Planigale, *P. novaeguineae*

3. Shared Australian/New Guinean carnivorous marsupials (2 species)

Order Dasyuromorphia

Family Dasyuridae

Carnivorous (predominantly insectivorous) species (2 species)

Subfamily Sminthopsinae [continued]

Sminthopsis (dunnarts) – 2 species: Chestnut Dunnart, *S. archeri*; Red-cheeked Dunnart, *S. virginiae*

4. American (predominantly South America, but also Central America and Mexico) endemic carnivorous marsupials (3 families; 13 genera; 58 species)

Order Microbiotheria

Family Microbiotheriidae (Monito del Montes) [1 genus; 3 species]

Dromiciops (monito del montes) – 3 species: Panchos's Monito del Monte, *D. bozinovici*; Mondaca's Monito del Monte, *D. mondaca*; Monito del Monte, *D. gliroides*

Order Paucituberculata

Family Caenolestidae (shrew-opossums, caenolestids) [3 genera; 7 species]

Caenolestes (caenolestids) – 5 species: Gray-bellied Shrew-opossum, *C. caniventer*; Condor Shrew-opossum, *C. condorensis*; Blackish Shrew-opossum, *C. convelatus*; Dusky Shrew-opossum, *C. fuliginosus*; Sangay Shrew-opossum, *C. sangay*

Lestoros – 1 species: Incan Shrew-opossum, *L. inca*

Rhyncholestes – 1 species: Long-nosed Shrew-opossum, *R. raphanurus*

Order Didelphimorphia

Family Didelphidae (opossums) [9 genera; 48 species]

Subfamily Didelphinae

Chacodelphys – 1 species: Pygmy Opossum, *C. formosa*

Chironectes – 1 species: Water Opossum, *C. minimus*

Lestodelphys – 1 species: Patagonian Opossum, *L. halli*

Lutreolina (lutrine opossums) – 2 species: Big Lutrine Opossum, *L. crassicaudata*; Massoia's Lutrine Opossum, *L. massoia*

Metachirus – 1 species: Brown Four-eyed Opossum, *M. nudicaudatus*

Monodelphis (short-tailed opossums) – 23 species: Sepia Short-tailed Opossum, *M. adusta*; Northern Three-striped Opossum, *M. americana*; Arlindo's Short-tailed Opossum, *M. arlindoi*; Guyanan Short-tailed Opossum, *M. brevicaudata*; Orange-sided Opossum, *M. dimidiata*; Gray Short-tailed Opossum, *M. domestica*; Emilie's Short-tailed Opossum, *M. emiliae*; Gardner's Short-tailed Opossum, *M. gardneri*; Amazonian Red-sided Opossum, *M. glirina*; Handley's Short-tailed Opossum, *M. handleyi*; Ihering's Three-striped Opossum, *M. iheringi*; Pygmy Short-tailed Opossum, *M. kunsi*; Osgood's Short-tailed Opossum, *M. osgoodi*; Hooded Red-sided Opossum, *M. palliolata*; Peruvian Short-tailed Opossum, *M. peruviana*; Pinocchio's Short-tailed Opossum, *M. pinocchio*; Reig's Opossum, *M. reigi*; Pine's Opossum, *M. ronaldi*; Saci's Short-tailed Opossum, *M. saci*; Santa Rosa Short-tailed Opossum, *M. sanctaerosae*; Tawny-headed Opossum, *M. scalops*; Touan Short-tailed Opossum, *M. touan*; Single-striped Opossum, *M. unistriata*

Philander (four-eyed opossums) – 7 species: Anderson's Four-eyed Opossum, *P. andersoni*; Orinoco Four-eyed Opossum, *P. deltae*; South-eastern Four-eyed Opossum, *P. frenatus*; McIlhenny's Four-eyed Opossum, *P. mcilhennyi*; Mondolfi's Four-eyed Opossum, *P. mondolfii*; Olrog's Four-eyed Opossum, *P. olrogi*; Gray Four-eyed Opossum, *P. opossum*

Thylamys (fat-tailed mouse opossums) – 11 species: Mesopotamian Fat-tailed Opossum, *T. citellus*; Elegant Fat-tailed Opossum, *T. elegans*; Karimi's Fat-tailed Opossum; *T. karimii*; Paraguayan Fat-tailed Opossum, *T. macrurus*; Pallid Fat-tailed Opossum, *T. pallidior*; Dry Chaco Fat-tailed Opossum;

T. pulchellus; Chacoan Fat-tailed Opossum, *T. pusillus*; Argentine Fat-tailed Opossum, *T. sponsorius*; Tate's Fat-tailed Opossum, *T. tatei*; Dwarf Fat-tailed Opossum, *T. velutinus*; Buff-bellied Fat-tailed Opossum, *T. venustus*

Tlacuatzin – 1 species: Tlacuatzin, *T. canescens*

References

Abbott I, Peacock D, Short J (2014) The new guard: the arrival and impacts of cats and foxes. In *Carnivores of Australia: Past, Present and Future.* (Eds AS Glen and CR Dickman) pp. 69–104. CSIRO Publishing, Melbourne.

Adams M, Cooper N, Armstrong J (2000) 'Revision of *Dasycercus* systematics'. Unpublished report to the South Australian Department for Environment and Heritage, Adelaide.

Adams-Hosking C, Grantham HS, Rhodes JR, McAlpine C, Moss PT (2011) Modelling climate-change induced shifts in the distribution of the Koala. *Wildlife Research* **38**, 122–130. doi:10.1071/WR10156

Agnolin F, Chimento N (2014) Morphological evidence supports Dryolestoid affinities for the living Australian marsupial mole *Notoryctes*. *PeerJ PrePrints* **2**, e755v1. doi.org/10.7287/peerj.preprints.755v1.

Aitken PF (1971) Rediscovery of the Large Desert Sminthopsis (*Sminthopsis psammophilus* Spencer) on Eyre Peninsula, South Australia. *Victorian Naturalist* **88**, 103–111.

Aitken PF (1972) *Sminthopsis murina* (Waterhouse) 1838. A new record from Kangaroo Island, South Australia. *South Australian Naturalist* **46**, 36–37.

Aland RC, Gosden E, Bradley AJ (2016) Seasonal morphometry of the vomeronasal organ in the marsupial mouse, *Antechinus subtropicus*. *Journal of Morphology* **277**, 1517–1530. doi:10.1002/jmor.20593

Albanese S, Ojeda RA (2012) Habitat use by a Neotropical desert marsupial (*Thylamys pallidior*): a multiscale approach. *Mammalian Biology* **77**, 237–243. doi:10.1016/j.mambio.2011.12.005

Albanese S, Dacar MA, Ojeda RA (2012) Unvarying diet of a Neotropical desert marsupial inhabiting a variable environment: the case of *Thylamys pallidior*. *Acta Theriologica* **57**, 185–188. doi:10.1007/s13364-011-0057-6

Albuja L, Patterson BD (1996) A new species of northern shrew-opossum (Paucituberculata: Caenolestidae) from the Cordillera del Cóndor, Ecuador. *Journal of Mammalogy* **77**, 41–53. doi:10.2307/1382707

Allen JA (1893a) Description of a new species of opossum from the Isthmus of Tehuantepec, Mexico. *Bulletin of the American Museum of Natural History* **5**, 235–236.

Allen JA (1893b) List of mammals and birds collected in northeastern Sonora and northwestern Chihuahua, Mexico, on the Lumholtz Archaeological Expedition, 1890-92. *Bulletin of the American Museum of Natural History* **5**, 3.

Allen JA, Buller AC (1897) Further notes on mammals collected in Mexico by Dr. Audley C. Buller, with descriptions of new species. *Bulletin of the American Museum of Natural History* **9**, 5.

Allison LM, Gibson LA, Aberton JG (2006) Dietary strategy of the Swamp Antechinus (*Antechinus minimus maritimus*) (Marsupialia: Dasyuridae) in coastal and inland heathland habitats. *Wildlife Research* **33**, 67–76. doi:10.1071/WR05038

Ameghino F (1891) Nuevos restos de mamíferos fósiles descubiertos por Carlos Ameghino en el Eoceno inferior de la Patagonia austral. Especies nuevas, adiciones y correciones. *Revista Argentina de Historia Natural* **1**, 289–328.

Andersen GE, Johnson CN, Jones ME (2016) Sympatric predator odour reveals a competitive relationship in size-structured mammalian carnivores. *Behavioral Ecology and Sociobiology* **70**, 1831–1841. doi:10.1007/s00265-016-2189-9

Anderson S (1997) Mammals of Bolivia: taxonomy and distribution. *Bulletin of the American Museum of Natural History* **231**, 1–652.

Andrew DL, Settle GA (1982) Observations on the behaviour of species of *Planigale* (Dasyuridae, Marsupialia) with particular reference to the Narrow-nosed Planigale (*Planigale tenuirostris*). In *Carnivorous Marsupials.* (Ed. M Archer) pp 311–324. Royal Zoological Society of New South Wales, Sydney.

Anon. (1923a) Australian fauna. *The Brisbane Courier* 9 April, p. 5.

Anon. (1923b) Off to the wilds – Australian explorers to search for fauna specimens for museums. *The Age*, 28 March, p. 5.

Anon. (1929a) Grief causes suicide. *Gloucester Journal*. 22 June. p. 20.

Anon. (1929b) Scientist's suicide. *The Yorkshire Post*. 19 June. p. 4.

Anon. (1929c) Well known scientist shoots himself. *Northern Daily Mail*. 18 June, p. 8.

Anon. (2015) *Martín Alonso Pinzón*. Wikipedia website, https://en.wikipedia.org/wiki/Mart%C3%ADn_Alonso_Pinz%C3%B3n

Aplin KP (2015) Family Notoryctidae (marsupial moles). In *Handbook of the Mammals of the World. Volume 5. Monotremes and Marsupials*. (Eds DE Wilson and RA Mittermeier) pp. 210–221. Lynx Edicions, Barcelona, Spain.

Archer M (1975) *Ningaui*, a new genus of tiny dasyurids (Marsupialia) and two new species, *N. timealeyi* and *N. ridei* from arid Western Australia. *Memoirs of the Queensland Museum* **17**, 237–249.

Archer M (1976) Revision of the marsupial genus *Planigale* Troughton (Dasyuridae). *Memoirs of the Queensland Museum* **17**, 341–365.

Archer M (1977) Revision of the dasyurid marsupial genus *Antechinomys* Krefft. *Memoirs of the Queensland Museum* **18**, 17–29.

Archer M (1979) Two new species of *Sminthopsis* Thomas (Dasyuridae: Marsupialia) from northern Australia, *S. butleri* and *S. douglasi*. *Australian Zoologist* **20**, 327–345.

Archer M (1981) Results of the Archbold Expeditions. No. 104. Systematic revision of the marsupial dasyurid genus *Sminthopsis* Thomas. *Bulletin of the American Museum of Natural History* **168**, 63–223.

Archer M (1982) Review of the dasyurid (Marsupialia) fossil record, integration of data bearing on phylogenetic interpretation and suprageneric classification. In *Carnivorous Marsupials*. (Ed. M Archer) pp. 397–443. Royal Zoological Society of New South Wales, Sydney.

Archer M (1984) The Australian marsupial radiation. In *Vertebrate Zoogeography and Evolution in Australia*. (Eds M Archer and G Clayton) pp. 633–708. Hesperian Press, Perth, WA.

Archer M (2008) The Australian mammal fauna. In *The Mammals of Australia*. 3rd edn. (Eds S Van Dyck and R Strahan) pp. 20–22. Reed New Holland, Sydney.

Archer M, Hand SJ (2006) The Australian marsupial radiation. In *Evolution and Biogeography of Australasian Vertebrates*. (Eds JR Merrick, M Archer, GM Hickey and MSY Lee) pp. 575–646. Auscipub, Sydney.

Archer M, Kirsch J (2006) The evolution and classification of marsupials. In *Marsupials*. (Eds PJ Armati, CR Dickman and ID Hume) pp. 1–21. Cambridge University Press, Cambridge, UK.

Archer M, Hand SJ, Godthelp H (1991) *Riversleigh*. Reed Books, Sydney.

Archer M, Beck R, Gott M, Hand S, Godthelp H, Black K (2011) Australia's first fossil marsupial mole (Notoryctemorphia) resolves controversies about their evolution and palaeoenvironmental origins. *Proceedings. Biological Sciences* **278**, 1498–1506. doi:10.1098/rspb.2010.1943

Arena DA, Black KH, Archer M, Hand SJ, Godthelp H, Creaser P (2014) Reconstructing a Miocene pitfall trap: recognition and interpretation of fossiliferous Cenozoic palaeokarst. *Sedimentary Geology* **304**, 28–43.

Argot C (2004a) Evolution of South American mammalian predators (Borhyaenoidea): anatomical and palaeobiological implications. *Zoological Journal of the Linnean Society* **140**, 487–521. doi:10.1111/j.1096-3642.2004.00110.x

Argot C (2004b) Functional-adaptive features and palaeobiologic implications of the postcranial skeleton of the late Miocene sabretooth borhyaenoid *Thylacosmilus atrox* (Metatheria). *Alcheringa* **28**, 229–266. doi:10.1080/03115510408619283

Argot C, Babot J (2011) Postcranial morphology, functional adaptations and palaeobiology of *Callistoe vincei*, a predaceous metatherian from the Eocene of Salta, north-western Argentina. *Palaeontology* **54**, 447–480. doi:10.1111/j.1475-4983.2011.01036.x

Ariel B (2013) The phylogeny and biogeography of the Monito del Monte (*Dromiciops gliroides*) and its relatives. BSc (Hons) thesis. Paper 117. Honors College, University of Maine, Orono ME, USA.

Armstrong LA, Krajewski C, Westerman M (1998) Phylogeny of the dasyurid marsupial genus *Antechinus* based on cytochrome-b, 12S-rRNA, and protamine-P1 genes. *Journal of Mammalogy* **79**, 1379–1389. doi:10.2307/1383028

Arnold J, Shield J (1970) Oxygen consumption and body temperature of the Chuditch (*Dasyurus geoffroii*). *Journal of Zoology* **160**, 391–404. doi:10.1111/j.1469-7998.1970.tb03089.x

Arrese CA, Hart NS, Thomas N, Beazley LD, Shand J (2002) Trichromacy in Australian marsupials. *Current Biology* **12**, 657–660. doi:10.1016/S0960-9822(02)00772-8

Arrese CA, Beazley LD, Neumeyer C (2006) Behavioural evidence for marsupial trichromacy. *Current Biology* **16**, R193–R194. doi:10.1016/j.cub.2006.02.036

Arthur AD, Catling PC, Reid A (2012) Relative influence of habitat structure, species interactions and rainfall on the post-fire population dynamics of ground-dwelling vertebrates. *Austral Ecology* **37**, 958–970. doi:10.1111/j.1442-9993.2011.02355.x

Aslin HJ (1974) The behaviour of *Dasyuroides byrnei* (Marsupialia) in captivity. *Zeitschrift für Tierpsychologie* **35**, 187–208. doi:10.1111/j.1439-0310.1974.tb00443.x

Aslin H (1975) Reproduction in *Antechinus maculatus* Gould (Dasyuridae). *Australian Wildlife Research* **2**, 77–80. doi:10.1071/WR9750077

Aslin HJ (1983) Reproduction in *Sminthopsis ooldea* (Marsupialia: Dasyuridae). *Australian Mammalogy* **6**, 93–95.

Astúa D (2015) Family Didelphidae (Opossums). In *Handbook of the Mammals of the World. Volume 5. Monotremes and Marsupials.* (Eds DE Wilson and RA Mittermeier) pp. 70–187. Lynx Edicions, Barcelona, Spain.

Astúa D, Carvalho RA, Maia PF, Magalhäes AR, Loretta D (2015) First evidence of gregarious denning in opossums (Didelphimorphia, Didelphidae), with notes on their social behaviour. *Biology Letters* **11**, 20150307. doi:10.1098/rsbl.2015.0307

Attard MRG, Chamoli U, Ferrara TL, Rogers TL, Wroe S (2011) Skull mechanics and implications for feeding behaviour in a large marsupial carnivore guild: the Thylacine, Tasmanian Devil and Spotted-tailed Quoll. *Journal of Zoology* **285**, 292–300. doi:10.1111/j.1469-7998.2011.00844.x

Attard MRG, Parr WCH, Wilson LAB, Archer M, Hand SJ, Rogers TL, *et al.* (2014) Virtual reconstruction and prey size preference in the mid Cenozoic thylacinid, *Nimbacinus dicksoni* (Thylacinidae, Marsupialia). *PLoS One* **9**(4), e93088. doi:10.1371/journal.pone.0093088

Attenborough D (2013) *A Curious Hoax? David Attenborough's Natural Curiosities Season 1, Episode 2*. UKTV, Cardiff, UK.

Bailey AM (1930) The obituary of Russell W. Hendee (with one illustration). *The Condor* **32**, 147–151. doi:10.2307/1363446

Baker AM (2012a) Science update: birth of a new species, the Buff-footed Antechinus, *Antechinus mysticus. Science Illustrated Magazine*, January–February, p. 20.

Baker AM (2012b) What's the difference? Describing a new dasyurid species. *Wildlife Australia Magazine*, Spring Edition, 18–21.

Baker AM (2013) Mysterious mammal! A new species of Australian marsupial. *Australian Geographic Magazine*, November–December.

Baker AM (2015) Family Dasyuridae (carnivorous marsupials). In *Handbook of the Mammals of the World*. Volume 5. *Monotremes and marsupials.* (Eds DE Wilson and RA Mittermeier) pp. 232–349. Lynx Edicions, Barcelona, Spain.

Baker AM, Van Dyck S (2013a) Monograph: taxonomy and redescription of the Yellow-footed Antechinus, *Antechinus flavipes* (Waterhouse) (Marsupialia: Dasyuridae). *Zootaxa* **3649**(1), 1–62. doi:10.11646/zootaxa.3649.1.

Baker AM, Van Dyck S (2013b) Taxonomy and redescription of the Atherton Antechinus, *Antechinus godmani* (Thomas) (Marsupialia: Dasyuridae). *Zootaxa* **3670**(4), 401–439. doi:10.11646/zootaxa.3670.4.1

Baker AM, Van Dyck S (2013c) Taxonomy and redescription of the Fawn Antechinus, *Antechinus bellus* (Thomas) (Marsupialia: Dasyuridae). *Zootaxa* **3613**(3), 201–228. doi:10.11646/zootaxa.3746.2.1

Baker AM, Van Dyck S (2015) Taxonomy and redescription of the Swamp Antechinus, *Antechinus minimus* (È. Geoffroy) (Marsupialia: Dasyuridae). *Memoirs of the Queensland Museum* **59**, 127–170.

Baker AM, Mutton TY, Van Dyck S (2012) A new dasyurid marsupial from eastern Queensland, Australia: the Buff-footed Antechinus, *Antechinus mysticus* sp. nov. (Marsupialia: Dasyuridae). *Zootaxa* **3515**, 1–37.

Baker AM, Mutton TY, Hines HB (2013) A new dasyurid marsupial from Kroombit Tops, south-east Queensland, Australia: the Silver-headed Antechinus, *Antechinus argentus* sp. nov. (Marsupialia: Dasyuridae). *Zootaxa* **3746**(2), 201–239. doi:10.11646/zootaxa.3746.2.1

Baker AM, Mutton TY, Hines HB, Van Dyck S (2014) The Black-tailed Antechinus, *Antechinus arktos* sp. nov.: a new species of carnivorous marsupial from montane regions of the Tweed Volcano caldera, eastern Australia. *Zootaxa* **3765**(2), 101–133. doi:10.11646/zootaxa.3765.2.1

Baker AM, Mutton TY, Mason ED, Gray EL (2015) A taxonomic assessment of the Australian Dusky Antechinus complex: a new species, the Tasman Peninsula Dusky Antechinus (*Antechinus vandycki* sp. nov.) and an elevation to species of the Mainland Dusky Antechinus (*Antechinus swainsonii mimetes* (Thomas)). *Memoirs of the Queensland Museum* **59**: 75-126.

Balazote Oliver A, Amico GC, Rivarola MD, Morales JM (2017) Population dynamics of *Dromiciops gliroides* (Microbiotheriidae) in an austral temperate forest. *Journal of Mammalogy* **98**, 1179–1184. doi:10.1093/jmammal/gyx051

Banks SC, Finlayson GR, Lawson SJ, Lindenmayer DB, Paetkau D, Ward SJ, *et al.* (2005a) The effects of habitat fragmentation due to forestry plantation establishment on demography and genetic variation in a marsupial carnivore, *Antechinus agilis. Biological Conservation* **122**, 581–597. doi:10.1016/j.biocon.2004.09.013

Banks SC, Lindenmayer DB, Ward SJ, Taylor AC (2005b) The effects of habitat fragmentation via forestry plantation establishment on spatial genotypic structure in the small marsupial carnivore, *Antechinus agilis. Molecular Ecology* **14**, 1667–1680. doi:10.1111/j.1365-294X.2005.02525.x

Banks PB, Bytheway JP, Carthey AJR, Hughes NK, Price CJ (2014) Olfaction and predator-prey interactions amongst mammals in Australia. In *Carnivores of Australia: Past, Present and Future.* (Eds AS Glen and CR Dickman) pp. 389–404. CSIRO Publishing, Melbourne.

Barkley LJ, Whitaker JO, Jr (1984) Confirmation of *Caenolestes* in Peru with information on diet. *Journal of Mammalogy* **65**, 328–330. doi:10.2307/1381173

Barry SJ (1984) Small mammals in a south-east Queensland rainforest: the effects of soil fertility and past logging disturbance. *Australian Wildlife Research* **11**, 31–39.

Baverstock PR, Archer M, Adams M, Richardson BJ (1982) Genetic relationships among 32 species of Australian dasyurid marsupials. In *Carnivorous Marsupials.* (Ed. M Archer) pp. 641–650. Royal Zoological Society of New South Wales, Sydney.

Baverstock PR, Adams M, Archer M, McKenzie NL, How R (1983) An electrophoretic and chromosomal study of the dasyurid marsupial genus *Ningaui* Archer. *Australian Journal of Zoology* **31**, 381–392. doi:10.1071/ZO9830381

Baverstock PR, Adams M, Archer M (1984) Electrophoretic resolution of species boundaries in the *Sminthopsis murina* complex (Dasyuridae). *Australian Journal of Zoology* **32**, 823–832. doi:10.1071/ZO9840823

Baynes A, Johnson KA (1996) The contributions of the Horn Expedition and cave deposits to knowledge of the original mammal fauna of central Australia. In *Exploring Central Australia: Society, the Environment and the 1894 Horn Expedition.* (Eds SR Morton and DJ Mulvaney) pp. 168–186. Surrey Beatty and Sons, Sydney.

Beck R (2008) A dated phylogeny of marsupials using a molecular supermatrix and multiple fossil constraints. *Journal of Mammalogy* **89**, 175–189. doi:10.1644/06-MAMM-A-437.1

Beck RMD, Travouillon KJ, Aplin K, Godthelp H, Archer M (2014) The osteology and systematics of the enigmatic Australian Oligo-Miocene metatherian *Yalkaparidon* (Yalkaparidontidae; Yalkaparidontia? Australidelphia; Marsupialia). *Journal of Mammalian Evolution* **21**, 127–172. doi:10.1007/s10914-013-9236-3

Begg RJ (1981) The small mammals of Little Nourlangie Rock, NT. 2. Ecology of the Sandstone Antechinus, *Antechinus bilarni* (Marsupialia: Dasyuridae). *Australian Wildlife Research* **8**, 57–72. doi:10.1071/WR9810057

Belcher CA (1995) Diet of the Tiger Quoll (*Dasyurus maculatus*). *Wildlife Research* **22**, 341–357. doi:10.1071/WR9950341

Belcher CA (2008) Response of the Tiger Quoll *Dasyurus maculatus* to disturbance from selective logging. *Australian Forestry* **71**, 280–286. doi:10.1080/00049158.2008.10675046

Belcher CA, Darrant JP (2004) Home range and spatial organisation of the marsupial carnivore, *Dasyurus maculatus maculatus* (Marsupialia: Dasyuridae) in south-eastern Australia. *Journal of Zoology* **262**, 271–280. doi:10.1017/S0952836903004631

Belcher CA, Darrant JP (2006) Habitat use by the Tiger Quoll (*Dasyurus maculatus*) (Marsupiala: Dasyuridae) in south-eastern Australia. *Journal of Zoology* **269**, 183–190.

Belcher CA, Burnett S, Jones M (2008) Spotted-tailed Quoll, *Dasyurus maculatus*. In *The Mammals of Australia*. 3rd edn. (Eds S Van Dyck and R Strahan) pp. 60–62. Reed New Holland, Sydney.

Bencini R, McCulloch C, Mills HR, Start AN (2001) Habitat and diet of the Dibbler (*Parantechinus apicalis*) on two islands in Jurien Bay, Western Australia. *Wildlife Research* **28**, 465–468. doi:10.1071/WR00014

Bennison K, Dickman CR, Godfree R (2013) Habitat use and ecological observations of the Ooldea Dunnart (*Sminthopsis ooldea*) at Uluru-Kata Tjuta National Park, Northern Territory. *Australian Mammalogy* **35**, 175–179. doi:10.1071/AM12048

Bennison K, Clayton J, Godfree R, Pavey C, Wilson M (2014) Surfacing behaviour and ecology of the marsupial mole (*Notoryctes typhlops*) at Uluru-Kata Tjuta National Park. *Australian Mammalogy* **36**, 184–188.

Benshemesh J, Johnson K (2003) Biology and conservation of marsupials moles (*Notoryctes*). In *Predators with Pouches: the Biology of Carnivorous Marsupials*. (Eds M Jones, C Dickman and M Archer) pp. 464–474. CSIRO Publishing, Melbourne.

Beolens B, Watkins M, Grayson M (2009) *The Eponym Dictionary of Mammals*. Johns Hopkins University Press, Baltimore MD, USA.

Bilney RJ, Cooke R, White JG (2011) Potential competition between two top-order predators following a dramatic contraction in the diversity of their prey base. *Animal Biology* **61**, 29–47. doi:10.1163/157075511X554400

Birney EC, Monjeau JA (2003) Latitudinal variation in South American marsupial biology. In *Predators with Pouches: the Biology of Carnivorous Marsupials*. (Eds M Jones, C Dickman and M Archer) pp. 297–317. CSIRO Publishing, Melbourne.

Birney EC, Monjeau JA, Phillips CJ, Sikes RS, Kim I (1996) *Lestodelphys halli*: new information of a poorly known Argentine marsupial. *Mastozoología Neotropical* **3**, 171–181.

Black KH, Archer M, Hand SJ, Godthelp H (2012) The rise of Australian marsupials: a synopsis of biostratigraphic, phylogenetic, palaeoecologic and palaeobiogeographic understanding. In *Earth and Life: Global Biodiversity, Extinction Intervals and Biogeographic Perturbations Through Time*. (Ed. JA Talent) pp. 983–1078. Springer, Dordecht, Netherlands.

Blacket MJ, Adams M, Krajewski C, Westerman M (2000) Genetic variation within the dasyurid marsupial genus *Planigale*. *Australian Journal of Zoology* **48**, 443–459. doi:10.1071/ZO00020

Blacket MJ, Adams M, Cooper SJB, Krajewski C, Westerman M (2001) Systematics and evolution of the dasyurid marsupial genus *Sminthopsis*: I. The macroura species group. *Journal of Mammalian Evolution* **8**, 149–170. doi:10.1023/A:1011322031747

Blacket MJ, Cooper SJB, Krajewski C, Westerman M (2006) Systematics and evolution of the dasyurid marsupial genus *Sminthopsis*: II. The murina species group. *Journal of Mammalian Evolution* **13**, 125–138. doi:10.1007/s10914-006-9008-4

Blacket MJ, Kemper C, Brandle R (2008) Planigales (Marsupialia : Dasyuridae) of eastern Australia's interior: a comparison of morphology, distributions and habitat preferences, with particular emphasis on South Australia. *Australian Journal of Zoology* **56**, 195–205. doi:10.1071/ZO08057

Blackhall S (1980) Diet of the Eastern Native-cat, *Dasyurus viverrinus* (Shaw), in southern Tasmania. *Australian Wildlife Research* **7**, 191–197. doi:10.1071/WR9800191

Blanco RE, Jones WW, Milne N (2013) Is the extant Southern Short-tailed Opossum a pigmy sabretooth predator? *Journal of Zoology* **291**, 100–110. doi:10.1111/jzo.12050

Bos DG (2001) Some observations on foraging behaviour in the Southern Ningaui, *Ningaui yvonneae. Australian Mammalogy* **23**, 59–61. doi:10.1071/AM01059

Bos DG, Carthew SM (2001) Population ecology of *Ningaui yvonneae* (Dasyuridae: Marsupialia) in the Middleback Ranges, Eyre Peninsula, South Australia. *Wildlife Research* **28**, 507–515. doi:10.1071/WR00021

Bos DG, Carthew SM (2003) The influence of behaviour and season on habitat selection by a small mammal. *Ecography* **26**, 810–820. doi:10.1111/j.0906-7590.2003.03584.x

Bos DG, Carthew SM (2007a) Patterns of movement in the small dasyurid *Ningaui yvonneae. Australian Journal of Zoology* **55**, 299–307. doi:10.1071/ZO06055

Bos DG, Carthew SM (2007b) Prey selection by the dasyurid *Ningaui yvonneae. Wildlife Research* **34**, 632–639. doi:10.1071/WR07001

Bos DG, Carthew SM, Lorimer MF (2002) Habitat selection by the small dasyurid *Ningaui yvonneae* (Marsupialia: Dasyuridae) in South Australia. *Austral Ecology* **27**, 103–109. doi:10.1046/j.1442-9993.2002.01163.x

Bowman DMJS, Woinarski JCZ (1994) Biogeography of Australian monsoon rainforest mammals: implications for the conservation of rainforest mammals. *Pacific Conservation Biology* **1**, 98–106. doi:10.1071/PC940098

Bozinovic F, Ruiz G, Rosenmann M (2004) Energetics and torpor of a South American "living fossil", the microbiotheriid *Dromiciops gliroides. Journal of Comparative Physiology. B, Biochemical, Systemic, and Environmental Physiology* **174**, 293–297. doi:10.1007/s00360-004-0414-8

Bozinovic F, Muñoz JLP, Naya DE, Cruz-Neto AP (2007) Adjusting energy expenditures to energy supply: food availability regulates torpor use and organ size in the Chilean Mouse-opossum *Thylamys elegans. Journal of Comparative Physiology. B, Biochemical, Systemic, and Environmental Physiology* **177**, 393–400. doi:10.1007/s00360-006-0137-0

Bradley AJ (1997) Reproduction and life history in the Red-tailed Phascogale, *Phascogale calura* (Marsupialia: Dasyuridae): the adaptive-stress senescence hypothesis. *Journal of Zoology* **241**, 739–755. doi:10.1111/j.1469-7998.1997.tb05745.x

Bradley AJ (2003) Stress, hormones and mortality in small carnivorous marsupials. In *Predators with Pouches: the Biology of Carnivorous Marsupials.* (Eds M Jones, C Dickman and M Archer) pp. 254–267. CSIRO Publishing, Melbourne.

Bradley AJ, Monamy V (1991) A physiological profile of male Dusky Antechinuses, *Antechinus swainsonii* (Marsupialia: Dasyuridae) surviving post-mating mortality. *Australian Mammalogy* **14**, 25–27.

Bradley AJ, Kemper CM, Kitchener DJ, Humphreys WF, How RA (1987) Small mammals of the Mitchell Plateau Region, Kimberley, Western Australia. *Australian Wildlife Research* **14**, 397–413. doi:10.1071/WR9870397

Bradley AJ, Foster WK, Taggart DA (2008) Red-tailed Phascogale, *Phascogale calura*. In *The Mammals of Australia.* 3rd edn. (Eds S Van Dyck and R Strahan) pp. 101–102. Reed New Holland, Sydney.

Braithewaite RW (1973) An ecological study of *Antechinus stuartii* (Marsupialia: Dasyuridae). MSc thesis. University of Queensland, Brisbane.

Braithwaite RW, Griffiths AD (1994) Demographic variation and range contraction in the Northern Quoll, *Dasyurus hallucatus* (Marsupialia: Dasyuridae). *Wildlife Research* **21**, 203–217. doi:10.1071/WR9940203

Braithwaite RW, Lonsdale WM (1987) The rarity of *Sminthopsis virginiae* (Marsupialia: Dasyuridae) in relation to natural and unnatural habitats. *Conservation Biology* **1**, 341–343. doi:10.1111/j.1523-1739.1987.tb00054.x

Braithwaite RW, Muller WJ (1997) Rainfall, groundwater and refuges: predicting extinctions of Australian tropical mammal species. *Austral Ecology* **22**, 57–67. doi:10.1111/j.1442-9993.1997.tb00641.x

Brass LJ (1959) Results of the Archbold Expeditions. No. 79. Summary of the fifth Archbold Expedition to New Guinea (1956-1957). *Bulletin of the American Museum of Natural History* **118**, 1–69.

Breed WG, Taggart DA, Moore HDM (2003) Sperm maturation and fertilisation in Australian and American insectivorous marsupials. In *Predators with Pouches: the Biology of Carnivorous Marsupials.* (Eds M Jones, C Dickman and M Archer) pp. 133–146. CSIRO Publishing, Melbourne.

British Museum (Natural History). Department of Zoology: Mammals Section Correspondence. DF200-DF299; DF232/20; DF232/7; DF232/1; DF232/5. NHM Archives.

Brooks SJ, Calver MC, Dickman CR, Meathrel CE, Bradley JS (1996) Does intraspecific variation in the energy value of a prey species to its predators matter in studies of ecological energetics? A case study using insectivorous vertebrates. *Ecoscience* **3**, 247–251. doi:10.1080/11956860.1996.11682338

Brown JS, Laundré JW, Gurung M (1999) The ecology of fear: optimal foraging, game theory, and trophic interactions. *Journal of Mammalogy* **80**, 385–399. doi:10.2307/1383287

Bruniche-Olsen A, Burridge CP, Austin JJ, Jones ME (2013) Disease induced changes in gene flow patterns among Tasmanian Devil populations. *Biological Conservation* **165**, 69–78. doi:10.1016/j.biocon.2013.05.014

Buckmaster AJ, Dickman CR (2013) Early onset of reproduction in the Agile Antechinus, *Antechinus agilis. Australian Mammalogy* **35**, 115–118. doi:10.1071/AM11046

Burbidge AA (1999) Conservation values and management of Australian islands for non-volant mammal conservation. *Australian Mammalogy* **21**, 67–74.

Burbidge A (2008) *Ningaui timealeyi.* In *IUCN Red List of Threatened Species. Version 2013.1.* IUCN, Gland, Switzerland.

Burbidge AA, Abbott I (2018) Mammals on Western Australian islands: occurrence and preliminary analysis. *Australian Journal of Zoology*, (In press).

Burbidge AA, Eisenberg JF (2006) Conservation and management. In *Marsupials.* (Eds PJ Armati, CR Dickman and ID Hume) pp. 299–330. Cambridge University Press, Cambridge, UK.

Burbidge AA, Manly BJF (2002) Mammal extinctions on Australian islands: causes and conservation implications. *Journal of Biogeography* **22**, 465–473.

Burbidge AA, McKenzie NL (1976) A further record of *Sminthopsis longicaudata* (Marsupialia, Dasyuridae). *Western Australian Naturalist (Perth)* **13**, 144–145.

Burbidge AA, McKenzie NL (1978) The islands of the north-west Kimberley Western Australia. *Wildlife Research Bulletin, Western Australia* **7**, 1–47.

Burbidge AA, McKenzie NL (1989) Patterns in the modern decline of Western Australia's vertebrate fauna: causes and conservation implications. *Biological Conservation* **50**, 143–198. doi:10.1016/0006-3207(89)90009-8

Burbidge A, Woinarski J (2008) *Pseudantechinus ningbing.* In *IUCN Red List of Threatened Species. Version 2013.1.* IUCN, Gland, Switzerland.

Burbidge AA, Johnson KA, Fuller PJ, Southgate RI (1988) Aboriginal knowledge of the mammals of the central deserts of Australia. *Australian Wildlife Research* **15**, 9–39. doi:10.1071/WR9880009

Burbidge AH, Rolfe JK, McKenzie NL, Roberts JD (2004) Biogeographic patterns in small ground-dwelling vertebrates of the Western Australian wheatbelt. *Records of the Western Australian Museum* **67**, 109–138. doi:10.18195/issn.0313-122x.67.2004.109-137

Burbidge A, Cooper N, Morris K (2008a) *Pseudantechinus roryi.* In *IUCN Red List of Threatened Species. Version 2013.1.* IUCN, Gland, Switzerland.

Burbidge A, Cooper N, Woolley P, Morris K (2008b) *Pseudantechinus woolleyae.* In *IUCN Red List of Threatened Species. Version 2013.1.* IUCN, Gland, Switzerland.

Burbidge AA, McKenzie NL, Fuller PJ (2008c) Long-tailed Dunnart, *Sminthopsis longicaudata.* In *The Mammals of Australia.* 3rd edn. (Eds S Van Dyck and R Strahan) pp. 148–150. Reed New Holland, Sydney.

Burbidge A, Robinson T, Ellis M, Dickman C, Menkhorst P, Woinarski J (2008d) *Sminthopsis crassicaudata*. In *IUCN Red List of Threatened Species. Version 2013.2.* IUCN, Gland, Switzerland.

Burbidge A, Robinson T, Woinarski J (2008e) *Sminthopsis ooldea*. In *IUCN Red List of Threatened Species. Version 2013.2.* IUCN, Gland, Switzerland.

Burbidge AA, McKenzie NL, Brennan KEC, Woinarski JCZ, Dickman CR, Baynes A, *et al.* (2008f) Conservation status and biogeography of Australia's terrestrial mammals. *Australian Journal of Zoology* **56**, 411–422. doi:10.1071/ZO08027

Burbidge AA, Fuller PJ (1979) Mammals of the Warburton region, Western Australia. *Records of the Western Australian Museum* **8**(1), 57–73.

Burmeister H (1854) *Systematische Uebersicht der Thiere Brasiliens, welche wahrend einer Reise durch die Provinzen von Rio de Janeiro und Minas Geraes gesammelt oder beobachtet wurden von Dr. Hermann Burmeister. Saugethiere (Mammalia).* Georg Reimer, Berlin, Germany.

Burnett S (1997) Colonizing Cane Toads cause population declines in native predators: reliable anecdotal information and management implications. *Pacific Conservation Biology* **3**, 65–72. doi:10.1071/PC970065

Burnett S (2008a) Atherton Antechinus *Antechinus godmani*. In *The Mammals of Australia.* 3rd edn. (Eds S Van Dyck and R Strahan) pp. 89–90. Reed New Holland, Sydney.

Burnett S (2008b) Common Planigale *Planigale maculata*. In *The Mammals of Australia.* 3rd edn. (Eds S Van Dyck and R Strahan) pp. 112–113. Reed New Holland, Sydney

Burnett S, Crowther MS (2008a) Rusty Antechinus *Antechinus adustus*. In *The Mammals of Australia.* 3rd edn. (Eds S Van Dyck and R Strahan) pp. 81–82. Reed New Holland, Sydney.

Burnett S, Crowther MS (2008b) Subtropical Antechinus *Antechinus subtropicus*. In *The Mammals of Australia.* 3rd edn. (Eds S Van Dyck and R Strahan) pp. 97–98. Reed New Holland, Sydney.

Burnett S, Dickman C (2008a) *Dasyurus maculatus*. In *IUCN Red List of Threatened Species. Version 2013.1.* IUCN, Gland, Switzerland.

Burnett S, Dickman C (2008b) *Planigale maculata*. In *IUCN Red List of Threatened Species. Version 2013.1.* IUCN, Gland, Switzerland.

Burnett S, Winter J (2008) *Sminthopsis douglasi*. In *IUCN Red List of Threatened Species. Version 2013.2.* IUCN, Gland, Switzerland.

Burwell CJ, Geiser F, Barritt M, May K, Pavey CR (2005) Additional records of prey of the Fat-tailed False Antechinus *Pseudantechinus macdonnellensis* from central Australia. *Australian Mammalogy* **27**, 227–229.

Busch M, Kravetz FO (1991) Diet composition of *Monodelphis dimidiata* (Marsupialia, Didelphidae). *Mammalia* **55**, 619–621.

Butler R (2013) Indonesia: palm oil giant to forgo development of New Guinea rainforest. *Climate Ark News Archive*, Mongabay.

Byrne M, Yeates DK, Joseph L, Kearney M, Bowler J, Williams MAJ, Cooper S, *et al.* (2008) Birth of a biome: insights into the assembly and maintenance of the Australian arid zone biota. *Molecular Ecology* **17**, 4398–4417. doi:10.1111/j.1365-294X.2008.03899.x

Cáceres NC, Ghizoni IR, Jr, Graipel ME (2002) Diet of two marsupials, *Lutreolina crassicaudata* and *Micoureus demerarae*, in a coastal Atlantic Forest island of Brazil. *Mammalia* **66**, 331–340. doi:10.1515/mamm.2002.66.3.331

Calaby JH (1960) Observations on the Banded Ant-eater *Myrmecobius f. fasciatus* Waterhouse (Marsupialia), with particular reference to its food habits. *Proceedings of the Zoological Society of London* **135**, 183–207. doi:10.1111/j.1469-7998.1960.tb05841.x

Calaby JH (1966) 'Mammals of the upper Richmond and Clarence Rivers, New South Wales'. Technical Paper 10. CSIRO Division of Wildlife Research, Canberra.

Calaby JH, Taylor JM (1981) Reproduction in two marsupial-mice *Antechinus bellus* and *Antechinus bilarni* (Dasyuridae), of tropical Australia. *Journal of Mammalogy* **62**, 329–341. doi:10.2307/1380709

Campos C, Ojeda R, Monge S, Dacar M (2001) Utilization of food resources by small and medium-sized mammals in the Monte Desert biome, Argentina. *Austral Ecology* **26**, 142–149. doi:10.1046/j.1442-9993.2001.01098.x

Canty PD, Brandle R (2008) 'Kowari *Dasycercus byrnei* distribution and monitoring in Sturts Stony Desert, South Australia, Spring 2007'. Report for the South Australia Arid Lands Natural Resources Management Board. Science and Conservation, South Australia Department of Environment and Heritage, Adelaide.

Cardoso MJ, Eldridge MDB, Oakwood M, Rankmore B, Sherwin WB, Firestone KB (2009) Effects of founder events on the genetic variation of translocated island populations: implications for conservation management of the Northern Quoll. *Conservation Genetics* **10**, 1719–1733. doi:10.1007/s10592-008-9774-z

Carmignotto AP, Monfort T (2006) Taxonomy and distribution of the Brazilian species of *Thylamys* (Didelphimorphia: Didelphidae). *Mammalia* **70**, 126–144. doi:10.1515/MAMM.2006.013

Carmignotto AP, de Vivo M, Langguth A (2012) Mammals of the Cerrado and Caatinga: distribution patterns of the tropical open biomes of central South America. In *Bones, Clones, and Biomes: the History and Geography of Recent Neotropical Mammals.* (Eds BD Patterson and LP Costa) pp. 307–350. University of Chicago Press, Chicago IL, USA.

Caro T (2013) The colours of extant mammals. *Seminars in Cell & Developmental Biology* **24**, 542–552. doi:10.1016/j.semcdb.2013.03.016

Carthew SM, Bos DG (2008) Southern Ningaui *Ningaui yvonneae*. In *The Mammals of Australia*. 3rd edn. (Eds S Van Dyck and R Strahan) pp. 120–121. Reed New Holland, Sydney.

Carthew SM, Goldingay RL (1997) Non-flying mammals as pollinators. *Trends in Ecology & Evolution* **12**, 104–108. doi:10.1016/S0169-5347(96)10067-7

Carthew SM, Keynes T (2000) Small mammals in a semi-arid community, with particular reference to *Ningaui yvonneae*. *Australian Mammalogy* **22**, 103–109.

Carvalho AB, Oliveira LF, Langguth A, Freygang CC, Ferraz RS, Mattevi MS (2011) Phylogenetic relationships and phylogeographic patterns in *Monodelphis* (Didelphimorphia: Didelphidae). *Journal of Mammalogy* **92**, 121–133. doi:10.1644/10-MAMM-A-075.1

Carvalho FMV, Pinheiro PS, Fernandez FAS, Nessimian JL (1999) Diet of small mammals in Atlantic Forest fragments in southeastern Brazil. *Revista Brasileira de Zoociências* **1**, 91–101.

Carvalho FMV, Fernandez FAS, Nessimian JL (2005) Food habits of sympatric opossums coexisting in small Atlantic Forest fragments in Brazil. *Mammalian Biology* **70**, 366–375. doi:10.1016/j.mambio.2005.08.003

Casella J, Cáceres NC (2006) Diet of four small mammal species from Atlantic Forest patches in south Brazil. *Neotropical Biology and Conservation* **1**, 5–11.

Castilheiro WFF, Filho MS (2013) Diet of *Monodelphis glirina* (Mammalia: Didelphidae) in forest fragments in southern Amazon. *Zoologia* **30**, 249–254. doi:10.1590/S1984-46702013000300001

Caughley G (1994) Directions in conservation biology. *Journal of Animal Ecology* **63**, 215–244. doi:10.2307/5542

Ceballos G (1990) Comparative natural history of small mammals from tropical forests in western Mexico. *Journal of Mammalogy* **71**, 263–266. doi:10.2307/1382182

Cerqueira R (1982) South American landscapes and their mammals. In *Mammalian Biology in South America.* (Eds MA Mares and HH Genoways) pp. 53–75. Pymatuning Laboratory of Ecology, University of Pittsburgh, Linesville PA, USA.

Cerqueira R, Tribe CJ (2008) Genus *Didelphis*. In *Mammals of South America, Volume 1: Marsupials, Xenarthrans, Shrews and Bats.* (Ed. AL Gardner) pp. 17–24. University of Chicago Press, Chicago IL, USA.

Chemisquy MA, Flores DA (2012) Taxonomy of the southernmost populations of *Philander* (Didelphimorphia, Didelphidae), with implications for the systematics of the genus. *Zootaxa* **3481**, 60–72.

Chen X, Dickman CR, Thompson MB (1998) Diet of the Mulgara, *Dasycercus cristicauda* (Marsupialia: Dasyuridae), in the Simpson Desert, central Australia. *Wildlife Research* **25**, 233–242. doi:10.1071/WR97087

Chen X, Dickman CR, Thompson MB (2004) Selective consumption by predators of different body regions of prey: is rate of energy intake important? *Journal of Zoology* **264**, 189–196. doi:10.1017/S0952836904005692

Chillo V, Ojeda RA (2012) Mammal functional diversity loss under human-induced disturbances in arid lands. *Journal of Arid Environments* **87**, 95–102. doi:10.1016/j.jaridenv.2012.06.016

Chimento NR, Agnolina FL, Novasa FE (2015) The bizarre 'metatherians' Groeberia and Patagonia, late surviving members of gondwanatherian mammals. *Historical Biology* **27**(5), 603–623. doi:10.1080/08912963.2014.903945

Chisholm AH (1966) Gould, Elizabeth (1804–1841). In *Australian Dictionary of Biography, National Centre of Biography*. Australian National University, Canberra.

Christensen P, Maisey K, Perry DH (1984) Radiotracking the Numbat, *Myrmecobius fasciatus*, in the Perup forest of Western Australia. *Australian Wildlife Research* **11**, 275–288. doi:10.1071/WR9840275

Churchill S (2001) *Recovery Plan for the Sandhill Dunnart (*Sminthopsis psammophila*)*. Biodiversity Conservation Program. Department for Environment and Heritage, Adelaide.

Claridge AW, Barry SC (2000) Factors influencing the distribution of medium-sized ground-dwelling mammals in southeastern mainland Australia. *Austral Ecology* **25**, 676–688. doi:10.1111/j.1442-9993.2000.tb00074.x

Claridge AW, Paull D, Dawson J, Mifsud G, Murray AJ, Poore R, *et al.* (2005) Home range of the Spotted-tailed Quoll (*Dasyurus maculatus*), a marsupial carnivore, in a rainshadow woodland. *Wildlife Research* **32**, 7–14. doi:10.1071/WR04031

Cockburn A, Lazenby-Cohen KA (1992) Use of nest trees by *Antechinus stuartii*, a semelparous lekking marsupial. *Journal of Zoology* **226**, 657–680. doi:10.1111/j.1469-7998.1992.tb07508.x

Cogger HJ, Ford H, Johnson C, Holman J, Butler D (2003) *Impacts of Land Clearing on Australian Wildlife in Queensland*. WWF-Australia, Sydney.

Cogger H, Dickman C, Ford H, Johnson C, Taylor MFJ (2017) *Australian Animals Lost to Bulldozers in Queensland 2013–15*. WWF-Australia Technical Report, WWF-Australia, Sydney.

Cole JR, Gibson DF (1991) Distribution of Stripe-faced Dunnarts *Sminthopsis macroura* and Desert Dunnarts *Sminthopsis youngsoni* (Marsupialia: Dasyuridae) in the Northern Territory. *Australian Mammalogy* **14**, 129–131.

Columbus C, De Las Casas B (1989) *Reprint of 1893 Version of the Journal of Christopher Columbus*. Translated by LC Jane. Bonanza Books, Modesto CA, USA.

Conde DA, Flesness N, Colchero F, Jones OR, Scheuerlein A (2011) An emerging role of zoos to conserve biodiversity. *Science* **331**, 1390–1391. doi:10.1126/science.1200674

Connell JH (1980) Diversity and the coevolution of competitors, or the ghost of competition past. *Oikos* **35**, 131–138. doi:10.2307/3544421

Conniff R (2010) *The Species Seekers: Heroes, Fools and the Mad Pursuit of Life on Earth*. W.W. Norton and Company, New York, USA.

Cooper CE, Withers PC (2004) Patterns of body temperature variation and torpor in the Numbat, *Myrmecobius fasciatus* (Marsupialia: Myrmecobiidae). *Journal of Thermal Biology* **29**, 277–284. doi:10.1016/j.jtherbio.2004.05.003

Cooper CE, Withers PC (2005) Physiological significance of the microclimate in night refuges of the Numbat *Myrmecobius fasciatus*. *Australian Mammalogy* **27**, 169–174.

Cooper CE, Withers PC (2010) Comparative physiology of Australian quolls (*Dasyurus*: Marsupialia). *Journal of Comparative Physiology. B, Biochemical, Systemic, and Environmental Physiology* **180**, 857–868. doi:10.1007/s00360-010-0452-3

Cooper CE, McAllan BM, Geiser F (2005) Effect of torpor on the water economy of an arid-zone marsupial, the Stripe-faced Dunnart (*Sminthopsis macroura*). *Journal of Comparative Physiology. B, Biochemical, Systemic, and Environmental Physiology* **175**, 323–328. doi:10.1007/s00360-005-0488-y

Cooper DW, Woolley PA (1983) Confirmation of a new species of small dasyurid marsupial by electrophoretic analysis of enzymes and proteins. *Australian Journal of Zoology* **31**, 743–751. doi:10.1071/ZO9830743

Cooper NK (2008) Rory's Pseudantechinus *Pseudantechinus roryi*. In *The mammals of Australia*. 3rd edn. (Eds S Van Dyck and R Strahan) pp. 75–76. Reed New Holland, Sydney.

Cooper NK, Aplin KP, Adams M (2000) A new species of false antechinus (Marsupialia: Dasyuromorphia: Dasyuridae) from the Pilbara region, Western Australia. *Records of the Western Australian Museum* **20**, 115–136.

Cooper NK, Withers PC, Stewart T (2005) Description of sternal glands in Western Australian *Sminthopsis* and *Ningaui* (Marsupialia: Dasyuridae). *Journal of the Royal Society of Western Australia* **88**, 45–49.

Cooper SJB, Adams M, Labrinidis A (2000) Phylogeography of the Australian dunnart *Sminthopsis crassicaudata* (Marsupialia: Dasyuridae). *Australian Journal of Zoology* **48**, 461–473. doi:10.1071/ZO00014

Corlett RT (2013) Special issue: Defaunation's impact in tropical terrestrial ecosystems. The shifted baseline: Prehistoric defaunation in the tropics and its consequences for biodiversity conservation. *Biological Conservation* **163**, 13–21. doi:10.1016/j.biocon.2012.11.012

Cornelio DL (2011) Status and conservation possibilities of Papua New Guinea's terrestrial mammals. *Jurnel Manajemen Hutan Tropika* **17**(2), 41–48.

Cortés PA, Franco M, Moreno-Gomez FN, Barrientos K, Nespolo RF (2014) Thermoregulatory capacities and torpor in the South American marsupial, *Dromiciops gliroides. Journal of Thermal Biology* **45**, 1–8. doi:10.1016/j.jtherbio.2014.07.003

Cortesão J (1937) The pre-Columbian discovery of America. *The Geographical Journal* **89**(1), 29–42. doi:10.2307/1786989

Costa LP, Leite YLR, Mendes SL, Ditchfield AD (2005) Mammal conservation in Brazil. *Conservation Biology* **19**, 672–679. doi:10.1111/j.1523-1739.2005.00666.x

Costanza R, de Groot R, Sutton P, van der Ploeg S, Anderson SJ, Kubiszewski I, *et al.* (2014) Changes in the global value of ecosystem services. *Global Environmental Change* **26**, 152–158. doi:10.1016/j.gloenvcha.2014.04.002

Covacevich JA, Roberts L, Storch D, Van Dyck S (1994) New *Phascogale tapoatafa* (Brush-tailed Phascogale) records from Cape York Peninsula, Australia. *Memoirs of the Queensland Museum* **37**(1), 82.

Cox MP, Dickman CR, Hunter JT (2003) Effects of rainforest fragmentation on non-flying mammals of the eastern Dorrigo Plateau, Australia. *Biological Conservation* **115**, 175–189. doi:10.1016/S0006-3207(03)00105-8

Creighton GK, Gardner AL (2008) Genus *Thylamys*. In *Mammals of South America. Volume 1. Marsupials, Xenarthrans, Shrews and Bats.* (Ed. AL Gardner) pp. 107–116. University of Chicago Press, Chicago IL, USA.

Crisp EA, Messer M, VandeBerg JL (1989) Changes in milk carbohydrates during lactation in a didelphid marsupial, *Monodelphis domestica. Physiological Zoology* **62**, 1117–1125. doi:10.1086/physzool.62.5.30156200

Croft DB (1982) Communication in the Dasyuridae: a review. In *Carnivorous Marsupials* (Ed. M. Archer) pp. 291–309. Royal Zoological Society of New South Wales, Sydney.

Croft DB (2003) Behaviour of carnivorous marsupials. In *Predators with Pouches: the Biology of Carnivorous Marsupials.* (Eds M Jones, C Dickman and M Archer) pp. 332–346. CSIRO Publishing, Melbourne.

Croft DB, Eisenberg JF (2006) Behaviour. In *Marsupials.* (Eds PJ Armati, CR Dickman and ID Hume) pp. 229–298. Cambridge University Press, Cambridge, UK.

Crowther MS (2002) Distributions of species of the *Antechinus stuartii–A. flavipes* complex as predicted by bioclimatic modelling. *Australian Journal of Zoology* **50**, 77–91. doi:10.1071/ZO01031

Crowther MS (2008) Yellow-footed Antechinus *Antechinus flavipes*. In *The Mammals of Australia.* 3rd edn. (Eds S Van Dyck and R Strahan) pp. 86–88. Reed New Holland, Sydney.

Crowther MS, Blacket MJ (2003) Biogeography and speciation in the Dasyuridae: why are there so many kinds of dasyurid? In *Predators with Pouches: the Biology of Carnivorous Marsupials.* (Eds M Jones, C Dickman and M Archer) pp. 124–130. CSIRO Publishing, Melbourne.

Crowther MS, Braithwaite RW (2008) Brown Antechinus *Antechinus stuartii*. In *The Mammals of Australia.* 3rd edn. (Eds S Van Dyck and R Strahan) pp. 94–96. Reed New Holland, Sydney.

Crowther MS, Dickman CR, Lynam AJ (1999) *Sminthopsis griseoventer boullangerensis* (Marsupialia : Dasyuridae), a new subspecies in the *S. murina* complex from Boullanger Island, Western Australia. *Australian Journal of Zoology* **47**(3), 215–243. doi:10.1071/ZO98062

Crowther MS, Spencer PBS, Alpers D, Dickman CR (2002) Taxonomic status of the Mardo, *Antechinus flavipes leucogaster* (Marsupialia: Dasyuridae): a morphological and molecular, reproductive and bioclimatic approach. *Australian Journal of Zoology* **50**, 627–647. doi:10.1071/ZO02030

Crowther MS, Sumner J, Dickman CR (2003) Speciation of *Antechinus stuartii* and *A. subtropicus* (Marsupialia: Dasyuridae) in eastern Australia: molecular and morphological evidence. *Australian Journal of Zoology* **51**, 443–462. doi:10.1071/ZO02009

Crowther MS, Lunney D, Lemon J, Stalenberg E, Wheeler R, Madani G, *et al.* (2014) Climate-mediated habitat selection in an arboreal folivore. *Ecography* **37**, 336–343. doi:10.1111/j.1600-0587.2013.00413.x

Cuaron AD, Emmons L, Helgen K, Reid F, Vazquez E (2008) *Tlacuatzin canescens*. In *IUCN Red List of Threatened Species. Version 2016.2.* IUCN, Gland, Switzerland.

Cunningham RB, Lindenmayer DB, MacGregor C, Barry S, Welsh A (2005) Effects of trap position, trap history, microhabitat and season on capture probabilities of small mammals in a wet eucalypt forest. *Wildlife Research* **32**, 657–671. doi:10.1071/WR04069

Cuthbert R (2010) Sustainability of hunting, population densities, intrinsic rates of increase and conservation of Papua New Guinean mammals: A quantitative review. *Biological Conservation* **143**, 1850–1859. doi:10.1016/j.biocon.2010.04.005

Cuttle P (1982a) Life history strategy of the dasyurid marsupial *Phascogale tapoatafa*. In *Carnivorous Marsupials.* (Ed. M Archer) pp. 13–22. Royal Zoological Society of New South Wales, Sydney.

Cuttle P (1982b) A preliminary report on aspects of the behaviour of the dasyurid marsupial *Phascogale tapoatafa*. In *Carnivorous Marsupials.* (Ed. M Archer) pp. 325–332. Royal Zoological Society of New South Wales, Sydney.

D'Elia G, Hurtado N, D'Anatro A (2016) Alpha taxonomy of *Dromiciops* (Microbiotheriidae) with the description of 2 new species of monito del monte. *Journal of Mammalogy* **97**(4), 1136–1152. doi:10.1093/jmammal/gyw068

da Fonseca GAB, Paglia AP, Sanderson J, Mittermeier RA (2003) Marsupials of the New World: status and conservation. In *Predators with Pouches: the Biology of Carnivorous Marsupials.* (Eds M Jones, C Dickman and M Archer) pp. 399–406. CSIRO Publishing, Melbourne.

Dahl K (1897) Biological notes on north-Australian mammalia. *The Zoologist* **671**, 189–216.

Davison A (1986) Dentition of Victorian *Antechinus* species. *Australian Mammalogy* **9**, 73–85.

Davison A (1991) A new subspecies of Dusky Antechinus, *Antechinus swainsonii* (Marsupialia:Dasyuridae) from western Victoria. *Australian Mammalogy* **14**, 103–113.

Dawson JP, Claridge AW, Triggs B, Paull DJ (2007) Diet of a native carnivore, the Spotted-tailed Quoll (*Dasyurus maculatus*), before and after an intensive wildfire. *Wildlife Research* **34**, 342–351. doi:10.1071/WR05101

Dawson TJ, Dawson L (2006) Evolution of arid Australia and consequences for vertebrates. In *Evolution and Biogeography of Australasian Vertebrates.* (Eds JR Merrick, M Archer, GM Hickey and MSY Lee) pp. 51–70. Auscipub, Sydney.

de Azara F (1801*) Essais sur l'histoire Naturelle des Quadrupe' des de la Province du Paraguay*. Charles Pougens, Paris, France.

De Las Casas B (1875) *Historia de las Indias*. Imprenta de Meguel Ginesta, Madrid, Spain.

de Moraes DA, Santori RT, Finotti R, Cerqueira R (2003) Nutritional and fibre contents of laboratory-established diets of Neotropical opossums (Didelphidae). In *Predators with Pouches: the Biology of Carnivorous Marsupials.* (Eds M Jones, C Dickman and M Archer) pp. 229–237. CSIRO Publishing, Melbourne.

de Muizon C (2011) Fresh light on southern early mammals. *Nature* **479**, 51–52. doi:10.1038/479051a

Degabriele R (1976) The forest mammals of Cape York Peninsula. *Habitat* **4**(4), 3–6.

Delciellos AC, Vieira MV (2009) Allometric, phylogenetic, and adaptive components of climbing performance in seven species of didelphid marsupials. *Journal of Mammalogy* **90**, 104–113. doi:10.1644/07-MAMM-A-115.1

Denny MJS (1975) Mammals of Sturt National Park, Tibooburra, New South Wales. *Australian Zoologist* **18**, 179–195.

Denny MJS (1982) Review of *Planigale* (Dasyuridae, Marsupialia) ecology. In *Carnivorous Marsupials.* (Ed. M Archer) pp. 131–138. Royal Zoological Society of New South Wales, Sydney.

Denny EA, Dickman CR (2010) *Review of Cat Ecology and Management Strategies in Australia*. Invasive Animals Cooperative Research Centre, Canberra.

Department of Environment (2015) *Threat Abatement Plan for Predation by Feral Cats*. Department of the Environment, Canberra.

Department of the Environment, Water, Heritage and the Arts (2008) *Threat Abatement Plan for Predation by the European Red Fox*. Department of the Environment, Water, Heritage and the Arts, Canberra.

Diaz GB, Ojeda RA, Dacar M (2001) Water conservation in the South American desert mouse opossum, *Thylamys pusilla* (Didelphimorphia, Didelphidae). *Comparative Biochemistry and Physiology Part A:Molecular & Integrative Physiology* **130**, 323–330. doi:10.1016/S1095-6433(01)00397-X

Diaz MM (2014) Marsupiales (Didelphimorphia: Didelphidae) de Iquitos y sus alrededores (Loret, Perú). *Therya* **5**, 111–151. doi:10.12933/therya-14-178

Diaz MM, Flores DA (2008) Early reproduction onset in four species of Didelphimorphia in the Peruvian Amazonia. *Mammalia* **72**, 126–130. doi:10.1515/MAMM.2008.012

Dickman CR (1980) Ecological studies of *Antechinus stuartii* and *Antechinus flavipes* (Marsupialia: Dasyuridae) in open-forest and woodland habitats. *Australian Zoologist* **20**, 433–446.

Dickman CR (1982) Some ecological aspects of seasonal breeding in *Antechinus* (Dasyuridae, Marsupialia). In *Carnivorous Marsupials.* (Ed. M Archer) pp. 139–150. Royal Zoological Society of New South Wales, Sydney.

Dickman CR (1984) Competition and coexistence among the small marsupials of Australia and New Guinea. *Acta Zoologica Fennica* **172**, 27–31.

Dickman CR (1985) Effects of photoperiod and endogenous control on timing of reproduction in the marsupial genus *Antechinus. Journal of Zoology* **206A**, 509–524.

Dickman CR (1986a) An experimental manipulation of the intensity of interspecific competition: effects on a small marsupial. *Oecologia* **70**, 536–543. doi:10.1007/BF00379900

Dickman CR (1986b) An experimental study of competition between two species of dasyurid marsupials. *Ecological Monographs* **56**, 221–241. doi:10.2307/2937075

Dickman CR (1986c) Niche compression: two tests of an hypothesis using narrowly sympatric predator species. *Australian Journal of Ecology* **11**, 121–134. doi:10.1111/j.1442-9993.1986.tb01383.x

Dickman CR (1987) Return of the phantom Dibbler. *Australian Natural History* **22**, 32–33.

Dickman CR (1988) Body size, prey size, and community structure in insectivorous mammals. *Ecology* **69**, 569–580. doi:10.2307/1941006

Dickman CR (1989a) Demographic responses of *Antechinus stuartii* (Marsupialia) to supplementary food. *Australian Journal of Ecology* **14**, 387–398. doi:10.1111/j.1442-9993.1989.tb01449.x

Dickman CR (1989b) Patterns in the structure and diversity of marsupial carnivore communities. In *Patterns in the Structure of Mammalian Communities.* (Eds DW Morris, Z Abramsky, BJ Fox and MR Willig) pp 241–251. Texas Tech University Press, Lubbock TX, USA.

Dickman CR (1991a) Mechanisms of competition among insectivorous mammals. *Oecologia* **85**, 464–471. doi:10.1007/BF00323757

Dickman CR (1991b) Use of trees by ground-dwelling mammals: implications for management. In *Conservation of Australia's Forest Fauna.* (Ed. D Lunney) pp. 125–136. Royal Zoological Society of New South Wales, Sydney.

Dickman CR (1992) Conservation of mammals in the Australasian region: the importance of islands. In *Australia and the Global Environmental Crisis.* (Eds

JN Coles and JM Drew) pp. 175–214. Academy Press, Canberra.

Dickman CR (1993) Evolution of semelparity in male dasyurid marsupials: a critique, and an hypothesis of sperm competition. In *The Biology and Management of Australasian Carnivorous Marsupials.* (Eds M Roberts, J Carnio, G Crawshaw and M Hutchins) pp. 25–38. Metropolitan Toronto Zoo, Ontario, and the American Association of Zoological Parks and Aquariums, Washington DC, USA.

Dickman CR (1996) Vagrants in the desert. *Nature Australia* **25**, 54–62.

Dickman CR (2003) Distributional ecology of dasyurid marsupials. In *Predators with Pouches: the Biology of Carnivorous Marsupials.* (Eds M Jones, C Dickman and M Archer) pp. 318–331. CSIRO Publishing, Melbourne.

Dickman CR (2005) Marsupials of the world: an introduction. In *Walker's Marsupials of the World.* (Ed. RM Nowak) pp. 1–67. The Johns Hopkins University Press, Baltimore MD, USA.

Dickman CR (2006) Species interactions: indirect effects. In *Ecology: an Australian Perspective.* (Eds P Attiwill and B Wilson) pp. 303–316. Oxford University Press, Melbourne.

Dickman CR (2007) *A Fragile Balance: the Extraordinary Story of Australian Marsupials.* University of Chicago Press, Chicago IL, USA.

Dickman CR (2008a) Agile Antechinus *Antechinus agilis.* In *The Mammals of Australia.* 3rd edn. (Eds S Van Dyck and R Strahan) pp. 83–84. Reed New Holland, Sydney

Dickman CR (2008b) Dusky Antechinus *Antechinus swainsonii.* In *The Mammals of Australia.* 3rd edn. (Eds S Van Dyck and R Strahan) pp. 99–100. Reed New Holland, Sydney.

Dickman CR (2008c) Grey-bellied Dunnart *Sminthopsis griseoventer.* In *The Mammals of Australia.* 3rd edn. (Eds S Van Dyck and R Strahan) pp. 141–143. Reed New Holland, Sydney.

Dickman CR (2012) Fences or ferals? Benefits and costs of conservation fencing in Australia. In *Fencing for Conservation: Restriction of Evolutionary Potential or a Riposte to Threatening Processes?* (Eds MJ Somers and MW Hayward) pp. 43–63. Springer, New York, USA.

Dickman CR (2014a) Measuring and managing the impacts of cats. In *Carnivores of Australia: Past, Present and Future.* (Eds AS Glen and CR Dickman) pp. 173–196. CSIRO Publishing, Melbourne.

Dickman CR (2014b) Micro-carnivores: the ecological role of small dasyurid predators in Australia. In *Carnivores of Australia: Past, Present and Future.* (Eds AS Glen and CR Dickman) pp. 241–262. CSIRO Publishing, Melbourne.

Dickman CR (2014c) Invasion ecology of Australasian marsupials. In *Invasion Biology and Ecological Theory: Insights from a Continent in Transformation.* (Eds HHT Prins and IJ Gordon) pp. 159–195. Cambridge University Press, Cambridge, UK.

Dickman CR (2015a) Family Peramelidae (bandicoots and echymiperas). In *Handbook of the Mammals of the World. Volume 5. Monotremes and Marsupials.* (Eds DE Wilson and RA Mittermeier) pp. 362–398. Lynx Edicions, Barcelona, Spain.

Dickman CR (2015b) *A Fragile Balance: the Extraordinary Story of Australian Marsupials.* 2nd edn, revised and updated. Australian Geographic, Sydney.

Dickman CR, Braithwaite RW (1992) Postmating mortality of males in the dasyurid marsupials, *Dasyurus* and *Parantechinus. Journal of Mammalogy* **73**(1), 143–147. doi:10.2307/1381875

Dickman CR, Read DG (1992) 'The biology and management of dasyurids of the arid zone in New South Wales'. Species Management Report No. 11. New South Wales National Parks and Wildlife Service, Sydney.

Dickman CR, Russell EM (2001) Marsupials. In *The New Encyclopedia of Mammals.* (Ed. DW Macdonald) pp. 802–807. Oxford University Press, Oxford, UK.

Dickman CR, Steeves TE (2004) Use of habitat by mammals in eastern Australian forests: are common species important in forest management? In *Conservation of Australia's Forest Fauna.* 2nd edn. (Ed. D Lunney) pp. 761–773. Royal Zoological Society of New South Wales, Sydney.

Dickman CR, Vieira EM (2006) Ecology and life histories. In *Marsupials.* (Eds PJ Armati, CR

Dickman and ID Hume) pp 199–228. Cambridge University Press, Cambridge, UK.

Dickman CR, Green K, Carron PL, Happold DCD, Osborne WS (1983) Coexistence, convergence and competition among *Antechinus* (Marsupialia) in the Australian high country. *Proceedings of the Ecological Society of Australia* **12**, 79–99.

Dickman CR, King DH, Adams M, Baverstock PR (1988) Electrophoretic identification of a new species of *Antechinus* (Marsupialia: Dasyuridae) in south-eastern Australia. *Australian Journal of Zoology* **36**, 455–463. doi:10.1071/ZO9880455

Dickman CR, Downey FJ, Predavec M (1993a) The Hairy-footed Dunnart *Sminthopsis hirtipes* (Marsupialia: Dasyuridae) in Queensland. *Australian Mammalogy* **16**, 69–72.

Dickman CR, Pressey RL, Lim L, Parnaby HE (1993b) Mammals of particular conservation concern in the Western Division of New South Wales. *Biological Conservation* **65**, 219–248. doi:10.1016/0006-3207(93)90056-7

Dickman CR, Predavec M, Downey FJ (1995) Long-range movements of small mammals in arid Australia: implications for land management. *Journal of Arid Environments* **31**, 441–452. doi:10.1016/S0140-1963(05)80127-2

Dickman CR, Parnaby HE, Crowther MS, King DH (1998) *Antechinus agilis* (Marsupialia:Dasyuridae), a new species from the *A. stuartii* complex in south-eastern Australia. *Australian Journal of Zoology* **46**, 1–26. doi:10.1071/ZO97036

Dickman CR, Haythornthwaite AS, McNaught GH, Mahon PS, Tamayo B, Letnic M (2001) Population dynamics of three species of dasyurid marsupials in arid central Australia: a 10-year study. *Wildlife Research* **28**, 493–506. doi:10.1071/WR00023

Dickman CR, Lunney D, Matthews A (2001) Ecological attributes and conservation of dasyurid marsupials in New South Wales, Australia. *Pacific Conservation Biology* **7**, 124–133. doi:10.1071/PC010124

Dickman C, Burnett S, McKenzie N (2008) *Sminthopsis murina*. In *IUCN Red List of Threatened Species. Version 2013.2.* IUCN, Gland, Switzerland.

Dickman CR, Greenville AC, Tamayo B, Wardle GM (2011) Spatial dynamics of small mammals in central Australian desert habitats: the role of drought refugia. *Journal of Mammalogy* **92**(6), 1193–1209. doi:10.1644/10-MAMM-S-329.1

Dobson A, Lyles A (2000) Black-footed Ferret recovery. *Science* **288**, 985–988. doi:10.1126/science.288.5468.985

Dobson FS (2013) Commentary: live fast, die young, and win the sperm competition. *Proceedings of the National Academy of Sciences of the United States of America* **110**(44), 17610–17611. doi:10.1073/pnas.1317368110

Dollman G (1930) On the mammals obtained by Mr. Shaw Mayer in New Guinea, and presented to the British Museum by Mr. J. Spedan Lewis F.Z.S. *Proceedings of the Zoological Society of London* **1930**, 429–435.

Duncan-Kemp AM (1933) *Our Sandhill Country: Nature and Man in South-western Queensland.* Angus and Robertson, Sydney.

Dunlop JN, Sawle M (1982) The habitat and life history of the Pilbara Ningaui *Ningaui timealeyi. Records of the Western Australian Museum* **10**(1), 47–52.

Dunlop JN, Cooper NK, Teale RJ (2008) Pilbara Ningaui *Ningaui timealeyi*. In *The Mammals of Australia.* 3rd edn. (Eds S Van Dyck and R Strahan) pp. 119–120. Reed New Holland, Sydney.

Dwyer PD (1977) Notes on *Antechinus* and *Cercartetus* (Marsupialia) in the New Guinea Highlands. *Proceedings of the Royal Society of Queensland* **88**, 69–73.

Dwyer PD, Hockings M, Wilmer J (1979) Mammals of Cooloola and Beerwah. *Proceedings of the Royal Society of Queensland* **90**, 65–84.

Echarri S, Prevosti FJ (2015) Differences in mandibular disparity between extant and extinct species of metatherian and placental carnivore clades. *Lethaia* **48**, 196–204. doi:10.1111/let.12099

Edwards GP (2013) Relative abundance of the Central Rock-rat, the Desert Mouse and the Fat-tailed Pseudantechinus at Ormiston Gorge in the West MacDonnell Ranges National Park, Northern Territory. *Australian Mammalogy* **35**, 144–148. doi:10.1071/AM12031

Eisenberg JF (1989) *Mammals of the Neotropics. Volume 1. The Northern Neotropics: Panama, Colombia,*

Venezuela, Guyana, Suriname, French Guiana. University of Chicago Press, Chicago IL, USA.

Eisenberg JF, Redford KH (1992) *Mammals of the Neotropics. Volume 2. The Southern Cone: Chile, Argentina, Uruguay, Paraguay.* University of Chicago Press, Chicago IL, USA.

Eisenberg JF, Redford KH (1999) *Mammals of the Neotropics. Volume 3. The Central Neotropics: Ecuador, Peru, Bolivia, Brazil.* University of Chicago Press, Chicago IL, USA.

Eisenberg JF, Collins LR, Wemmer C (1975) Communication in the Tasmanian Devil (*Sarcophilus harrisii*) and a survey of auditory communication in the Marsupialia. *Zeitschrift für Tierpsychologie* **37**, 379–399. doi:10.1111/j.1439-0310.1975.tb00887.x

Elith J, Leathwick JR (2009) Species distribution models: ecological explanation and prediction across space and time. *Annual Review of Ecology, Evolution and Systematics* **40**, 677–697. doi:10.1146/annurev.ecolsys.110308.120159

Ellis M, Menkhorst P, van Weenen J, Burbidge A (2008a) *Ningaui yvonneae*. In *IUCN Red List of Threatened Species. Version 2013.1.* IUCN, Gland, Switzerland.

Ellis M, van Weenen J, Burnett S (2008b) *Planigale tenuirostris*. In *IUCN Red List of Threatened Species. Version 2013.1.* IUCN, Gland, Switzerland.

Ellis M, van Weenen J, Pennay M (2008c) *Planigale gilesi*. In *IUCN Red List of Threatened Species. Version 2013.1.* IUCN, Gland, Switzerland.

Emmons LH (1999) *Neotropical Rainforest Mammals: a Field Guide.* 2nd edn. University of Chicago Press, Chicago IL, USA.

Epstein B, Jones M, Hamede R, Hendricks S, McCallum H, Murchison EP, *et al.* (2016) Rapid evolutionary response to a transmissible cancer in Tasmanian Devils. *Nature Communications* **7**, 12684. doi:10.1038/ncomms12684

Estbergs JA, Braithwaite RW (1985) The diet of the Rufous Owl *Ninox rufa* near Cooinda in the Northern Territory. *Emu* **85**, 202–205. doi:10.1071/MU9850202

Esvelt K, Church G, Lunshof J (2014) "Gene drives" and CRISPR could revolutionize ecosystem management. *Scientific American*, blog. <http://blogs.scientificamerican.com/guest-blog/gene-drives-and-crispr-could-revolutionize-ecosystem-management>.

Ewer RF (1968a) *Ethology of Mammals*. Elek Science, London, UK.

Ewer RF (1968b) A preliminary survey of the behaviour in captivity of the dasyurid marsupial, *Sminthopsis crassicaudata* (Gould). *Zeitschrift für Tierpsychologie* **25**, 319–365. doi:10.1111/j.1439-0310.1968.tb00019.x

Ewer RF (1969) Some observations on the killing and eating of prey by two dasyurid marsupials: the Mulgara *Dasycercus cristicauda*, and the Tasmanian Devil *Sarcophilus harrisii*. *Zeitschrift für Tierpsychologie* **26**, 23–38. doi:10.1111/j.1439-0310.1969.tb01935.x

Fadem BH, Kraus DB, Sheffet RH (1986) Nest-building in Gray Short-tailed Opossums: temperature effects and sex differences. *Physiology & Behavior* **36**, 667–670. doi:10.1016/0031-9384(86)90351-3

Fairbridge D, Anderson R, Wilkes T, Pell G (2003) Bait uptake by free living Brush-tailed Phascogales *Phascogale tapoatafa* and other non-target mammals during simulated buried fox baiting. *Australian Mammalogy* **25**, 31–40.

Fancourt BA, Hawkins CE, Nicol SC (2013) Evidence of rapid population decline of the Eastern Quoll (*Dasyurus viverrinus*) in Tasmania. *Australian Mammalogy* **35**, 195–205. doi:10.1071/AM13004

Fanning FD (1982) Reproduction, growth and development in *Ningaui* sp. (Dasyuridae, Marsupialia) from the Northern Territory. In *Carnivorous Marsupials.* (Ed. M Archer) pp. 23–37 Royal Zoological Society of New South Wales, Sydney.

Farina RA, Vizcaino SF, De Iuliis G (2013) *Megafauna: Giant Beasts of Pleistocene South America*. Indiana University Press, Bloomington IN, USA.

Fernandez FAS, Galliez M, Leite MS, Queiroz TL, Palmerin AF (2015) Natural history of the Water Opossum *Chironectes minimus*: a review. *Oecologia Australis* **19**, 47–62. doi:10.4257/oeco.2015.1901.04

Fernandez Duro C (1892) *Pinzón en el Descubrimiento de las Indias, con Noticias Críticas de Algunas Obras Recientes Relacionadas con el Mismo Descubrimiento* [Reprint 2017]. *Forgotten Books,* Madrid, Spain.

Ferreira MS, Vieira MV, Cerqueira R, Dickman CR (2016) Seasonal dynamics with compensatory effects regulate populations of tropical forest marsupials: a 16-year study. *Oecologia* **182**, 1095–1106. doi:10.1007/s00442-016-3735-x

Finlayson HH (1958) A case of duplex convergence in Australian mammals, with a review of some aspects of the morphology of *Phascogale (Antechinus) swainsonii* Waterhouse and *Phascogale (Antechinus) flavipes* Waterhouse. *Proceedings of the Royal Society of South Australia* **81**, 141–151.

Firestone K (2000) Phylogenetic relationships among quolls revisited: the mtDNA control region as a useful tool. *Journal of Mammalian Evolution* **7**, 1–22. doi:10.1023/A:1009425815903

Firestone KB (2003) The application of genetic research to conservation management in carnivorous marsupials with special emphasis on dasyurids. In *Predators with Pouches: the Biology of Carnivorous Marsupials.* (Eds M Jones, C Dickman and M Archer) pp. 475-486. CSIRO Publishing, Melbourne.

Firestone KB, Elphinstone MS, Sherwin WB, Houlden BA (1999) Phylogeographical population structure of Tiger Quolls *Dasyurus maculatus* (Dasyuridae: Marsupialia), an endangered carnivorous marsupial. *Molecular Ecology* **8**, 1613–1625. doi:10.1046/j.1365-294x.1999.00745.x

Firth RSC, Woinarski JCZ, Brennan KG, Hempel G (2006) Environmental relationships of the Brush-tailed Rabbit-rat, *Conilurus penicillatus*, and other small mammals on the Tiwi Islands, northern Australia. *Journal of Biogeography* **33**, 1820–1837. doi:10.1111/j.1365-2699.2006.01543.x

Fisher A, Woinarski JCZ, Churchill S, Trainor C, Griffiths AD, Palmer C, *et al.* (2000) Distribution of the rock-dwelling dasyurids *Pseudantechinus bilarni* and *P. ningbing* in the Northern Territory. *Northern Territory Naturalist* **16**, 1–13.

Fisher DO (2005) Population density and presence of the mother are related to natal dispersal in male and female *Antechinus stuartii. Australian Journal of Zoology* **53**, 103–110. doi:10.1071/ZO04068

Fisher DO, Blomberg SP (2011) Costs of reproduction and terminal investment by females in a semelparous marsupial. *PLoS One* **6**, e15226. doi:10.1371/journal.pone.0015226

Fisher DO, Dickman CR (1993a) The body size–prey size relationship in dasyurid marsupials: tests of three hypotheses. *Ecology* **74**, 1871–1883. doi:10.2307/1939944

Fisher DO, Dickman CR (1993b) Diets of insectivorous marsupials in arid Australia: selection for prey type, size or hardness? *Journal of Arid Environments* **25**, 397–410. doi:10.1006/jare.1993.1072

Fisher DO, Double MC, Blomberg SP, Jennions MD, Cockburn A (2006) Post-mating sexual selection increases lifetime fitness of polyandrous females in the wild. *Nature* **444**, 89–92. doi:10.1038/nature05206

Fisher DO, Nuske S, Green S, Seddon JM, McDonald B (2011) The evolution of sociality in small, carnivorous marsupials: the lek hypothesis revisited. *Behavioral Ecology and Sociobiology* **65**, 593–605. doi:10.1007/s00265-010-1060-7

Fisher DO, Dickman CR, Jones ME, Blomberg SP (2013) Sperm competition drives the evolution of suicidal reproduction in mammals. *Proceedings of the National Academy of Sciences of the United States of America* **110**(44), 17910–17914. doi:10.1073/pnas.1310691110

Fisher DO, Johnson CN, Lawes MJ, Fritz SA, McCallum H, Blomberg SP, *et al.* (2014) The current decline of tropical marsupials in Australia: is history repeating? *Global Ecology and Biogeography* **23**, 181–190. doi:10.1111/geb.12088

Fitzherbert EB, Struebig MJ, Morel A, Danielsen F, Bruhl CA, Donald PF, Phalan B (2008) How will oil palm expansion affect biodiversity? *Trends in Ecology & Evolution* **23**(10), 538–545. doi:10.1016/j.tree.2008.06.012

Flannery TF (1995) *The Mammals of New Guinea.* 2nd edn. Reed Books, Sydney.

Fleay D (1929) The Brush-tailed Phascogale. *Victorian Naturalist* **46**, 135–136.

Fleay D (1935) Breeding of *Dasyurus viverrinus* and general observations on the species. *Journal of Mammalogy* **16**, 10–16. doi:10.2307/1374524

Fleay D (1947) *Gliders of the Gum Trees.* Bread and Cheese Club, Melbourne.

Fleay D (1950) Experiences with Australia's Brush-tailed Tuan. *Animal Kingdom* **53**, 152–157.

Fleming PJS, Allen BL, Allen LR, Ballard G-A, Bengsen A, Gentle MN, *et al.* (2014) Management of wild canids in Australia: free-ranging dogs and red foxes. In *Carnivores of Australia: Past, Present and Future.* (Eds AS Glen and CR Dickman) pp. 105–149. CSIRO Publishing, Melbourne.

Fletcher TP (1985) Aspects of reproduction in the male Eastern Quoll, *Dasyurus viverrinus* (Shaw) (Marsupialia : Dasyuridae), with notes on polyoestry in the female. *Australian Journal of Zoology* **33**, 101–110. doi:10.1071/ZO9850101

Flores DA, Giannini N, Abdala F (2006) Comparative postnatal ontogeny of the skull in the Australidelphian Metatherian *Dasyurus albopunctatus* (Marsupialia: Dasyuromorpha: Dasyuridae). *Journal of Morphology* **267**, 426–440. doi:10.1002/jmor.10420

Flores DA, Diaz MM, Barquez RM (2007) Systematics and distribution of marsupials in Argentina: a review. In *The Quintessential Naturalist: Honoring the Life and Legacy of Oliver P. Pearson.* (Eds DA Kelt, EP Lessa, J Salazar-Bravo and JL Patton) pp. 579–669. University of California Publications in Zoology, Oakland CA, USA.

Flores DA, Barquez RM, Diaz MM (2008) A new species of *Philander* Brisson, 1762 (Didelphimorphia, Didelphidae). *Mammalian Biology* **73**, 14–24. doi:10.1016/j.mambio.2007.04.002

Fonseca GAB, Kierulff MC (1989) Biology and natural history of Brazilian Atlantic Forest small mammals. *Bulletin of the Florida State Museum Biological Sciences* **34**, 99–152.

Fontúrbel FE, Silva-Rodriguez EA, Cardenas NH, Jimenez JE (2010) Spatial ecology of Monito del Monte (*Dromiciops gliroides*) in a fragmented landscape of southern Chile. *Mammalian Biology* **75**, 1–9. doi:10.1016/j.mambio.2009.08.004

Fontúrbel FE, Candia AB, Salazar DA, Malebran J, Gonzalez-Browne C, Botto-Mahan C (2014) How forest marsupials are affected by habitat degradation and fragmentation? A meta-analysis. *Naturwissenschaften* **101**, 599–602. doi:10.1007/s00114-014-1193-z

Forasiepi AM, Goin FJ, Alejandra Abello M, Cerdeño E (2014) A unique, Late Oligocene shrew-like marsupial from western Argentina and the evolution of dental morphology. *Journal of Systematic Palaeontology* **12**(5), 549–564. doi:10.1080/14772019.2013.799611

Ford F (2014) *John Gould's Extinct and Endangered Mammals of Australia.* National Library of Australia, Canberra.

Formoso AE, Martin GM, Teta P, Carbajo AE, Sauthier DEU, Pardiñas UFJ (2015) Regional extinctions and quaternary shifts in the geographic range of *Lestodelphys halli*, the southernmost living marsupial: clues for its conservation. *PLoS One* **10**(7), e0132130. doi:10.1371/journal.pone.0132130

Foster WK, Taggart DA (2008) Gender and parental influences on the growth of a sexually dimorphic carnivorous marsupial. *Journal of Zoology* **275**, 221–228. doi:10.1111/j.1469-7998.2008.00429.x

Foster WK, Bradley AJ, Caton W, Taggart DA (2006) Comparison of growth and development of the Red-tailed Phascogale (*Phascogale calura*) in three captive colonies. *Australian Journal of Zoology* **54**, 343–352. doi:10.1071/ZO06033

Foulkes JN (2008) Ooldea Dunnart *Sminthopsis ooldea.* In *The Mammals of Australia.* 3rd edn. (Eds S Van Dyck and R Strahan) pp. 154–155. Reed New Holland, Sydney.

Fox BJ (1982a) A review of dasyurid ecology and speculation on the role of limiting similarity in community organization. In *Carnivorous Marsupials.* (Ed. M Archer) pp. 97–116. Royal Zoological Society of New South Wales, Sydney.

Fox BJ (1982b) Ecological separation and coexistence of *Sminthopsis murina* and *Antechinus stuartii* (Dasyuridae, Marsupialia): a regeneration niche? In *Carnivorous Marsupials.* (Ed. M Archer) pp. 187–197. Royal Zoological Society of New South Wales, Sydney.

Fox BJ (1989) Small mammal community pattern in Australian heathland: a taxonomic rule for species assembly. In *Patterns in the Structure of Mammalian Communities.* (Eds DW Morris, Z Abramsky, BJ Fox and MR Willig) pp. 91–103. Texas Tech University Press, Lubbock TX, USA.

Fox BJ (2008) Common Dunnart *Sminthopsis murina.* In *The Mammals of Australia.* 3rd edn. (Eds S Van Dyck and R Strahan) pp. 152–154. Reed New Holland, Sydney.

Fox BJ, Archer E (1984) The diets of *Sminthopsis murina* and *Antechinus stuartii* (Marsupialia: Dasyuridae)

in sympatry. *Australian Wildlife Research* **11**, 235–248. doi:10.1071/WR9840235

Fox BJ, Whitford D (1982) Polyoestry in a predictable coastal environment: reproduction, growth and development in *Sminthopsis murina* (Dasyuridae, Marsupialia). In *Carnivorous Marsupials.* (Ed. M Archer) pp. 39–48. Royal Zoological Society of New South Wales, Sydney.

Franco M, Quijano A, Soto-Gamboa M (2011) Communal nesting, activity patterns, and population characteristics in the near-threatened Monito del Monte, *Dromiciops gliroides. Journal of Mammalogy* **92**(5), 994–1004. doi:10.1644/10-MAMM-A-256.1

Franco M, Contreras C, Cortés P, Chappell MA, Soto-Gamboa M, Nespolo RF (2012) Aerobic power, huddling and the efficiency of torpor in the South American marsupial, *Dromiciops gliroides. Biology Open* **1**, 1178–1184. doi:10.1242/bio.20122790

Frank A, Soderquist T (2005) The importance of refuge habitat in the local conservation of Stripe-faced Dunnarts *Sminthopsis macroura* on arid rangelands. *Australian Mammalogy* **27**, 75–79.

Frankham GJ, Thompson S, Ingleby S, Soderquist T, Eldridge MDB (2017) Does the 'extinct' Eastern Quoll (*Dasyurus viverrinus*) persist in Barrington Tops, New South Wales? *Australian Mammalogy* **39**, 243–247. doi:10.1071/AM16029

Friend GR (1985) Ecological studies of a population of *Antechinus bellus* (Marsupialia: Dasyuridae) in tropical northern Australia. *Australian Wildlife Research* **12**, 151–162. doi:10.1071/WR9850151

Friend GR, Pearson DJ (2008) Little Long-tailed Dunnart *Sminthopsis dolichura*. In *The Mammals of Australia.* 3rd edn. (Eds S Van Dyck and R Strahan) pp. 134–135. Reed New Holland, Sydney.

Friend GR, Johnson BW, Mitchell DS, Smith GT (1997) Breeding, population dynamics and habitat relationships of *Sminthopsis dolichura* (Marsupialia: Dasyuridae) in semiarid shrublands of Western Australia. *Wildlife Research* **24**, 245–262. doi:10.1071/WR96070

Friend JA (1989) Myrmecobiidae. In *Fauna of Australia. Volume 1B. Mammalia.* (Eds DW Walton and BJ Richardson) pp. 583–590. Australian Government Publishing Service, Canberra.

Friend JA (1990) The Numbat *Myrmecobius fasciatus* (Myrmecobiidae): history of decline and potential recovery. *Proceedings of the Ecological Society of Australia* **16**, 369–377.

Friend JA (2004) *Dibbler Recovery Plan: June 2003–June 2013.* Western Australian Wildlife Management Program No. 38. Department of Conservation and Land Management, Western Australia Threatened Species and Communities Unit, Albany, Western Australia.

Friend JA (2015) Family Myrmecobiidae (Numbat). In *Handbook of the Mammals of the World. Volume 5. Monotremes and Marsupials.* (Eds DE Wilson and RA Mittermeier) pp. 222–231. Lynx Edicions, Barcelona, Spain.

Friend JA, Thomas ND (2003) Conservation of the Numbat (*Myrmecobius fasciatus*). In *Predators with Pouches: the Biology of Carnivorous Marsupials.* (Eds M Jones, C Dickman and M Archer) pp. 452–463. CSIRO Publishing, Melbourne.

Friend T (2008) *Phascogale calura.* In *IUCN Red List of Threatened Species. Version 2013.1.* IUCN, Gland, Switzerland.

Friend T, Burbidge A, Morris K (2008) *Parantechinus apicalis.* In *IUCN Red List of Threatened Species. Version 2013.1.* IUCN, Gland, Switzerland.

Frigo L, Woolley PA (1997) Growth and development of pouch young of the Stripe-faced Dunnart, *Sminthopsis macroura* (Marsupialia: Dasyuridae), in captivity. *Australian Journal of Zoology* **45**, 157–170. doi:10.1071/ZO97002

Frost C (1987) *A History of British Taxidermy.* Sudbury, UK.

Galliez M, de Souza Leite M, Queiroz TL, dos Santos Fernandez FA (2009) Ecology of the Water Opossum *Chironectes minimus* in Atlantic Forest streams of southeastern Brazil. *Journal of Mammalogy* **90**, 93–103. doi:10.1644/07-MAMM-A-397.1

Gardner AL (2005) Order Didelphimorphia. In *Mammal Species of the World: a Taxonomic and Geographic Reference.* 3rd edn. (Eds DE Wilson and DM Reeder) pp. 3–18. Johns Hopkins University Press, Baltimore MD, USA.

Gardner AL (2008) Genus *Chacodelphys.* In *Mammals of South America. Volume 1. Marsupials, Xenarthrans,*

Shrews and Bats. (Ed. AL Gardner) pp. 39–40. University of Chicago Press, Chicago IL, USA.

Gardner AL, Dagosto M (2008) Genus *Metachirus.* In *Mammals of South America. Volume 1. Marsupials, Xenarthrans, Shrews and Bats.* (Ed. AL Gardner) pp. 35–38. University of Chicago Press, Chicago IL, USA.

Gates JA (2001a) *Recovery Plan for the Kangaroo Island Dunnart* Sminthopsis aitkeni*: 2002–2006.* Report to the Threatened Species and Communities Section, Environment Australia. National Parks and Wildlife South Australia, Adelaide.

Gates JA (2001b) *The Distribution and Ecology of the Kangaroo Island Dunnart* Sminthopsis aitkeni. Endangered Species Project Number 006499, Endangered Species Program, Environment Australia, Canberra.

Geiser F (1986) Thermoregulation and torpor in the Kultarr, *Antechinomys laniger* (Marsupialia: Dasyuridae). *Journal of Comparative Physiology. B, Biochemical, Systemic, and Environmental Physiology* **156**, 751–757. doi:10.1007/BF00692755

Geiser F (2003) Thermal biology and energetics of carnivorous marsupials. In *Predators with Pouches: the Biology of Carnivorous Marsupials.* (Eds M Jones, C Dickman and M Archer) pp. 238–253. CSIRO Publishing, Melbourne.

Geiser F (2004) The role of torpor in the life of Australian arid zone mammals. *Australian Mammalogy* **26**, 125–134.

Geiser F (2013) Primer: hibernation. *Current Biology* **23**(5), R188–R193. doi:10.1016/j.cub.2013.01.062

Geiser F, Baudinette RV (1988) Daily torpor and thermoregulation in the small dasyurid marsupials *Planigale gilesi* and *Ningaui yvonneae. Australian Journal of Zoology* **36**, 473–481. doi:10.1071/ZO9880473

Geiser F, Körtner G (2010) Hibernation and daily torpor in Australian mammals. *Australian Zoologist* **35**(2), 204–215. doi:10.7882/AZ.2010.009

Geiser F, Martin GM (2013) Torpor in the Patagonian Opossum (*Lestodelphys halli*): implications for the evolution of daily torpor and hibernation. *Naturwissenschaften* **100**, 975–981. doi:10.1007/s00114-013-1098-2

Geiser F, Pavey CR (2007) Basking and torpor in a rock-dwelling desert marsupial: survival strategies in a resource-poor environment. *Journal of Comparative Physiology. B, Biochemical, Systemic, and Environmental Physiology* **177**, 885–892. doi:10.1007/s00360-007-0186-z

Geoffroy [Saint Hilaire] È (1803) Note sur les espece du genre dasyure. *Bulletin des Sciences par la Societe Philomathique de Paris* **3**(81), 258–259.

Giarla TC, Voss RS, Jansa SA (2010) Species limits and phylogenetic relationships in the didelphid marsupial genus *Thylamys* based on mitochondrial DNA sequences and morphology. *Bulletin of the American Museum of Natural History* **346**, 1–67. doi:10.1206/716.1

Gibbons P, Lindenmayer D (2002) *Tree Hollows and Wildlife Conservation in Australia.* CSIRO Publishing, Melbourne.

Gibbs LS (1917) *Dutch NW New Guinea. A contribution to the Phytogeography and Flora of the Arfak Mountains and Surrounds.* Taylor and Francis, London, UK.

Gibson DF, Cole JR (1992) Aspects of the ecology of the Mulgara, *Dasycercus cristicauda*, (Marsupialia: Dasyuridae) in the Northern Territory. *Australian Mammalogy* **15**, 105–112.

Gibson LA, Wilson BA, Cahill DM, Hill J (2004) Modelling habitat suitability of the Swamp Antechinus (*Antechinus minimus maritimus*) in the coastal heathlands of southern Victoria, Australia. *Biological Conservation* **117**, 143–150. doi:10.1016/S0006-3207(03)00288-X

Gilfillan SL (2001a) An ecological study of a population of *Pseudantechinus macdonnellensis* (Marsupialia: Dasyuridae) in central Australia. I. Invertebrate food supply, diet and reproductive strategy. *Wildlife Research* **28**, 469–480. doi:10.1071/WR99062

Gilfillan SL (2001b) An ecological study of a population of *Pseudantechinus macdonnellensis* (Marsupialia: Dasyuridae) in central Australia. II. Population dynamics and movements. *Wildlife Research* **28**, 481–492. doi:10.1071/WR99063

Glen AS, Dickman CR (2006a) Diet of the Spotted-tailed Quoll (*Dasyurus maculatus*) in eastern

Australia: effects of season, sex and size. *Journal of Zoology* **269**, 241–248.

Glen AS, Dickman CR (2006b) Home range, denning behaviour and microhabitat use of the carnivorous marsupial *Dasyurus maculatus* in eastern Australia. *Journal of Zoology* **268**, 347–354. doi:10.1111/j.1469-7998.2006.00064.x

Glen AS, Dickman CR (2011) Why are there so many Spotted-tailed Quolls *Dasyurus maculatus* in parts of north-eastern New South Wales? *Australian Zoologist* **35**, 711–718. doi:10.7882/AZ.2011.023

Glen AS, Dickman CR (2013) Population viability analysis shows Spotted-tailed Quolls may be vulnerable to competition. *Australian Mammalogy* **35**, 180–183. doi:10.1071/AM12045

Glen AS, Dickman CR (2014) The importance of predators. In *Carnivores of Australia: Past, Present and Future.* (Eds AS Glen and CR Dickman) pp. 1–12. CSIRO Publishing, Melbourne.

Glen AS, Dickman CR, Soulé ME, Mackey BG (2007) Evaluating the role of the Dingo as a trophic regulator in Australian ecosystems. *Austral Ecology* **32**, 492–501. doi:10.1111/j.1442-9993.2007.01721.x

Glen AS, de Tores PJ, Sutherland DR, Morris KD (2009a) Interactions between Chuditch (*Dasyurus geoffroii*) and introduced predators: a review. *Australian Journal of Zoology* **57**, 347–356. doi:10.1071/ZO09041

Glen AS, Cardoso MJ, Dickman CR, Firestone KB (2009b) Who's your daddy? Paternity testing reveals promiscuity and multiple paternity in the carnivorous marsupial *Dasyurus maculatus* (Marsupialia: Dasyuridae). *Biological Journal of the Linnean Society. Linnean Society of London* **96**, 1–7. doi:10.1111/j.1095-8312.2008.01094.x

Glen AS, Wayne A, Maxwell M, Cruz J (2010a) Comparative diets of the Chuditch, a threatened marsupial carnivore, in the northern and southern jarrah forests, Western Australia. *Journal of Zoology* **282**, 276–283. doi:10.1111/j.1469-7998.2010.00738.x

Glen AS, Berry O, Sutherland DR, Garretson S, Robinson T, de Tores PJ (2010b) Forensic DNA confirms intraguild killing of a Chuditch (*Dasyurus geoffroii*) by a feral cat (*Felis catus*). *Conservation Genetics* **11**, 1099–1101. doi:10.1007/s10592-009-9888-y

Glen AS, Pennay M, Dickman CR, Wintle BA, Firestone KB (2011) Diets of sympatric native and introduced carnivores in the Barrington Tops, eastern Australia. *Austral Ecology* **36**, 290–296. doi:10.1111/j.1442-9993.2010.02149.x

Glen AS, Dickman CR, Letnic M (2014) Carnivore communities: challenges and opportunities for conservation. In *Carnivores of Australia: Past, Present and Future.* (Eds AS Glen and CR Dickman) pp. 405–415. CSIRO Publishing, Melbourne.

Godsell J (1982) The population ecology of the Eastern Quoll *Dasyurus viverrinus* (Dasyuridae, Marsupialia), in southern Tasmania. In *Carnivorous Marsupials.* (Ed. M Archer) pp. 199–207. Royal Zoological Society of New South Wales, Sydney.

Golding BG (1979) Use of artificial hollows by mammals and birds in the Wombat Forest, Daylesford, Victoria. MSc thesis. Monash University, Melbourne.

Goldingay RL (2000) Small dasyurid marsupials – are they effective pollinators? *Australian Journal of Zoology* **48**, 597–606. doi:10.1071/ZO00015

González EM, Claramunt S (2000) Behaviors of captive Short-tailed Opossums, *Monodelphis dimidiata* (Wagner, 1847) (Didelphimorphia, Didelphidae). *Mammalia* **64**, 271–285. doi:10.1515/mamm.2000.64.3.271

Goswami A, Milne N, Wroe S (2011) Biting through constraints: cranial morphology, disparity and convergence across living and fossil carnivorous mammals. *Proceedings. Biological Sciences* **278**, 1831–1839. doi:10.1098/rspb.2010.2031

Gould J (1841) Observations on *Dasyurus maugei* and *D. viverrinus* of Geoffroy, and description of a new species. *Proceedings of the Zoological Society of London* **1840**, 151.

Gould J (1842) Characters of a new species of *Perameles*, and a new species of *Dasyurus*. *Proceedings of the Zoological Society of London* **1842**, 41–42.

Gould J (1845-1863) *The Mammals of Australia.* John Gould, London, UK.

Gray JE (1821) On the natural arrangement of vertebrose animals. *London Medical Repository* **15**, 308.

Gray JE (1843) *List of Specimens of Mammalia in the Collection of the British Museum of London*. British Museum (Natural History), London, UK.

Gray E (2013) A field study of breeding biology in a new species of carnivorous marsupial, the Buff-footed Antechinus, *Antechinus mysticus*. BSc (Hons) thesis. Queenland University of Technology, Brisbane.

Gray EL, Baker AM, Firn J (2016) Autecology of a new species of carnivorous marsupial, the endangered Black-tailed Dusky Antechinus (*Antechinus arktos*), compared to a sympatric congener, the Brown Antechinus (*Antechinus stuartii*). *Mammal Research* **62**, 47–63. doi:10.1007/s13364-016-0281-1

Green B (1984) Composition of milk and energetics of growth in marsupials. *Symposia of the Zoological Society of London* **51**, 369–387.

Green B, Merchant J, Newgrain K (1987) Milk composition in the Eastern Quoll, *Dasyurus viverrinus* (Marsupialia: Dasyuridae). *Australian Journal of Biological Sciences* **40**, 379–387. doi:10.1071/BI9870379

Green B, Krause WJ, Newgrain K (1996) Milk composition in the North American Opossum (*Didelphis virginiana*). *Comparative Biochemistry and Physiology B* **113**, 619–623. doi:10.1016/0305-0491(95)02034-9

Green B, Merchant J, Newgrain K (1997) Lactational energetics of a marsupial carnivore, the Eastern Quoll (*Dasyurus viverrinus*). *Australian Journal of Zoology* **45**, 295–306. doi:10.1071/ZO97003

Green K (1989) Altitudinal and seasonal differences in the diets of *Antechinus swainsonii* and *A. stuartii* (Marsupialia: Dasyuridae) in relation to the availability of prey in the Snowy Mountains. *Australian Wildlife Research* **16**, 581–592. doi:10.1071/WR9890581

Green K (2001) Autumnal body mass reduction in *Antechinus swainsonii* (Dasyuridae) in the Snowy Mountains. *Australian Mammalogy* **23**, 31–36. doi:10.1071/AM01031

Green RH (1967) Notes on the Devil (*Sarcophilus harrisii*) and the Quoll (*Dasyurus viverrinus*) in north-eastern Tasmania. *Records of the Queen Victoria Museum* **27**, 1–13.

Green RH (1972) The murids and small dasyurids in Tasmania. Parts 5, 6 and 7. *Records of the Queen Victoria Museum* **46**, 1–34.

Green RH, Scarborough TJ (1990) The Spotted-tailed Quoll *Dasyurus maculatus* (Dasyuridae, Marsupialia) in Tasmania. *Tasmanian Naturalist* **100**, 1–15.

Greenville AC, Wardle GM, Dickman CR (2012) Extreme climatic events drive mammal irruptions: regression analysis of 100-year trends in desert rainfall and temperature. *Ecology and Evolution* **2**(11), 2645–2658. doi:10.1002/ece3.377

Greenville AC, Wardle GM, Nguyen V, Dickman CR (2016) Population dynamics of desert mammals: similarities and contrasts within a multispecies assemblage. *Ecosphere* **7**, e01343. doi:10.1002/ecs2.1343

Greenville AC, Wardle GM, Dickman CR (2017) Desert mammal populations are limited by introduced predators rather than future climate change. *Royal Society Open Science* **4**, 170384. doi:10.1098/rsos.170384

Gressitt JL (Ed.) (1981) *Biogeography and Ecology of New Guinea*. Springer, Dordrecht, Netherlands.

Grossek GL, Woolley PA, Menzies JL (2010) The diet of some New Guinean dasyurid marsupials. *Australian Mammalogy* **32**, 145–155. doi:10.1071/AM10002

Groves CP (2005) Order Dasyuromorphia. In *Mammal Species of the World*. (Eds DE Wilson and DM Reeder) pp. 23–37. The Johns Hopkins University Press, Baltimore MD, USA.

Guerrero E, Sguerra S, eds. (2009) *Protected Areas and Development in Latin America: from Santa Marta 1997 to Bariloche 2007 and Perspectives for a new Decade*. Fundación Natura, Comité Colombiano UICN y Parques Nacionales Naturales Colombia, Bogotá, Colombia.

Guiler ER (1970a) Observations on the Tasmanian Devil, *Sarcophilus harrisii* (Marsupialia: Dasyuridae). I. Numbers, home range, movements, and food in two populations. *Australian Journal of Zoology* **18**, 49–62. doi:10.1071/ZO9700049

Guiler ER (1970b) Observations on the Tasmanian Devil, *Sarcophilus harrisii* (Marsupialia: Dasyuridae). II. Reproduction, breeding, and growth of pouch young. *Australian Journal of Zoology* **18**, 63–70. doi:10.1071/ZO9700063

Guiler ER (1982) Temporal and spatial distribution of the Tasmanian Devil, *Sarcophilus harrisii* (Dasyuridae: Marsupialia). *Papers and Proceedings of the Royal Society of Tasmania* **116,** 153–163.

Gurgel-Gonçalves R, Galváo C, Costa J, Peterson AT (2012) Geographic distribution of Chagas disease vectors in Brazil based on ecological niche modeling. *Journal of Tropical Medicine* **2012**, 705326. doi:10.1155/2012/705326

Haby NA, Foulkes J, Brook BW (2013) Using climate variables to predict small mammal occurrence in hot, dry environments. *Landscape Ecology* **28**, 741–753. doi:10.1007/s10980-013-9859-z

Hall R (2011) Australia–SE Asia collision: plate tectonics and crustal flow. In *The SE Asian Gateway: History and Tectonics of the Australia–Asia Collision.* (Eds R Hall, MA Cottam and MEJ Wilson) pp. 75–109. Special Publication 355. Geological Society, London, UK.

Hall S (1980) The diets of two coexisting species of *Antechinus* (Marsupialia: Dasyuridae). *Australian Wildlife Research* **7**, 365–378. doi:10.1071/WR9800365

Hamede R, McCallum H, Jones M (2008) Seasonal, demographic and density-related patterns of contact between Tasmanian Devils (*Sarcophilus harrisii*): implications for transmission of devil facial tumour disease. *Austral Ecology* **33**, 614–622. doi:10.1111/j.1442-9993.2007.01827.x

Hamede RK, Bashford J, McCallum H, Jones M (2009) Contact networks in a wild Tasmanian Devil (*Sarcophilus harrisii*) population: using social network analysis to reveal seasonal variability in social behaviour and its implications for tramsmission of devil facial tumour disease. *Ecology Letters* **12**, 1147–1157. doi:10.1111/j.1461-0248.2009.01370.x

Hamede R, Lachish S, Belov K, Woods G, Kreiss A, Pearse A-M, *et al.* (2012) Reduced effect of Tasmanian Devil facial tumor disease at the disease front. *Conservation Biology* **26**(1), 124–134. doi:10.1111/j.1523-1739.2011.01747.x

Happold M (1972) Maternal and juvenile behaviour in the marsupial jerboa *Antechinomys spenceri* (Dasyuridae). *Australian Mammalogy* **1**, 27–37.

Harder JD, Fleck DW (1997) Reproductive ecology of New World marsupials. In *Marsupial Biology: Recent Research, New Perspectives.* (Eds NR Saunders and LA Hinds) pp. 175–203. University of New South Wales Press, Sydney.

Hardy CM, Hinds LA, Kerr PJ, Lloyd ML, Redwood AJ, Shellam GR, *et al.* (2006) Biological control of vertebrate pests using virally vectored immunocontraception. *Journal of Reproductive Immunology* **71**, 102–111. doi:10.1016/j.jri.2006.04.006

Hart RP, Kitchener DJ (1986) First record of *Sminthopsis psammophila* (Marsupialia: Dasyuridae) from Western Australia. *Records of the Western Australian Museum* **13**, 139–144.

Hartman CG (1921) Breeding habits, development, and birth of the Opossum. *Annual Report of the Smithsonian Institution* **1921**, 347–364.

Hartman CG (1925a) Hysterectomy and the oestrous cycle in the Opossum. *The American Journal of Anatomy* **35**, 25–29. doi:10.1002/aja.1000350103

Hartman CG (1925b) The interruption of pregnancy by ovariectomy in the aplacental Opossum: a study in the physiology of implantation. *The American Journal of Physiology* **71**, 436–454.

Hawkins CE, Baars C, Hesterman H, Hocking GJ, Jones ME, Lazenby B, *et al.* (2006) Emerging disease and population decline of an island endemic, the Tasmanian Devil *Sarcophilus harrisii. Biological Conservation* **131**, 307–324. doi:10.1016/j.biocon.2006.04.010

Hawkins CE, McCallum H, Mooney N, Jones M, Holdsworth M (2008) *Sarcophilus harrisii. In IUCN Red List of Threatened Species. Version 2013.2.* IUCN, Gland, Switzerland.

Haythornthwaite AS (2005a) Microhabitat use and foraging behaviour of *Sminthopsis youngsoni* (Marsupialia: Dasyuridae) in arid central Australia. *Wildlife Research* **32**, 609–615. doi:10.1071/WR04126

Haythornthwaite AS, Dickman CR (2000) Foraging strategies of an insectivorous marsupial, *Sminthopsis youngsoni* (Marsupialia: Dasyuridae), in Australian sandridge desert. *Austral Ecology* **25**, 193–198. doi:10.1046/j.1442-9993.2000.01037.x

Haythornthwaite AS, Dickman CR (2006a) Distribution, abundance, and individual strategies: a multi-scale analysis of dasyurid marsupials in arid central Australia. *Ecography* **29**, 285–300. doi:10.1111/j.2006.0906-7590.04307.x

Haythornthwaite AS, Dickman CR (2006b) Long-distance movements by a small carnivorous marsupial: how *Sminthopsis youngsoni* (Marsupialia: Dasyuridae) uses habitat in an Australian sandridge desert. *Journal of Zoology* **270**, 543–549. doi:10.1111/j.1469-7998.2006.00186.x

Hayward MW, Moseby K, Read JL (2014) The role of predator exclosures in the conservation of Australian fauna. In *Carnivores of Australia: Past, Present and Future.* (Eds AS Glen and CR Dickman) pp. 355–371. CSIRO Publishing, Melbourne.

Heinsohn T, Hope G (2006) The Torresian connections: zoogeography of New Guinea. In *Evolution and Biogeography of Australasian Vertebrates.* (Eds JR Merrick, M Archer, GM Hickey and MSY Lee) pp. 71–94. Auscipub, Sydney.

Helgen KM (2007a) A taxonomic and geographic overview of the mammals of Papua. In *The Ecology of Papua (Ecology of Indonesia Series Volume VI).* (Eds AJ Marshall and B Beehler) pp. 689–749. Periplus Editions, Singapore.

Helgen KM (2007b) The mammal fauna of the Kaijende Highlands, Enga Province, Papua New Guinea. In *A Rapid Biodiversity Assessment of the Kaijende Highlands, Enga Province, Papua New Guinea.* pp. 52–68. Conservation International, Arlington VA, USA.

Helgen KM, Oliver PM (2004) A review of the mammal fauna of the Trans-Fly ecoregion. WWF South Pacifc Program, Papua New Guinea.

Helgen KM, Opiang MD (2011) Notes on the high elevation mammal fauna of the Kaijende Highlands, Enga Province, Papua New Guinea. *Rapid Assessment Program Bulletin of Biological Assessment* **60**, 235–245.

Helgen KM, Veatch EG (2015) Recently extinct Australian marsupials and monotremes. In *Handbook of the Mammals of the World. Volume 5. Monotremes and Marsupials.* (Eds DE Wilson and RA Mittermeier) pp. 17–33. Lynx Edicions, Barcelona, Spain.

Helgen K, Dickman C, Lunde D, Burnett S, Woinarski J, Woolley P (2008) *Sminthopsis virginiae*. In . *IUCN Red List of Threatened Species. Version 2013.2.* IUCN, Gland, Switzerland.

Herbert A (1996) *Distribution, Habitat Preference and Status of the Kangaroo Island Dunnart (Sminthopsis aitkeni).* Nature Conservation Society of South Australia, Adelaide.

Hershkovitz P (1997) Composition of the family Didelphidae Gray, 1821 (Didelphoidea: Marsupialia), with a review of the morphology and behavior of the included four eyed opossums of the genus *Philander* Tiedemann, 1808. *Fieldiana. Zoology* **86**, 1–103.

Hesterman H, Jones SM, Schwarzenberger F (2008) Pouch appearance is a reliable indicator of the reproductive status in the Tasmanian Devil and the Spotted-tailed Quoll. *Journal of Zoology* **275**, 130–138. doi:10.1111/j.1469-7998.2008.00419.x

Hill JE (1990) A memoir and bibliography of Michael Rogers Oldfield Thomas, F.R.S. *Bulletin of the British Museum (Natural History). Historical Series* **18**(1), 25–113.

Hill JP (1900) On the foetal membranes, placentation and parturition, of the Native Cat (*Dasyurus viverrinus*). *Anatomischer Anzeiger* **18**, 364–373.

Hill JP (1910) Contributions to the embryology of the Marsupialia. 4. The early development of the Marsupialia with special reference to the Native Cat (*Dasyurus viverrinus*). *The Quarterly Journal of Microscopical Science* **56**, 1–134.

Hill JP, Hill WCO (1953) The growth-stages of the pouch-young of the Native Cat (*Dasyurus viverrinus*) together with observations on the anatomy of the new-born young. *Transactions of the Zoological Society of London* **28**, 349–452.

Hill JP, O'Donoghue CH (1913) The reproductive cycle in the marsupial *Dasyurus viverrinus. The Quarterly Journal of Microscopical Science* **59**, 133–174.

Hindmarsh R, Majer JD (1977) Food requirements of Mardo (*Antechinus flavipes* (Waterhouse)) and the effect of fire on Mardo abundance. Research Paper Number 31. Forests Department of Western Australia, Perth WA.

Hinton MAC (1929) Mr M.R.Oldfield Thomas. *Nature* **124**(3116), 101–102. doi:10.1038/124101a0

Hohnen R, Ashby J, Tuft K, McGregor H (2013) Individual identification of Northern Quolls (*Dasyurus hallucatus*) using remote cameras. *Australian Mammalogy* **35**, 131–135. doi:10.1071/AM12015

Holdaway RN, Worthy TH (2006) Evolution of New Zealand and its vertebrates. In *Evolution and Biogeography of Australasian Vertebrates.* (Eds JR Merrick, M Archer, GM Hickey. and MSY Lee) pp. 111–128. Auscipub, Sydney.

Holleley CE, Dickman CR, Crowther MS, Oldroyd BP (2006) Size breeds success: multiple paternity, multivariate selection and male semelparity in a small marsupial, *Antechinus stuartii. Molecular Ecology* **15**, 3439–3448. doi:10.1111/j.1365-294X.2006.03001.x

Holloway JC, Geiser F (1996) Reproductive status and torpor of the marsupial *Sminthopsis crassicaudata*: effect of photoperiod. *Journal of Thermal Biology* **21**, 373–380. doi:10.1016/S0306-4565(96)00023-X

Horner BE, Taylor JM (1959) Results of the Archbold Expeditions. No. 80. Observations on the biology of the Yellow-footed Marsupial Mouse *Antechinus flavipes flavipes. American Museum Novitates Number* **1972**, 1–24.

How RA, Cooper NK (2002) Terrestrial small mammals of the Abydos Plain in the north-eastern Pilbara, Western Australia. *Journal of the Royal Society of Western Australia* **85**, 71–82.

How RA, Dell J, Muir BG (1988) Vertebrate fauna. In: The biological survey of the eastern goldfields of Western Australia: Part 4. Lake Johnston–Hyden study area. *Records of the Western Australian Museum* **30**, 44–83.

How RA, Dell J, Cooper NK (1991) Vertebrate fauna of the Abydos-Woodstock Reserve, northeast Pilbara. *Records of the Western Australian Museum* **37**, 78–125.

How RA, Cooper NK, Girardi L, Bow BG (2002) The Mardo: an examination of geographic variation in morphology and reproductive potential of *Antechinus flavipes* in southwestern Australia. *Records of the Western Australian Museum* **20**, 441–447.

How RA, Spencer PBS, Schmitt LH (2009) Island populations have high conservation value for northern Australia's top marsupial predator ahead of a threatening process. *Journal of Zoology* **278**, 206–217. doi:10.1111/j.1469-7998.2009.00569.x

Huang C (1987) Detection of prey by captive *Ningaui ridei* and *Sminthopsis griseoventer* (Marsupialia: Dasyuridae). *Australian Mammalogy* **10**, 23–26.

Hughes C, Gaffney R, Dickman CR (2011) A preliminary study assessing risk to Tasmanian Devils from poisoning for Red Foxes. *The Journal of Wildlife Management* **75**(2), 385–392. doi:10.1002/jwmg.47

Hughes RL (1982) Reproduction in the Tasmanian Devil *Sarcophilus harrisii* (Dasyuridae, Marsupialia). In *Carnivorous Marsupials.* (Ed. M Archer) pp. 49–63. Royal Zoological Society of New South Wales, Sydney.

Hume ID (2003) Nutrition of carnivorous marsupials. In *Predators with Pouches: the Biology of Carnivorous Marsupials.* (Eds M Jones, C Dickman and M Archer) pp. 221–228. CSIRO Publishing, Melbourne.

Hume ID (2006) Nutrition and digestion. In *Marsupials.* (Eds PJ Armati, CR Dickman and ID Hume) pp. 137–158. Cambridge University Press, Cambridge, UK.

Hume ID, Smith C, Woolley PA (2000) Anatomy and physiology of the gastrointestinal tract of the Julia Creek Dunnart, *Sminthopsis douglasi* (Marsupialia: Dasyuridae). *Australian Journal of Zoology* **48**, 475–485. doi:10.1071/ZO00016

Hunsaker D (1977) Ecology of New World marsupials. In *The Biology of Marsupials.* (Ed. D Hunsaker) pp. 95–156. Academic Press, New York, USA.

Hunsaker D, Shupe D (1977) Behavior of New World marsupials. In *The Biology of Marsupials.* (Ed. D Hunsaker) pp. 279–344. Academic Press, New York, USA.

Hunt DM, Chan J, Carvalho LS, Hokoc JN, Ferguson MC, Arrese CA, *et al.* (2009) Cone visual pigments in two species of South American marsupials. *Gene* **433**, 50–55. doi:10.1016/j.gene.2008.12.006

Hunter DO, Britz T, Jones M, Letnic M (2015) Reintroduction of Tasmanian Devils to mainland Australia can restore top-down control in ecosystems where Dingoes have been extirpated. *Biological Conservation* **191**, 428–435. doi:10.1016/j.biocon.2015.07.030

Hunter ML, Jr, Lindenmayer DB, Calhoun AJK (2007) *Saving the Earth as a Career: Advice on Becoming a Conservation Professional*. Blackwell Publishing, Oxford, UK.

Hutchinson GE (1957) Concluding remarks. *Cold Spring Harbor Symposia on Quantitative Biology* **22**, 415–427. doi:10.1101/SQB.1957.022.01.039

Hutson GD (1975) Sequences of prey-catching behaviour in the brush-tailed marsupial rat (*Dasyuroides byrnei*). *Zeitschrift für Tierpsychologie* **39**, 39–60. doi:10.1111/j.1439-0310.1975.tb00899.x

Ibarra-Cerdeña CN, Sánchez-Cordero V, Ibarra-López P, Iñiguez-Dávalos LI (2007) Noteworthy record of *Musonycteris harrisoni* and *Tlacuatzin canescens* pollinating a columnar cactus in west-central Mexico. *International Journal of Zoological Research* **3**, 223–226. doi:10.3923/ijzr.2007.223.226

Ibbett M, Woinarski JCZ, Oakwood M (2017) Declines in the mammal assemblage of a rugged sandstone environment in Kakadu National Park, Northern Territory, Australia. *Australian Mammalogy*, (In press). doi:10.1071/AM17011

Inns RW (1976) Some seasonal changes in *Antechinus flavipes* (Marsupialia: Dasyuridae). *Australian Journal of Zoology* **24**, 523–531. doi:10.1071/ZO9760523

Iredale T, Troughton E le G (1934) A checklist of the mammals recorded from Australia. *Memoirs of the Australian Museum* **6**, 1–122. doi:10.3853/j.0067-1967.6.1934.516

IUCN (2012) *IUCN Red List Categories and Criteria*. 2nd edn. Version 3.1. IUCN, Gland, Switzerland.

Jackson S, Groves C (2015) *Taxonomy of Australian Mammals*. CSIRO Publishing, Melbourne.

Jansa SA, Barker FK, Voss RS (2014) The early diversification history of didelphid marsupials: a window into South America's 'Splendid Isolation.' *Evolution* **68**(3), 684–695. doi:10.1111/evo.12290

Janssens PA, Messer M (1988) Changes in nutritional metabolism during weaning. In *The Developing Marsupial: Models for Biomedical Research*. (Eds CH Tyndale-Biscoe and PA Janssens) pp. 162–175. Springer, Berlin, Germany.

Jentink FA (1911) New and interesting mammals of the Dutch New Guinea Expedition to the Snow-Mountains. *Notes from the Leyden Museum* **33**, 233–238.

Johnson CN (2006) *Australia's Mammal Extinctions: a 50,000 Year History*. Cambridge University Press, Melbourne.

Johnson CN (2014) The rise and fall of large marsupial carnivores. In *Carnivores of Australia: Past, Present and Future*. (Eds AS Glen and CR Dickman) pp. 13–26. CSIRO Publishing, Melbourne.

Johnson CN, Wroe S (2003) Causes of extinction of vertebrates during the Holocene of mainland Australia: arrival of the Dingo, or human impact? *The Holocene* **13**, 941–948. doi:10.1191/09596836 03hl682fa

Johnson KA, Roff AD (1980) Discovery of *Ningauis* (*Ningaui* sp: Dasyuridae: Marsupialia) in the Northern Territory, Australia. *Australian Mammalogy* **3**, 127–129.

Johnson KA, Woinarski JCZ, Langford DG (2008) Carpentarian Pseudantechinus *Pseudantechinus mimulus*. In *The Mammals of Australia*. 3rd edn. (Eds S Van Dyck and R Strahan) pp. 71–72. Reed New Holland, Sydney.

Johnson KA, McQuillan PB, Kirkpatrick JB (2011) Nocturnal mammals, diurnal lizards, and the pollination ecology of the cryptic flowering *Acrotriche serrulata* (Ericaceae). *International Journal of Plant Sciences* **172**, 173–182. doi:10.1086/657280

Johnstone CP, Reina RD, Lill A (2010) Impact of anthropogenic habitat fragmentation on population health in a small, carnivorous marsupial. *Journal of Mammalogy* **91**(6), 1332–1341. doi:10.1644/10-MAMM-A-034.1

Jones FW (1949) The study of a generalized marsupial (*Dasycercus cristicauda* Krefft). *Transactions of the Zoological Society of London* **26**, 409–501. doi:10.1111/j.1096-3642.1949.tb00225.x

Jones ME (1998) Function of vigilance in sympatric marsupial carnivores: the Eastern Quoll and the Tasmanian Devil. *Animal Behaviour* **56**, 1279–1284. doi:10.1006/anbe.1998.0893

Jones M (2008a) Eastern Quoll *Dasyurus viverrinus*. In *The Mammals of Australia*. 3rd edn. (Eds S Van

Dyck and R Strahan) pp. 62–64. Reed New Holland, Australia.

Jones M (2008b) Tasmanian Devil *Sarcophilus harrisii*. In *The Mammals of Australia*. 3rd edn. (Eds S Van Dyck and R Strahan) pp. 78–80. Reed New Holland, Sydney.

Jones ME, Barmuta LA (1998) Diet overlap and relative abundance of sympatric dasyurid carnivores: a hypothesis of competition. *Journal of Animal Ecology* **67**, 410–421. doi:10.1046/j.1365-2656.1998.00203.x

Jones ME, Barmuta LA (2000) Niche differentiation among sympatric Australian dasyurid carnivores. *Journal of Mammalogy* **81**, 434–447. doi:10.1644/1545-1542(2000)081<0434:NDASAD>2.0.CO;2

Jones ME, Rose RK (2001) Eastern Quoll, *Dasyurus viverrinus*. *Mammalian Species* **677**, 1–9. doi:10.1644/1545-1410(2001)677<0001:DV>2.0.CO;2

Jones ME, Grigg GC, Beard LA (1997) Body temperatures and activity patterns of Tasmanian Devils (*Sarcophilus harrisii*) and Eastern Quolls (*Dasyurus viverrinus*) through a subalpine winter. *Physiological Zoology* **70**(1), 53–60. doi:10.1086/639541

Jones ME, Oakwood M, Belcher CA, Morris K, Murray AJ, Woolley PA, *et al.* (2003) Carnivore concerns: problems, issues and solutions for conserving Australasia's marsupial carnivores. In *Predators with Pouches: the Biology of Carnivorous Marsupials*. (Eds M Jones, C Dickman and M Archer) pp. 422–434. CSIRO Publishing, Melbourne.

Jones ME, Paetkou D, Geffen E, Moritz C (2004a) Genetic diversity and population structure of Tasmanian Devils, the largest marsupial carnivore. *Molecular Ecology* **13**, 2197–2209. doi:10.1111/j.1365-294X.2004.02239.x

Jones ME, Smith GC, Jones SM (2004b) Is anti-predator behaviour in Tasmanian Eastern Quolls (*Dasyurus viverrinus*) effective against introduced predators? *Animal Conservation* **7**, 155–160. doi:10.1017/S136794300400126X

Jones ME, Cockburn A, Hawkins C, Hesterman H, Lachish S, Mann D, *et al.* (2008) Life history change in disease-ravaged Tasmanian Devil populations. *Proceedings of the National Academy of Sciences of the United States of America* **105**(29) 10023–10027.

Jones ME, Burnett S, Claridge AW, Fancourt B, Körtner G, Morris K, *et al.* (2014) Australia's surviving marsupial carnivores: threats and conservation. In *Carnivores of Australia: Past, Present and Future*. (Eds AS Glen and CR Dickman) pp. 197–240. CSIRO Publishing, Melbourne.

Jones ME, Apfelbach R, Banks PB, Cameron EZ, Dickman CR, Frank A, *et al.* (2016) A nose for death: integrating trophic and informational networks for conservation and management. *Frontiers in Ecology and Evolution* **4**, 124. doi:10.3389/fevo.2016.00124

Keast A (1977) Historical biogeography of the marsupials. In *The Biology of Marsupials*. (Eds B Stonehouse and B Gilmore) pp. 69–95. Macmillan Press, London, UK.

Keeley T, O'Brien JK, Fanson BG, Masters K, McGreevy PD (2012) The reproductive cycle of the Tasmanian Devil (*Sarcophilus harrisii*) and factors associated with reproductive success in captivity. *General and Comparative Endocrinology* **176**, 182–191. doi:10.1016/j.ygcen.2012.01.011

Keighery GJ, Halse SA, Harvey MS, McKenzie NL (Eds) (2004) A biodiversity survey of the Western Australian Agricultural Zone. *Records of the Western Australian Museum* **67**, 1–384.

Kelly LT, Nimmo DG, Spence-Bailey LM, Clarke MF, Bennett AF (2010) The short-term responses of small mammals to wildfire in semiarid mallee shrubland. *Wildlife Research* **37**, 293–300. doi:10.1071/WR10016

Kelly LT, Nimmo DG, Spence-Bailey LM, Taylor RS, Watson SJ, Clarke MF, *et al.* (2012) Managing fire mosaics for small mammal conservation: a landscape perspective. *Journal of Applied Ecology* **49**, 412–421. doi:10.1111/j.1365-2664.2012.02124.x

Kelly LT, Dayman R, Nimmo DG, Clarke MF, Bennett AF (2013) Spatial and temporal drivers of small mammal distributions in a semi-arid environment: the role of rainfall, vegetation and life-history. *Austral Ecology* **38**, 786–797. doi:10.1111/aec.12018

Kelt DA, Martínez DR (1989) Notes on distribution and ecology of two marsupials endemic to the Valdivian forests of southern South America. *Journal of Mammalogy* **70**, 220–224. doi:10.2307/1381695

Kelt DA, Meserve PL (2014) Status and challenges for conservation of small mammal assemblages in South America. *Biological Reviews of the Cambridge Philosophical Society* **89**, 705–722. doi:10.1111/brv.12080

Kemper CM, Cooper SJB, Medlin GC, Adams M, Stemmer D, Saint KM, *et al.* (2011) Cryptic Grey-bellied Dunnart (*Sminthopsis griseoventer*) discovered in South Australia: genetic, morphological and subfossil analyses show the value of collecting voucher material. *Australian Journal of Zoology* **59**, 127–144. doi:10.1071/ZO11037

Kennedy ML, Schnell GD, Romero-Almarez ML, Malakouti BS, Sánchez-Hernández C, Best TL, *et al.* (2013) Demographic features, distribution, and habitat selection of the Gray Mouse Opossum (*Tlacuatzin canescens*) in Colima, Mexico. *Acta Theriologica* **58**, 285–298. doi:10.1007/s13364-012-0117-6

Kerr JB, Hedger MP (1983) Spontaneous spermatogenic failure in the marsupial mouse *Antechinus stuartii* Macleay (Dasyuridae: Marsupialia). *Australian Journal of Zoology* **31**, 445–466. doi:10.1071/ZO9830445

Kinnear JE (2018) Mammal conservation and invasive species control in Australia: harnessing a potential extinction machine. *Australian Mammalogy*, (In press). doi.org/10.1071/AM1702

Kinnear JE, Onus ML, Bromilow RN (1988) Fox control and rock-wallaby population dynamics. *Australian Wildlife Research* **15**, 435–450. doi:10.1071/WR9880435

Kinnear JE, Onus ML, Sumner NR (1998) Fox control and rock-wallaby population dynamics – II. An update. *Wildlife Research* **25**, 81–88. doi:10.1071/WR96072

Kinnear JE, Sumner NR, Onus ML (2002) The red fox in Australia – an exotic predator turned biocontrol agent. *Biological Conservation* **108**, 335–359. doi:10.1016/S0006-3207(02)00116-7

Kinnear JE, Krebs CJ, Pentland C, Orell P, Holme C, Karvinen R (2010) Predator-baiting experiments for the conservation of rock-wallabies in Western Australia: a 25-year review with recent advances. *Wildlife Research* **37**, 57–67. doi:10.1071/WR09046

Kinnear JE, Pentland C, Moore N, Krebs CJ (2017) Fox control and 1080 baiting conundrums: time to prepare for a CRISPR solution. *Australian Mammalogy* **39**, 127–136. doi:10.1071/AM16020

Kirsch JAW, Waller PF (1979) Notes on the trapping and behavior of the Caenolestidae (Marsupialia). *Journal of Mammalogy* **60**, 390–395. doi:10.2307/1379811

Kitchener DJ (1981) Breeding, diet and habitat preference of *Phascogale calura* (Gould 1844) (Marsupalia: Dasyuridae) in the southern wheatbelt, Western Australia. *Records of the Western Australian Museum* **9**(2), 173–186.

Kitchener DJ (1988) A new species of false antechinus (Marsupialia: Dasyuridae) from the Kimberley, Western Australia. *Records of the Western Australian Museum* **14**(1), 61–71.

Kitchener DJ (1991) *Pseudantechinus mimulus* (Thomas, 1906) (Marsupialia, Dasyuridae): rediscovery and redescription. *Records of the Western Australian Museum* **15**(1), 191–202.

Kitchener DJ, Caputi N (1988) A new species of false antechinus (Marsupialia: Dasyuridae) from Western Australia, with remarks on the generic classification within the Parantechini. *Records of the Western Australian Museum* **14**(1), 35–59.

Kitchener DJ, Chapman A (1978) Mammals of Dongolocking Nature Reserve. *Records of the Western Australian Museum* **6**, 53–58.

Kitchener DJ, Stoddart J, Henry J (1983) A taxonomic appraisal of the genus *Ningaui* Archer (Marsupialia: Dasyuridae), including description of a new species. *Australian Journal of Zoology* **31**, 361–379. doi:10.1071/ZO9830361

Kitchener DJ, Stoddart J, Henry J (1984) A taxonomic revision of the *Sminthopsis murina* complex (Marsupialia, Dasyuridae) in Australia, including descriptions of four new species. *Records of the Western Australian Museum* **11**, 201–248.

Kitchener DJ, Cooper N, Bradley A (1986) Reproduction in male *Ningaui* (Marsupialia: Dasyuridae). *Australian Wildlife Research* **13**, 13–25. doi:10.1071/WR9860013

Knight EH, Fox BJ (2000) Does habitat structure mediate the effects of forest fragmentation and human-induced disturbance on the abundance of *Antechinus stuartii? Australian Journal of Zoology* **48**, 577–595. doi:10.1071/ZO00018

Kohl KD, Coogan SCP, Raubenheimer D (2015) Do wild carnivores forage for prey or for nutrients? Evidence for nutrient-specific foraging in vertebrate predators. *BioEssays* **37**, 701–709. doi:10.1002/bies.201400171

Körtner G (2007) 1080 aerial baiting for the control of wild dogs and its impact on Spotted-tailed Quoll (*Dasyurus maculatus*) populations in eastern Australia. *Wildlife Research* **34**, 48–53. doi:10.1071/WR06076

Körtner G, Geiser F (2011) Activity and torpor in two sympatric Australian desert marsupials. *Journal of Zoology* **283**, 249–256. doi:10.1111/j.1469-7998.2010.00766.x

Körtner G, Gresser S, Mott B, Tamayo B, Pisanu P, Bayne P, Harden R (2004) Diet of the Spotted-tailed Quoll (*Dasyurus maculatus*) in eastern Australia: effects of season, sex and size. *Wildlife Research* **31**, 474–484.

Körtner G, Pavey CR, Geiser F (2008) Thermal biology, torpor, and activity in free-living mulgaras in arid zone Australia during the winter reproductive season. *Physiological and Biochemical Zoology* **81**, 442–451. doi:10.1086/589545

Körtner G, Rojas AD, Geiser F (2010) Thermal biology, torpor use and activity patterns of a small diurnal marsupial from a tropical desert: sexual differences. *Journal of Comparative Physiology. B, Biochemical, Systemic, and Environmental Physiology* **180**, 869–876. doi:10.1007/s00360-010-0459-9

Körtner G, Riek A, Pavey CR, Geiser F (2016) Activity patterns and torpor in two free-ranging carnivorous marsupials in arid Australia in relation to precipitation, reproduction, and ground cover. *Journal of Mammalogy* **97**, 1555–1564. doi:10.1093/jmammal/gyw113

Kraaijeveld K, Kraaijeveld-Smit FJL, Adcock GJ (2003) Does female mortality drive male semelparity in dasyurid marsupials? *Proceedings. Biological Sciences* **270**(supplement), S251–S253. doi:10.1098/rsbl.2003.0082

Kraaijeveld-Smit FJL, Ward SJ, Temple-Smith PD (2003) Paternity success and the direction of sexual selection in a field population of a semelparous marsupial, *Antechinus agilis*. *Molecular Ecology* **12**, 475–484. doi:10.1046/j.1365-294X.2003.01745.x

Krajewski C, Westerman M (2003) Molecular systematics of Dasyuromorphia. In *Predators with Pouches: the Biology of Carnivorous Marsupials*. (Eds M Jones, C Dickman and M Archer) pp. 3–20. CSIRO Publishing, Melbourne.

Krajewski C, Blacket M, Buckley L, Westerman M (1997a) A multigene assessment of phylogenetic relationships within the dasyurid marsupial subfamily Sminthopsinae. *Molecular Phylogenetics and Evolution* **8**, 236–248. doi:10.1006/mpev.1997.0421

Krajewski C, Young J, Buckley L, Woolley PA, Westerman M (1997b) Reconstructing the evolutionary radiation of dasyurine marsupials with cytochrome *b*, 12S rRNA, and protamine P1 gene trees. *Journal of Mammalian Evolution* **4**(3), 217–236. doi:10.1023/A:1027349725642

Krajewski C, Wroe S, Westerman M (2000a) Molecular evidence for the pattern and timing of cladogenesis in dasyurid marsupials. *Zoological Journal of the Linnean Society* **130**, 375–404. doi:10.1111/j.1096-3642.2000.tb01635.x

Krajewski C, Woolley P, Westerman M (2000b) The evolution of reproductive strategies in dasyurid marsupials: implications of molecular phylogeny. *Biological Journal of the Linnean Society. Linnean Society of London* **71**, 417–435. doi:10.1111/j.1095-8312.2000.tb01267.x

Krajewski C, Moyer GR, Sipiorski JT, Fain MG, Westerman M (2004) Molecular systematics of the enigmatic 'phascolosoricine' marsupials of New Guinea. *Australian Journal of Zoology* **52**, 389–415. doi:10.1071/ZO04020

Krajewski C, Torunskey R, Sipiorski JT, Westerman M (2007) Phylogenetic relationships of the dasyurid marsupial genus *Murexia*. *Journal of Mammalogy* **88**(3), 696–705. doi:10.1644/06-MAMM-A-310R.1

Krajewski C, Anderson FE, Woolley PA, Westerman M (2012) Molecular evidence for a deep clade of dunnarts (Marsupialia: Dasyuridae: Sminthopsis). *Journal of Mammalian Evolution* **19**(4), 265–276. doi:10.1007/s10914-012-9204-3

Krefft G (1866) On the vertebrated animals of the lower Murray and Darling, their habits, economy, and geographical distribution. *Transactions of the Philosophical Society of New South Wales* **1862-1865**, 1–33.

Krefft G (1867) On the classification of the small Dasyuridae of Australia, with descriptions of two new genera and one new species. *Proceedings of the Zoological Society of London* **1866**, 431–435.

Krefft G (1871) *The Mammals of Australia, Illustrated by Miss Harriett Scott, and Mrs Helena Forde, for the Council of Education; with a Short Account of the Species Hitherto Described*. Thomas Richards, Government Printer, Sydney.

Krockenberger A (2006) Lactation. In *Marsupials.* (Eds PJ Armati, CR Dickman and ID Hume) pp. 108–136. Cambridge University Press, Cambridge, UK.

Kuhnen VV, Martinez de Aguiar MA, Gonçalves AZ, Setz EZF (2017) Realized trophic niche driven by apparent competition: an example with marsupials. *Biotropica*, **49**(6), 832–837. doi:10.1111/btp.12468

Kutt AS (2003) New records of the Julia Creek Dunnart *Sminthopsis douglasi* in central-north Queensland. *Australian Zoologist* **32**, 257–260. doi:10.7882/AZ.2003.008

Kutt AS (2008) Chestnut Dunnart *Sminthopsis archeri*. In *The Mammals of Australia.* 3rd edn. (Eds S Van Dyck and R Strahan) 127–128. Reed New Holland, Sydney.

Kutt AS, Van Dyck S, Christie S (2005) A significant range extension for the Chestnut Dunnart *Sminthopsis archeri* (Marsupialia: Dasyuridae) in north Queensland. *Australian Zoologist* **33**(2), 265–268. doi:10.7882/AZ.2005.024

Lachish S, McCallum H, Jones M (2009) Demography, disease and the devil: life-history changes in a disease-affected population of Tasmanian Devils (*Sarcophilus harrisii*). *Journal of Animal Ecology* **78**, 427–436. doi:10.1111/j.1365-2656.2008.01494.x

Laidlaw WS, Hutchings S, Newell GR (1996) Home range and movement patterns of *Sminthopsis leucopus* (Marsupialia: Dasyuridae) in coastal dry heathland, Victoria. *Australian Mammalogy* **19**, 1–9.

Lambert C, Gaikhorst G, Matson P (2011) Captive breeding of the Sandhill Dunnart, *Sminthopsis psammophila* (Marsupialia: Dasyuridae): reproduction, husbandry and growth and development. *Australian Mammalogy* **33**, 21–27. doi:10.1071/AM10004

Laundré JW, Hernández L, Ripple WJ (2010) The landscape of fear: ecological implications of being afraid. *The Open Ecology Journal* **3**, 1–7. doi:10.2174/1874213001003030001

Laurance WF (1990) Distributional records for two 'relict' dasyurid marsupials in north Queensland rainforest. *Australian Mammalogy* **13**, 215–218.

Laurance WF (1993) The pre-European and present distributions of *Antechinus godmani* (Marsupialia: Dasyuridae), a restricted rainforest endemic. *Australian Mammalogy* **16**(1), 23–27.

Laurie EMO, Hill JE (1954) *List of Land Mammals of New Guinea, Celebes and Adjacent Islands 1758-1952.* British Museum (Natural History), London, UK.

Lavergne S, Mouquet N, Thuiller W, Ronce O (2010) Biodiversity and climate change: integrating evolutionary and ecological responses of species and communities. *Annual Review of Ecology, Evolution and Systematics* **41**, 321–350. doi:10.1146/annurev-ecolsys-102209-144628

Lazenby BT, Dickman CR (2013) Patterns of detection and capture are associated with cohabiting predators and prey. *PLoS One* **8**, e59846. doi:10.1371/journal.pone.0059846

Lazenby-Cohen KA, Cockburn A (1988) Lek promiscuity in a semelparous mammal, *Antechinus stuartii* (Marsupialia: Dasyuridae)? *Behavioral Ecology and Sociobiology* **22**, 195–202. doi:10.1007/BF00300569

Lazenby-Cohen KA, Cockburn A (1993) Intense predation by owls on lekking Brown Antechinus *Antechinus stuartii*. In *Australian Raptor Studies.* (Ed. P Olsen) pp. 175–180. Royal Australasian Ornithologists Union, Melbourne.

Le Souef AS, Burrell H (1926) *The Wild Animals of Australasia Embracing the Mammals of New Guinea and the Nearer Pacific Islands.* G.G. Harrap, Sydney.

Leary T, Seri L, Flannery T, Wright D, Hamilton S, Helgen K, *et al.* (2008a) *Dasyurus spartacus.* In *IUCN Red List of Threatened Species. Version 2013.* IUCN, Gland, Switzerland.

Leary T, Seri L, Wright D, Hamilton S, Helgen K, Singadan R, *et al.* (2008b) *Murexia habbema.* In *IUCN Red List of Threatened Species. Version 2013.2.* IUCN, Gland, Switzerland.

Leary T, Seri L, Wright D, Hamilton S, Helgen K, Singadan R, *et al.* (2008c) *Murexia melanurus*. In *IUCN Red List of Threatened Species. Version 2013.2.* IUCN, Gland, Switzerland.

Leary T, Seri L, Wright D, Hamilton S, Helgen K, Singadan R, *et al.* (2008d) *Murexia naso*. In *IUCN Red List of Threatened Species. Version 2013.2.* IUCN, Gland, Switzerland.

Leary T, Seri L, Wright D, Hamilton S, Helgen K, Singadan R, *et al.* 2008e) *Murexia rothschildi*. In *IUCN Red List of Threatened Species. Version 2013.2.* IUCN, Gland, Switzerland.

Leary T, Seri L, Wright D, Hamilton S, Helgen K, Singadan R *et al.* (2008f) *Myoictis melas*. In *IUCN Red List of Threatened Species. Version 2013.1.* IUCN, Gland, Switzerland.

Leary T, Seri L, Wright D, Hamilton S, Helgen K, Singadan R, *et al.* (2008g) *Myoictis wallacei*. In *IUCN Red List of Threatened Species. Version 2013.1.* IUCN, Gland, Switzerland.

Leary T, Seri L, Flannery T, Wright D, Hamilton S, Helgen K, *et al.* (2008h) *Planigale novaeguineae*. In *IUCN Red List of Threatened Species. Version 2013.1.* IUCN, Gland, Switzerland.

Lee AK, Cockburn A (1985) *Evolutionary Ecology of Marsupials*. Cambridge University Press, Cambridge, UK.

Lee AK, McDonald IR (1985) Stress and population regulation in small mammals. *Oxford Reviews of Reproductive Biology* **7**, 261–304.

Lee AK, Bradley AJ, Braithwaite RW (1977) Corticosteroid levels and male mortality in *Antechinus stuartii*. In *The Biology of Marsupials*. (Eds B Stonehouse and D Gilmore) pp. 209–220. Macmillan Press, London, UK.

Lee AK, Woolley P, Braithwaite RW (1982) Life history strategies of dasyurid marsupials. In *Carnivorous Marsupials*. (Ed. M Archer) pp. 1–11. Royal Zoological Society of New South Wales, Sydney.

Leiner NO, Setz EZF, Silva WR (2008) Semelparity and factors affecting the reproductive activity of the Brazilian Slender Opossum (*Marmosops paulensis*) in southeastern Brazil. *Journal of Mammalogy* **89**, 153–158. doi:10.1644/07-MAMM-A-083.1

Leite YLR, Costa LP, Stallings JR (1996) Diet and vertical space use of three sympatric opossums in a Brazilian Atlantic Forest reserve. *Journal of Tropical Ecology* **12**, 435–440. doi:10.1017/S0266467400009640

Leslie AM, Stewart M, Price E, Munn AJ (2015) Daily changes in food availability, but not long-term unpredictability, determine daily torpor-bout occurrences and frequency in Stripe-faced Dunnarts (*Sminthopsis macroura*). *Australian Journal of Zoology* **63**, 12–17. doi:10.1071/ZO14058

Lessa LG, da Costa FN (2010) Diet and seed dispersal by five marsupials (Didelphimorphia: Didelphidae) in a Brazilian cerrado reserve. *Mammalian Biology* **75**, 10–16. doi:10.1016/j.mambio.2008.11.002

Lessa ICM, Ferreguetti AC, Kajin M, Dickman CR, Bergallo HG (2017) You can't run but you can hide: the negative influence of human presence on mid-sized mammals on an Atlantic island. *Journal of Coastal Conservation* **21**, 829–836. doi:10.1007/s11852-017-0544-2

Letnic M (2001) Long distance movements and the use of fire mosaics by small mammals in the Simpson Desert, central Australia. *Australian Mammalogy* **23**, 125–134. doi:10.1071/AM01125

Letnic M, Koch F, Gordon C, Crowther MS, Dickman CR (2009) Keystone effects of an alien top-predator stem extinctions of native mammals. *Proceedings. Biological Sciences* **276**, 3249–3256. doi:10.1098/rspb.2009.0574

Letnic M, Fillios M, Crowther MS (2012) Could direct killing by larger Dingoes have caused the extinction of the Thylacine from mainland Australia? *PLoS One* **7**, e34877. doi:10.1371/journal.pone.0034877

Letnic M, Fillios M, Crowther MS (2014) The arrival and impacts of the Dingo. In *Carnivores of Australia: Past, Present and Future*. (Eds AS Glen and CR Dickman) pp. 53–68. CSIRO Publishing, Melbourne.

Leung LK-P (1999) Ecology of Australian tropical rainforest mammals. I. The Cape York Antechinus, *Antechinus leo* (Dasyuridae : Marsupialia). *Wildlife Research* **26**, 287–306. doi:10.1071/WR96042

Leung LK-P (2008) Cinnamon Antechinus *Antechinus leo*. In *The Mammals of Australia*. 3rd edn. (Eds S Van Dyck and R Strahan) pp. 91–92. Reed New Holland, Sydney.

Lew D, Perez-Hernandez R, Ventura J (2006) Two new species of *Philander* (Didelphimorphia, Didelphidae) from northern South America. *Journal of Mammalogy* **87**, 224–237. doi:10.1644/05-MAMM-A-065R2.1

Lidicker WZ, Marlow BJ (1970) A review of the dasyurid marsupial genus *Antechinomys* Krefft. *Mammalia* **34**, 212–227. doi:10.1515/mamm.1970.34.2.212

Lim BK, Engstrom MD, Patton JC, Bickham JW (2010) Molecular phylogenetics of Reig's Short-tailed Opossum (*Monodelphis reigi*) and its distributional range extension into Guyana. *Mammalian Biology* **75**, 287–293. doi:10.1016/j.mambio.2009.03.009

Lim L (1992) *The Recovery Plan for the Kowari.* Australian National Parks and Wildlife Service and South Australian Department of Environment and Planning, Adelaide.

Lim L (1998) *Kowari Recovery Plan Project No.186 Final Report Research Phase.* Australian Nature Conservancy Agency, Canberra and Department of Environment and Heritage, Brisbane.

Lim L (2008) Kowari *Dasyuroides byrnei*. In *The Mammals of Australia.* 3rd edn. (Eds S Van Dyck and R Strahan) pp. 52–54. Reed New Holland, Sydney.

Lima M, Stenseth NC, Yoccoz NG, Jaksic FM (2001) Demography and population dynamics of the mouse opossum (*Thylamys elegans*) in semi-arid Chile: seasonality, feedback structure and climate. *Proceedings. Biological Sciences* **268**, 2053–2064. doi:10.1098/rspb.2001.1735

Lindenmayer DB, Cunningham RB, Donnelly CF, Triggs BJ, Belvedere M (1994) The diversity, abundance and microhabitat requirements of terrestrial mammals in contiguous forests and retained linear strips in the montane ash forests of the central highlands of Victoria. *Forest Ecology and Management* **67**, 113–133. doi:10.1016/0378-1127(94)90011-6

Llewelyn J, Webb JK, Schwarzkopf L, Alford R, Shine R (2010) Behavioural responses of carnivorous marsupials (*Planigale maculata*) to toxic invasive Cane Toads (*Bufo marinus*). *Austral Ecology* **35**, 560–567. doi:10.1111/j.1442-9993.2009.02067.x

Lloyd P, Sanders M, Reis T, Abbott A (2013) Targeted trapping surveys shed new light on the distribution and habitat characteristics of the Carpentarian Pseudantechinus (*Pseudantechinus mimulus*), a threatened dasyurid marsupial. *Australian Mammalogy* **35**, 220–223. doi:10.1071/AM12027

Long J, Archer M, Flannery T, Hand S (2002) *Prehistoric Mammals of Australia and New Guinea: One Hundred Million Years of Evolution.* University of New South Wales Press, Sydney.

López-Aguirre C, Archer M, Hand SJ, Laffan SW (2017) Extinctions of South American sparassodontans (Metatheria): environmental fluctuations or complex ecological processes? *Palaeontology* **60**, 91–115. doi:10.1111/pala.12272

Loyola RD, Lemes P, Faleiro FV, Trindade-Filho J, Machado RB (2012) Severe loss of suitable climatic conditions for marsupial species in Brazil: challenges and opportunities for conservation. *PLoS One* **7**, e46257. doi:10.1371/journal.pone.0046257

Luckett WP, Woolley PA (1996) Ontogeny and homology of the dentition in dasyurid marsupials: development in *Sminthopsis virginiae. Journal of Mammalian Evolution* **3**, 327–364. doi:10.1007/BF02077449

Lundie-Jenkins G, Payne A (2000) *Recovery Plan for the Julia Creek Dunnart (*Sminthopsis douglasi*) 2000–2004.* Queensland Parks and Wildlife Service, Brisbane.

Lunney D (2008) White-footed Dunnart *Sminthopsis leucopus*. In *The Mammals of Australia.* 3rd edn. (Eds S Van Dyck and R Strahan) pp. 145–148. Reed New Holland, Sydney.

Lunney D, Ashby E (1987) Population changes in *Sminthopsis leucopus* (Gray) (Marsupialia: Dasyuridae), and other small mammal species, in forest regenerating from logging and fire near Bega, New South Wales. *Australian Wildlife Research* **14**, 275–284. doi:10.1071/WR9870275

Lunney D, Leary T (1989) Movement patterns of the White-footed Dunnart *Sminthopsis leucopus* (Marsupialia: Dasyuridae) in a logged, burnt forest near Bega on the south coast of New South Wales. *Australian Wildlife Research* **16**, 207–215. doi:10.1071/WR9890207

Lunney D, Matthews A (2001) The contribution of the community to defining the distribution of a vulnerable species, the Spotted-tailed Quoll,

Dasyurus maculatus. *Wildlife Research* **28**, 537–545. doi:10.1071/WR00018

Lunney D, Ashby E, Grigg J, O'Connell M (1986) Food availability and habitat selection of *Sminthopsis leucopus* (Gray) (Marsupialia: Dasyuridae) in logged forest on the south coast of New South Wales. *Australian Mammalogy* **9**, 105–110.

Lunney D, O'Connell M, Sanders J, Forbes S (1989) Habitat of the White-Footed Dunnart *Sminthopsis leucopus* (Gray) (Dasyuridae: Marsupialia) in a logged, burnt forest near Bega, New South Wales. *Australian Journal of Ecology* **14**, 335–344. doi:10.1111/j.1442-9993.1989.tb01442.x

Lunney D, Matthews A, Grigg J (2001) The diet of *Antechinus agilis* and *A. swainsonii* in unlogged and regenerating sites in Mumbulla State Forest, south-eastern New South Wales. *Wildlife Research* **28**, 459–464. doi:10.1071/WR00015

Lunney D, Menkhorst P, Burnett S (2008) *Sminthopsis leucopus*. In *IUCN Red List of Threatened Species. Version 2013.2*. IUCN, Gland, Switzerland.

Lydekker R (1896) *A Hand-Book to the Marsupialia and Monotremata*. Edward Lloyd, London, UK.

Macleay WS (1842) Additional particulars respecting *Antechinus stuartii*, a new Marsupial Quadruped. *Annals & Magazine of Natural History* **8**, 337–338. doi:10.1080/03745484209442765

Magnusdottir R, Wilson BA, Hersteinsson P (2008) Dispersal and the influence of rainfall on a population of the carnivorous marsupial Swamp Antechinus (*Antechinus minimus maritimus*). *Wildlife Research* **35**, 446–454. doi:10.1071/WR06156

Mahon PS (1999) Predation by feral cats and red foxes and the dynamics of small mammal populations in arid Australia. PhD thesis. University of Sydney, Sydney.

Mahoney JA, Ride WDL (1988) Dasyuridae. In *Zoological Catalogue of Australia. Volume 5. Mammalia*. Pp. 14–33 Australian Government Publishing Service, Canberra.

Majer JD (1978) 'Further notes on the food requirements of the Mardo (*Antechinus flavipes* (Waterhouse)'. Research Paper Number 49. Forests Department of Western Australia, Perth WA.

Malcolm JR (1995) Forest structure and the abundance and diversity of Neotropical small mammals. In *Forest Canopies*. (Eds MD Lowman and NM Nadkarni) pp. 179–197. Academic Press, San Diego CA, USA.

Mangan JA, McKenzie C (2008) Publicist and proselytizer: the officer-hunter as scientist and naturalist. *The International Journal of the History of Sport* **25**(9), 1189–1217. doi:10.1080/09523360802166154

Mangan JA, McKenzie CC (2010) *Militarism, Hunting, Imperialism: Blooding the Martial Male*. Routledge, London, UK.

Marchesan D, Carthew SM (2004) Autecology of the Yellow-footed Antechinus (*Antechinus flavipes*) in a fragmented landscape in southern Australia. *Wildlife Research* **31**, 273–282. doi:10.1071/WR02038

Marden L (1986) The first landfall of Columbus. *National Geographic* **170**(November), 572–577.

Marlow B (1961) Reproductive behaviour of the marsupial mouse, *Antechinus flavipes* (Waterhouse) (Marsupialia) and the development of the pouch young. *Australian Journal of Zoology* **9**(2), 203–220. doi:10.1071/ZO9610203

Marlow BJ (1962) *Marsupials of Australia*. Jacaranda Press, Brisbane.

Marshall LG (1978) *Chironectes minimus*. *Mammalian Species* **109**, 1–6.

Martin GM (2011) Geographic distribution of *Rhyncholestes raphanurus* Osgood, 1924 (Paucituberculata: Caenolestidae), an endemic marsupial of the Valdivian temperate rainforest. *Australian Journal of Zoology* **59**, 118–126. doi:10.1071/ZO11038

Martin GM, González-Chávez B (2016) Observations on the behavior of *Caenolestes fuliginosus* (Tomes, 1863) (Marsupialia, Paucituberculata, Caenolestidae) in captivity. *Journal of Mammalogy* **97**, 568–575. doi:10.1093/jmammal/gyv203

Martin GM, Udrizar Sauthier DE (2011) Observations on the captive behavior of the rare Patagonian Opossum *Lestodelphys halli* (Thomas, 1921) (Marsupialia, Didelphimorphia, Didelphidae). *Mammalia* **75**, 281–286. doi:10.1515/mamm.2011.032

Martinez-Lanfranco JA, Flores D, Jayat JP, D'Elia G (2014) A new species of lutrine opossum, genus *Lutreolina* Thomas (Didelphidae), from the South

American Yungas. *Journal of Mammalogy* **95**(2), 225–240. doi:10.1644/13-MAMM-A-246

Mason ED, Firn J, Hines HB, Baker AM (2017) Breeding biology and growth in a new, threatened carnivorous marsupial. *Mammal Research* **62**, 179–187. doi:10.1007/s13364-016-0303-z

Masters P (1993) The effects of fire-driven succession and rainfall on small mammals in spinifex grasslands at Uluru National Park, Northern Territory. *Wildlife Research* **20**, 803–813. doi:10.1071/WR9930803

Masters P (1998) The Mulgara *Dasycercus cristicauda* (Marsupialia: Dasyuridae) at Uluru National Park, Northern Territory. *Australian Mammalogy* **20**, 403–407.

Masters P (2003) Movement patterns and spatial organisation of the Mulgara, *Dasycercus cristicauda*, (Masupialia: Dasyuridae) in central Australia. *Wildlife Research* **30**, 339–344. doi:10.1071/WR01089

Masters P (2008) Crest-tailed Mulgara *Dasycercus cristicauda*. In *The Mammals of Australia*. 3rd edn. (Eds S Van Dyck and R Strahan) pp. 49–50. Reed New Holland, Sydney.

Masters P, Dickman CR (2012) Population dynamics of *Dasycercus blythi* (Marsupialia: Dasyuridae) in central Australia: how does the Mulgara persist? *Wildlife Research* **39**, 419–428. doi:10.1071/WR11156

Masters P, Dickman CR, Crowther MS (2003) The effects of cover reduction on Mulgara (*Dasycercus cristicauda*), rodent and invertebrate populations in central Australia: implications for management. *Austral Ecology* **28**, 658–665. doi:10.1046/j.1442-9993.2003.01327.x

Maxwell S, Burbidge AA, Morris K (1996) *The 1996 Action Plan for Australian Marsupials and Monotremes*. Australasian Marsupial and Monotreme Specialist Group, IUCN Species Survival Commission, Gland, Switzerland.

May-Collado LJ, Kilpatrick CW, Agnarsson I (2015) Mammals from 'down under': a multi-gene species-level phylogeny of marsupial mammals (Mammalia, Metatheria). *PeerJ* **3**, e805. doi:10.7717/peerj.805

Maynard D, Gordon T (2014) *Tasmanian Tiger: Precious Little Remains*. Queen Victoria Museum and Art Gallery, Launceston, Tasmania.

McAllan B (2003) Timing of reproduction in carnivorous marsupials. In *Predators with Pouches: the Biology of Carnivorous Marsupials*. (Eds M Jones, C Dickman and M Archer) pp. 147–168. CSIRO Publishing, Melbourne.

McAllan BM (2009) Reproductive parameters of surviving 'die-off' male *Antechinus flavipes* and *Antechinus stuartii* (Dasyuridae: Marsupialia) in their second year of life. *Australian Mammalogy* **31**, 17–23. doi:10.1071/AM08109

McAllan BM, Dickman CR (1986) The role of photoperiod in the timing of reproduction in the dasyurid marsupial *Antechinus stuartii*. *Oecologia* **68**, 259–264. doi:10.1007/BF00384797

McAllan BM, Geiser F (2014) Torpor during reproduction in mammals and birds: dealing with an energetic conundrum. *Integrative and Comparative Biology* **56**, 1–17.

McAllan BM, Dickman CR, Crowther MS (2006) Photoperiod as a reproductive cue in the marsupial genus *Antechinus*: ecological and evolutionary consequences. *Biological Journal of the Linnean Society. Linnean Society of London* **87**, 365–379. doi:10.1111/j.1095-8312.2006.00571.x

McAllan BM, Feay N, Bradley AJ, Geiser F (2012) The influence of reproductive hormones on the torpor patterns of the marsupial *Sminthopsis macroura*: bet-hedging in an unpredictable environment. *General and Comparative Endocrinology* **179**, 265–276. doi:10.1016/j.ygcen.2012.08.024

McDonald KR (1991) New distribution records for *Antechinus godmani* (Thomas), a restricted rainforest endemic. *Memoirs of the Queensland Museum* **30**(3), 487–491.

McElroy JW (1941) The ocean navigation of Columbus on his first voyage. *The American Neptune* **1**, 209–240.

McFarland D (1992) *Fauna of the Channel Country Biogeographic Region, South West Queensland*. Queensland National Parks and Wildlife Service, Department of Environment and Heritage, Brisbane.

McKenzie N (2008a) *Sminthopsis gilberti*. In *IUCN Red List of Threatened Species. Version 2013.2*. IUCN, Gland, Switzerland.

McKenzie N (2008b) *Sminthopsis granulipes*. In *IUCN Red List of Threatened Species. Version 2013.2*. IUCN, Gland, Switzerland.

McKenzie NL, Archer M (1982) *Sminthopsis youngsoni* (Marsupialia: Dasyuridae) the Lesser Hairy-footed Dunnart, a new species from arid Australia. *Australian Mammalogy* **5**, 267–279.

McKenzie NL, Cole JR (2008) Lesser Hairy-footed Dunnart *Sminthopsis youngsoni*. In *The Mammals of Australia*. 3rd edn. (Eds S Van Dyck and R Strahan) pp. 160–161. Reed New Holland, Sydney.

McKenzie N, Dickman C (2008a) *Sminthopsis hirtipes*. In *IUCN Red List of Threatened Species. Version 2013.2*. IUCN, Gland, Switzerland.

McKenzie N, Dickman C (2008b) *Sminthopsis youngsoni*. In *IUCN Red List of Threatened Species. Version 2013.2*. IUCN, Gland, Switzerland.

McKenzie NL, Dickman CR (2008c) Wongai Ningaui *Ningaui ridei*. In *The Mammals of Australia*. 3rd edn. (Eds S Van Dyck and R Strahan) pp. 117–118. Reed New Holland, Sydney.

McKenzie NL, Hall NJ (Eds) (1992) The biological survey of the eastern goldfields of Western Australia. Part 8, Kurnalpi-Kalgoorlie study area. *Records of the Western Australian Museum* **41** 1–125.

McKenzie N, Kemper C (2008) *Sminthopsis griseoventer*. In *IUCN Red List of Threatened Species. Version 2013.2*. IUCN, Gland, Switzerland.

McKenzie NL, Kitchener DJ (2008) White-tailed Dunnart *Sminthopsis granulipes*. In *The Mammals of Australia*. 3rd edn. (Eds S Van Dyck and R Strahan) pp. 140–141. Reed New Holland, Sydney.

McKenzie NL, Robinson AC (1987) *A Biological Survey of the Nullarbor Region: South and Western Australia in 1984*. Department of Environment and Planning, Adelaide.

McKenzie NL, Hall N, Muir WP (2000) Nonvolant mammals of the southern Carnarvon Basin, Western Australia. *Records of the Western Australian Museum* **61**, 479–510. doi:10.18195/issn.0313-122x.61.2000.479-510

McKenzie NL, Gibson N, Keighery GJ, Rolfe JK (2004) Patterns in the biodiversity of terrestrial environments in the Western Australian wheatbelt. *Records of the Western Australian Museum* **67**, 293–336. doi:10.18195/issn.0313-122x.67.2004.293-335

McKenzie NL, Burbidge AA, Baynes A, Brereton RN, Dickman CR, Gordon G, *et al.* (2007) Analysis of factors implicated in the recent decline of Australia's mammal fauna. *Journal of Biogeography* **34**, 597–611. doi:10.1111/j.1365-2699.2006.01639.x

McKenzie N, van Weenen J, Kemper K (2008a) *Sminthopsis dolichura*. In *IUCN Red List of Threatened Species. Version 2013.2*. IUCN, Gland, Switzerland.

McKenzie N, Woinarski J, Burbidge A (2008b) *Sminthopsis longicaudata*. In *IUCN Red List of Threatened Species. Version 2013.2*. IUCN, Gland, Switzerland.

McKnight M (2008a) *Dasyurus viverrinus*. In *IUCN Red List of Threatened Species. Version 2013*. IUCN, Gland, Switzerland.

McKnight M (2008b) *Sminthopsis butleri*. In *IUCN Red List of Threatened Species. Version 2013.2*. IUCN, Gland, Switzerland.

McLean AL (2015) Conservation biology of an endangered semi-arid marsupial, the Sandhill Dunnart (*Sminthopsis psammophila*). PhD thesis. University of Adelaide, Adelaide.

Meißner K, Gansloßer U (1985) Development of young in the Kowari, *Dasyuroides byrnei* Spencer, 1896. *Zoo Biology* **4**(4), 351–359. doi:10.1002/zoo.1430040405

Meng J (2014) Special Topic: Paleontology in China Mesozoic mammals of China: implications for phylogeny and early evolution of mammals. *National Science Review* **1**, 521–542. doi:10.1093/nsr/nwu070

Menkhorst P (1995) *Mammals of Victoria*. Oxford University Press, Melbourne.

Menkhorst PW, Beardsell CM (1982) Mammals of southwestern Victoria from the Little Desert to the coast. *Proceedings of the Royal Society of Victoria* **94**, 221–247.

Menkhorst PW, Emison WB, Marcius JM (1984) *A Survey of Mammals and Birds in the Whitfield Area of North-eastern Victoria*. Technical Report Series Number 14. Arthur Rylah Institute for Environmental Research Department of Conservation, Forests & Lands, Melbourne.

Menkhorst P, Rhind S, Ellis M (2008) *Phascogale tapoatafa*. In *IUCN Red List of Threatened Species. Version 2013.1*. IUCN, Gland, Switzerland.

Menzies BR, Renfree MB, Heider T, Mayer F, Hildebrandt TB, Pask AJ (2012) Limited genetic diversity preceded extinction of the Tasmanian Tiger. *PLoS One* **7**(4), e35433. doi:10.1371/journal.pone.0035433

Menzies JI (1972) The relative abundance of *Planigale novaeguineae* and other small mammals in the south Papuan savannas. *Mammalia* **36**(3), 400–405. doi:10.1515/mamm.1972.36.3.400

Menzies JI (1991) *A Handbook of New Guinea Marsupials and Monotremes.* Kristen Press, Madang, Papua New Guinea.

Menzies J (2011) *A Handbook of New Guinea's Marsupials and Monotremes.* University of Papua New Guinea Press, Port Moresby, Papua New Guinea.

Merchant JC, Newgrain K, Green B (1984) Growth of the Eastern Quoll, *Dasyurus viverrinus* (Shaw), (Marsupialia) in captivity. *Australian Wildlife Research* **11**, 21–29. doi:10.1071/WR9840021

Meredith RW, Westerman M, Case JA, Springer MS (2008) A phylogeny and timescale for marsupial evolution based on sequences for five nuclear genes. *Journal of Mammalian Evolution* **15**, 1–36. doi:10.1007/s10914-007-9062-6

Meredith RW, Krajewski C, Westerman M, Springer MS (2009) Relationships and divergence times among the orders and families of Marsupialia. *Museum of Northern Arizona Bulletin* **65**, 383–406.

Meserve PL, Lang BK, Patterson BD (1988) Trophic relationships of small mammals in a Chilean temperate rainforest. *Journal of Mammalogy* **69**, 721–730. doi:10.2307/1381627

Mifsud G, Woolley PA (2012) Predation of the Julia Creek dunnart (*Sminthopsis douglasi*) and other native fauna by cats and foxes on Mitchell grass downs in Queensland. *Australian Mammalogy* **34**, 188–195. doi:10.1071/AM11035

Miles MA, de Souza AA, Póvoa MM (1981) Mammal tracking and nest location in Brazilian forest with an improved spool-and-line device. *Journal of Zoology* **195**, 331–347. doi:10.1111/j.1469-7998.1981.tb03469.x

Millar RP, Glover TD (1973) Regulation of seasonal sexual activity in a small mammal, the Rock Hyrax, *Procavia capensis. Journal of Reproduction and Fertility. Supplement* **19**, 203–220.

Miller S, Bencini R, Mills H, Moro D (2003) Food availability for the Dibbler (*Parantechinus apicalis*) on Boullanger and Whitlock Islands, Western Australia. *Wildlife Research* **30**, 649–654. doi:10.1071/WR01082

Miller W, Hayes VM, Ratan A, Petersen DC, Wittekindt NE, Miller J, *et al.* (2011) Genetic diversity and population structure of the endangered marsupial *Sarcophilus harrisii* (Tasmanian Devil). *Proceedings of the National Academy of Sciences of the United States of America* **108**(30), 12348–12353. doi:10.1073/pnas.1102838108

Millis AL, Taggart DA, Bradley AJ, Phelan J, Temple-smith PD (1999) Reproductive biology of the Brush-tailed Phascogale, *Phascogale tapoatafa* (Marsupialia: Dasyuridae). *Journal of Zoology* **248**, 325–335. doi:10.1111/j.1469-7998.1999.tb01032.x

Mills HR, Bencini R (2000) New evidence for facultative male die-off in island populations of Dibblers, *Parantechinus apicalis. Australian Journal of Zoology* **48**, 501–510. doi:10.1071/ZO00025

Mills HR, Moro D, Spencer PBS (2004) Conservation significance of island versus mainland populations: a case study of Dibblers (*Parantechinus apicalis*) in Western Australia. *Animal Conservation* **7**, 387–395. doi:10.1017/S1367943004001568

Mitchell KJ, Pratt RC, Watson LN, Gibb GC, Llamas B, Kasper M, *et al.* (2014) Molecular phylogeny, biogeography, and habitat preference evolution of marsupials. *Molecular Biology and Evolution* **31**(9), 2322–2330. doi:10.1093/molbev/msu176

Monamy V, Fox BJ (2005) Differential habitat use by a local population of subadult Common Dunnarts, *Sminthopsis murina*, following wildfire in coastal wet heath, New South Wales, Australia. *Wildlife Research* **32**, 617–624. doi:10.1071/WR04105

Monastersky R (2014) Life: a status report. *Nature* **516**, 158–161. doi:10.1038/516158a

Morán-Ordóñez A, Briscoe NJ, Wintle BA (2017) Modelling species responses to extreme weather provides new insights into constraints on range and likely climate change impacts for Australian mammals. *Ecography*, (In press). doi:10.1111/ecog.02850

Moro D (2003) Translocation of captive-bred Dibblers *Parantechinus apicalis* (Marsupialia: Dasyuridae) to Escape Island, Western Australia. *Biological*

Conservation **111**, 305–315. doi:10.1016/S0006-3207(02)00296-3

Morris KD, McKenzie NL (2008) Gilbert's Dunnart *Sminthopsis gilberti*. In *The Mammals of Australia.* 3rd edn. (Eds S Van Dyck and R Strahan) pp. 138–139. Reed New Holland, Sydney.

Morris K, Johnson B, Orell P, Gaikhorst G, Wayne A, Moro D (2003) Recovery of the threatened Chuditch (*Dasyurus geoffroii*): a case study. In *Predators with Pouches: the Biology of Carnivorous Marsupials.* (Eds M Jones, C Dickman and M Archer) pp. 435–451. CSIRO Publishing, Melbourne.

Morris K, Woinarski J, Ellis M, Robinson T, Copley P (2008) *Antechinomys laniger.* In *IUCN Red List of Threatened Species. Version 2013.2.* IUCN, Gland, Switzerland.

Morris K, Austin JJ, Belov K (2013) Low major histocompatibility complex diversity in the Tasmanian Devil predates European settlement and may explain susceptibility to disease epidemics. *Biology Letters* **9**, 20120900. doi:10.1098/rsbl.2012.0900

Morton SR (1978a) An ecological study of *Sminthopsis crassicaudata* (Marsupialia: Dasyuridae) I. Distribution, study areas and methods. *Australian Wildlife Research* **5**, 151–162. doi:10.1071/WR9780151

Morton SR (1978b) An ecological study of *Sminthopsis crassicaudata* (Marsupialia: Dasyuridae) II. Behaviour and social organization. *Australian Wildlife Research* **5**, 163–182. doi:10.1071/WR9780163

Morton SR (1978c) An ecological study of *Sminthopsis crassicaudata* (Marsupialia: Dasyuridae) III. Reproduction and life history. *Australian Wildlife Research* **5**, 183–211. doi:10.1071/WR9780183

Morton SR (1978d) Torpor and nest-sharing in free-living *Sminthopsis crassicaudata* (Marsupialia) and *Mus musculus* (Rodentia). *Journal of Mammalogy* **59**, 569–575. doi:10.2307/1380234

Morton SR (1980) Ecological correlates of caudal fat storage in small mammals. *Australian Mammalogy* **3**, 81–86.

Morton SR (1982) Dasyurid marsupials of the Australian arid zone: an ecological review. In *Carnivorous Marsupials.* (Ed. M Archer) pp. 117–130 Royal Zoological Society of New South Wales, Sydney.

Morton SR (1993) Determinants of diversity in animal communities of arid Australia. In *Species Diversity in Ecological Communities.* (Eds RE Ricklefs and D Schluter) pp. 159–169. The University of Chicago Press, Chicago IL, USA.

Morton SR, Dickman CR (2008a) Fat-tailed Dunnart *Sminthopsis crassicaudata*. In *The Mammals of Australia.* 3rd edn. (Eds S Van Dyck and R Strahan) pp. 132–133. Reed New Holland, Sydney.

Morton SR, Dickman CR (2008b) Stripe-faced Dunnart *Sminthopsis macroura*. In *The Mammals of Australia.* 3rd edn. (Eds S Van Dyck and R Strahan) pp. 150–152. Reed New Holland, Sydney.

Morton SR, Denny MJS, Read DG (1983) Habitat preferences and diets of sympatric *Sminthopsis crassicaudata* and *S. macroura* (Marsupialia: Dasyuridae). *Australian Mammalogy* **6**, 29–34.

Morton SR, Dickman CR, Fletcher TP (1989) Dasyuridae. In *Fauna of Australia. Volume 1B. Mammalia.* (Eds DW Walton and BJ Richardson) pp. 560–582. Australian Government Publishing Service, Canberra.

Moss GL, Croft DB (1988) Behavioural mechanisms of microhabitat selection and competition among three species of arid zone dasyurid marsupial. *Australian Journal of Ecology* **13**, 485–493. doi:10.1111/j.1442-9993.1988.tb00997.x

Munks MW (2012) Progress in development of immunocontraceptive vaccines for permanent non-surgical sterilization of cats and dogs. *Reproduction in Domestic Animals* **47**, 223–227. doi:10.1111/j.1439-0531.2012.02079.x

Munn AJ, Kern P, McAllan BM (2010) Coping with chaos: unpredictable food supplies intensify torpor use in an arid-zone marsupial, the Fat-tailed Dunnart (*Sminthopsis crassicaudata*). *Naturwissenschaften* **97**, 601–605. doi:10.1007/s00114-010-0670-2

Mutton TY (2011) Systematics of the dasyurid marsupial genus *Antechinus*. BSc (Hons) thesis. Queensland University of Technology, Brisbane.

Mutton TY, Gray EL, Fuller SJ, Baker AM (2017) Life history, breeding biology and movement in a new species of carnivorous marsupial, the Buff-footed

Antechinus (*Antechinus mysticus*) and a sympatric congener, the Subtropical Antechinus (*Antechinus subtropicus*). *Mammal Research* **62**, 373–385. doi:10.1007/s13364-017-0325-1

Myers P, Patton JL (2008) Genus *Lestoros*. In *Mammals of South America. Volume 1. Marsupials, Xenarthrans, Shrews and Bats.* (Ed. AL Gardner) pp. 124–125. University of Chicago Press, Chicago IL, USA.

Naish D (2012) Marsupial 'dogs', 'bears', 'sabre-tooths' and 'weasels' of island South America: meet the borhyaenoids. *Scientific American*, blog, <http://blogs.scientificamerican.com/tetrapod-zoology/meet-the-borhyaenoids-2012/>.

National Reserve System Task Group (2009) *Australia's Strategy for the National Reserve System 2009–2030.* Australian Government, Canberra.

Naylor R, Richardson SJ, McAllan BM (2008) Boom and bust: a review of the physiology of the marsupial genus *Antechinus. Journal of Comparative Physiology. B, Biochemical, Systemic, and Environmental Physiology* **178**, 545–562. doi:10.1007/s00360-007-0250-8

Nelson JE, Gemmell RT (2003) Birth in the Northern Quoll *Dasyurus hallucatus* (Marsupialia: Dasyuridae). *Australian Journal of Zoology* **51**, 187–198. doi:10.1071/ZO02016

Nespolo RF, Verdugo C, Cortes PA, Bacigalupe LD (2010) Bioenergetics of torpor in the microbiotherid marsupial, Monito del Monte (*Dromiciops gliroides*): the role of temperature and food availablility. *Journal of Comparative Physiology. B, Biochemical, Systemic, and Environmental Physiology* **180**, 767–773. doi:10.1007/s00360-010-0449-y

Nova PV, Avilla LS, Oliveira EV (2015) Didelphidae marsupials (Mammalia, Didelphimorphia) from the Late Pleistocene deposit of the Gruta dos Moura Cave, northern Brazil. *Anais da Academia Brasileira de Ciencias* **87**(1), 193–208. doi:10.1590/0001-3765201520140229

O'Donnell S, Webb JK, Shine R (2010) Conditioned taste aversion enhances the survival of an endangered predator imperilled by a toxic invader. *Journal of Applied Ecology* **47**, 558–565. doi:10.1111/j.1365-2664.2010.01802.x

Oakwood M (1997) The ecology of the Northern Quoll, *Dasyurus hallucatus*. PhD thesis. Australian National University, Canberra.

Oakwood M (2000) Reproduction and demography of the Northern Quoll, *Dasyurus hallucatus*, in the lowland savanna of northern Australia. *Australian Journal of Zoology* **48**, 519–539. doi:10.1071/ZO00028

Oakwood M (2002) Spatial and social organisation of a carnivorous marsupial *Dasyurus hallucatus* (Marsupialia: Dasyuridae). *Journal of Zoology* **257**, 237–248. doi:10.1017/S0952836902000833

Oakwood M (2004) Death after sex. *Biologist (Columbus, Ohio)* **51**(1), 5–8.

Oakwood M (2008) Northern Quoll *Dasyurus hallucatus*. In *The Mammals of Australia.* 3rd edn. (Eds S Van Dyck and R Strahan) pp. 57–59. Reed New Holland, Sydney.

Oakwood M, Bradley AJ, Cockburn A (2001) Semelparity in a large marsupial. *Proceedings of the Royal Society of London Series B* **268**, 407–411. doi:10.1098/rspb.2000.1369

Oehser PH (1980) In memorium: Alexander Wetmore. *The Auk* **97**, 608–615.

Ojala-Barbour R, Pinto CM, Brito J, Albuja L, Lee TE, Patterson BD (2013) A new species of shrew-opossum (Paucituberculata: Caenolestidae) with a phylogeny of extant caenolestids. *Journal of Mammalogy* **94**, 967–982. doi:10.1644/13-MAMM-A-018.1

Ojeda RA, Mares MA (1982) Conservation of South American mammals: Argentina as a paradigm. In *Mammalian Biology in South America.* (Eds MA Mares and HH Genoways) pp. 505–521. Pymatuning Laboratory of Ecology, University of Pittsburgh, Linesville PA, USA.

Olsen P (2010) *Upside Down World: Early European Impressions of Australia's Curious Animals.* National Library of Australia, Canberra.

Osgood WH (1921) A monographic study of the American marsupial *Caenolestes*. PhD thesis. University of Chicago, Chicago IL, USA.

Osgood WH (1924) Review of living caenolestids with description of a new genus from Chile. *Field Museum of Natural History Zoological Series No. 222* **14**(2), 165–173.

Paddle RN (2000) *The Last Tasmanian Tiger.* Cambridge University Press, Cambridge, UK.

Painter J, Krajewski C, Westerman M (1995) A molecular phylogeny of planigales. *Journal of Mammalogy* **76**, 406–413. doi:10.2307/1382351

Palacios AG, Bozinovic F, Vielma A, Arrese CA, Hunt DM, Peichl L (2010) Retinal photoreceptor arrangement, SWS 1 and LWS opsin sequence, and electroretinography in the South American marsupial *Thylamys elegans* (Waterhouse, 1839). *The Journal of Comparative Neurology* **518**, 1589–1602. doi:10.1002/cne.22292

Palma RE, Valladares-Gomez A (2015) Family Microbiotheriidae (Monito del Monte). In *Handbook of the Mammals of the World. Volume 5. Monotremes and Marsupials.* (Eds DE Wilson and RA Mittermeier) pp. 200–209. Lynx Edicions: Barcelona, Spain.

Parrott ML, Ward SJ, Temple-Smith PD, Selwood L (2007) Effects of drought on weight, survival and breeding success of Agile Antechinus (*Antechinus agilis*), Dusky Antechinus (*A. swainsonii*) and Bush Rats (*Rattus fuscipes*). *Wildlife Research* **34**, 437–442. doi:10.1071/WR07071

Pastro LA (2013) The effects of wildfire on small mammals and lizards in the Simpson Desert, central Australia. PhD thesis. University of Sydney, Sydney.

Patterson BD (2008a) Genus *Rhyncholestes.* In *Mammals of South America. Volume 1 Marsupials, Xenarthrans, Shrews and Bats.* (Ed. AL Gardner) p. 126. University of Chicago Press, Chicago IL, USA.

Patterson BD (2008b) Order Paucituberculata. In *Mammals of South America. Volume 1. Marsupials, Xenarthrans, Shrews and Bats.* (Ed. AL Gardner). p. 119. University of Chicago Press, Chicago IL, USA.

Patterson BD (2015) Family Caenolestidae (Shrew-opossums). In *Handbook of the Mammals of the World. Volume 5. Monotremes and marsupials.* (Eds DE Wilson and RA Mittermeier) pp. 188–199. Lynx Edicions, Barcelona, Spain.

Patterson BD, Rogers MA (2008) Order Microbiotheria. In *Mammals of South America. Volume 1. Marsupials, Xenarthrans, Shrews and Bats.* (Ed. AL Gardner) p. 117. University of Chicago Press, Chicago IL, USA.

Patterson BD, Meserve PL, Lang BK (1990) Quantitative habitat associations of small mammals along an elevational transect in temperate rainforests of Chile. *Journal of Mammalogy* **71**, 620–633. doi:10.2307/1381803

Patton JL, da Silva MNF (2008) Genus *Philander.* In *Mammals of South America. Volume 1. Marsupials, Xenarthrans, Shrews and Bats.* (Ed. AL Gardner) pp. 27–34. University of Chicago Press, Chicago IL, USA.

Patton JL, Berlin B, Berlin EA (1982) Aboriginal perspectives of a mammal community in Amazonian Perú: knowledge and utilization patterns among the Aguaruna Jívaro. In *Mammalian Biology in South America.* (Eds MA Mares and HH Genoways) pp. 111–128. Pymatuning Laboratory of Ecology, University of Pittsburgh, Linesville PA, USA.

Paull DC (2013) Refuge sites, activity and torpor in wild Common Dunnarts (*Sminthopsis murina*) in a temperate heathland. *Australian Mammalogy* **35**, 153–159. doi:10.1071/AM12016

Pavan SE (2015) A new species of *Monodelphis* (Didelphimorphia: Didelphidae) from the Brazilian Atlantic Forest. *American Museum Novitates* **3832**, 1–15. doi:10.1206/3832.1

Pavan SE, Jansa SA, Voss RS (2014) Molecular phylogeny of short-tailed opossums (Didelphidae: *Monodelphis*): taxonomic implications and tests of evolutionary hypotheses. *Molecular Phylogenetics and Evolution* **79**, 199–214. doi:10.1016/j.ympev.2014.05.029

Pavan SE, Mendes-Oliveira AC, Voss RS (2017) A new species of *Monodelphis* (Didelphimorphia: Didelphidae) from the Brazilian Amazon. *American Museum Novitates* **3872**, 1–20. doi:10.1206/3872.1

Pavey C (2002) Long-tailed Dunnart: *Sminthopsis longicaudata.* Threatened Species of the Northern Territory.

Pavey CR, Geiser F (2008) Basking and diurnal foraging in the dasyurid marsupial *Pseudantechinus macdonnellensis. Australian Journal of Zoology* **56**, 129–135. doi:10.1071/ZO08032

Pavey CR, Goodship N, Geiser F (2003) Home range and spatial organisation of the rock-dwelling carnivorous marsupial, *Pseudantechinus macdonnellensis. Wildlife Research* **30**, 135–142. doi:10.1071/WR03005

Pavey CR, Burwell CJ, Körtner G, Geiser F (2009) Vertebrate diet decreases winter torpor use in a desert marsupial. *Naturwissenschaften* **96**, 679–683. doi:10.1007/s00114-009-0516-y

Pavey CR, Nano CEM, Cooper SJB, Cole JR, McDonald PJ (2011) Habitat use, population dynamics and species identification of mulgara, *Dasycercus blythi* and *D. cristicauda*, in a zone of sympatry in central Australia. *Australian Journal of Zoology* **59**, 156–169. doi:10.1071/ZO11052

Pavey CR, Burwell CJ, Benshemesh J (2012) Diet and prey selection of the Southern Marsupial Mole: an enigma from Australia's sand deserts. *Journal of Zoology* **287**, 115–123. doi:10.1111/j.1469-7998.2011.00889.x

Pavey CR, Burwell CJ, Körtner G, Geiser F (2016) Why is the marsupial Kaluta, *Dasykaluta rosamondae*, diurnally active in winter: foraging advantages or predator avoidance in arid northern Australia? *Journal of Arid Environments* **133**, 25–28. doi:10.1016/j.jaridenv.2016.05.006

Pearson DJ, Churchill SK (2008) Sandhill Dunnart *Sminthopsis psammophila*. In *The Mammals of Australia*. 3rd edn. (Eds S Van Dyck and R Strahan) pp. 156–158. Reed New Holland, Sydney.

Pearson DJ, McKenzie NL (2008) Hairy-footed Dunnart *Sminthopsis hirtipes*. In *The Mammals of Australia*. 3rd edn. (Eds S Van Dyck and R Strahan) pp. 143–145. Reed New Holland, Sydney.

Pearson OP (2008) Genus *Lestodelphys*. In *Mammals of South America. Volume 1. Marsupials, Xenarthrans, Shrews and Bats*. (Ed. A Gardner) p. 50. University of Chicago Press, Chicago IL, USA.

Pearson OP, Pearson AK (1982) Ecology and biogeography of the southern rainforests of Argentina. In *Mammalian Biology in South America*. (Eds MA Mares and HH Genoways) pp. 129–142. Pymatuning Laboratory of Ecology, University of Pittsburgh, Linesville PA, USA.

Pearson DJ, Robinson AC (1990) New records of the Sandhill Dunnart, *Sminthopsis psammophila* (Marsupialia: Dasyuridae) in South and Western Australia. *Australian Mammalogy* **13**, 57–59.

Peck DT (1993) *Cristoforo Colombo, God's Navigator*. Columbian Publishers, Columbus WI, USA.

Pellis SM, Officer RCE (1987) An analysis of some predatory behaviour patterns in four species of carnivorous marsupials (Dasyuridae), with comparative notes on the eutherian cat *Felis catus*. *Ethology* **75**, 177–196. doi:10.1111/j.1439-0310.1987.tb00652.x

Pellis SM, Vanderlely R, Nelson JE (1992) The roles of vision and vibrissae in the predatory behaviour of Northern Quolls *Dasyurus hallucatus* (Marsupialia: Dasyuridae). *Australian Mammalogy* **15**, 55–60.

Pemberton D, Renouf D (1993) A field study of communication and social behaviour of the Tasmanian Devil at feeding sites. *Australian Journal of Zoology* **41**, 507–526. doi:10.1071/ZO9930507

Pethers H (2012) Oldfield Thomas and his mammalian legacy. Natural History Museum Library and Archives Blog Post, http://www.nhm.ac.uk/natureplus/community/library/blog/2012/11/27/oldfield-thomas-and-his-nhm-legacy

Pianka ER (1986) *Ecology and Natural History of Desert Lizards*. Princeton University Press, Princeton NJ, USA.

Pine RH, Handley CO, Jr (2008) Genus *Monodelphis*. In *Mammals of South America. Volume 1. Marsupials, Xenarthrans, Shrews and Bats*. (Ed. AL Gardner) pp. 82–106. University of Chicago Press, Chicago IL, USA.

Pine RH, Dalby PL, Matson JO (1985) Ecology, postnatal development, morphometrics, and taxonomic status of the Short-tailed Opossum, *Monodelphis dimidiata*, an apparently semelparous annual marsupial. *Annals of the Carnegie Museum* **54**, 195–231.

Pine RH, Flores DA, Bauer K (2013) The second known specimen of *Monodelphis unistriata* (Wagner) (Mammalia: Didelphimorphia), with redescription of the species and phylogenetic analysis. *Zootaxa* **3640**, 425–441. doi:10.11646/zootaxa.3640.3.6

Pinotti BT, Naxara L, Pardini R (2011) Diet and food selection by small mammals in an old-growth Atlantic Forest of south-eastern Brazil. *Studies on Neotropical Fauna and Environment* **46**, 1–9. doi:10.1080/01650521.2010.535250

Pocock RI (1930) Obituary notice of Michael Rogers Oldfield Thomas. (With portrait). *Proceedings of the Royal Society of London (B)* **106**, i–v, 1 pl.

Poran NS (1998) Vomeronasal organ and its associated structures in the opossum *Monodelphis domestica*. *Microscopy Research and Technique* **43**, 500–510. doi:10.1002/(SICI)1097-0029(19981215)43:6<500::AID-JEMT3>3.0.CO;2-H

Possingham H, Jarman P, Kearns A (2004) Independent review of Western Shield – February 2003. *Conservation Science Western Australia* **5**, 2–18.

Potter LC (2017) Camera trap constraints in focus: assessing detectability and identification of small mammals in camera trap studies. BSc (Hons) thesis, Charles Darwin University, Darwin.

Potter TI (2017) Who killed the wolf spider? A who-dunnart: intraguild predation among taxonomically disparate micro-carnivores. BSc (Hons) thesis. University of Sydney, Sydney.

Prado JL, Martinez-Maza C, Alberdi MT (2015) Megafauna extinction in South America: A new chronology for the Argentine Pampas. *Palaeogeography, Palaeoclimatology, Palaeoecology* **425**, 41–49. doi:10.1016/j.palaeo.2015.02.026

Predavec M (1990) Predation and prey community structure: a study of the impact of *Antechinus stuartii* (Marsupialia: Dasyuridae) on leaf-litter invertebrate communities. BSc (Hons) thesis. University of Sydney, Sydney.

Prevosti FJ, Pereira JA (2014) Community structure of South American carnivores in the past and present. *Journal of Mammalian Evolution* **21**, 363–368. doi:10.1007/s10914-014-9273-6

Prins HHT, Weyerhaeuser FJ (1987) Epidemics in populations of wild ruminants: anthrax and impala, rinderpest and buffalo in Lake Manyara National Park, Tanzania. *Oikos* **49**, 28–38. doi:10.2307/3565551

Queensland Herbarium (2013) *Regional Ecosystem Description Database (REDD). Version 6.1.* Queensland Department of Science, Information Technology, Innovation and the Arts, Brisbane.

Queensland Museum Archives. Correspondence, manuscripts and related material concerning vertebrate zoology, 1887–1929. (1887/2267; 1888/2403; 1892/3684; 16/70; 19/329; 22/203; 23/159; 23/139A; 23/388; 23/461; 23/416; 24/191; 24/170; 24/215; 24/410; 24/632; 25/276; 26/10; 26/11; 26/210; 27/570).

Rayner K, Chambers B, Johnson B, Morris KD, Mills HR (2012) Spatial and dietary requirements of the Chuditch (*Dasyurus geoffroii*) in a semiarid climatic zone. *Australian Mammalogy* **34**, 59–67. doi:10.1071/AM10045

Read DG (1984a) Movements and home ranges of three sympatric dasyurids, *Sminthopsis crassicaudata, Planigale gilesi* and *P. tenuirostris* (Marsupialia), in semiarid western New South Wales. *Australian Wildlife Research* **11**, 223–234. doi:10.1071/WR9840223

Read DG (1984b) Reproduction and breeding season of *Planigale gilesi* and *P. tenuirostris* (Marsupialia: Dasyuridae). *Australian Mammalogy* **7**, 161–173.

Read DG (1985) Development and growth of *Planigale tenuirostris* (Marsupialia: Dasyuridae) in the laboratory. *Australian Mammalogy* **8**, 69–78.

Read DG (1987a) A comparison of growth rates in dependent juveniles of *Planigale gilesi* and *P. tenuirostris* (Marsupialia : Dasyuridae). *Australian Journal of Zoology* **35**, 161–171. doi:10.1071/ZO9870161

Read DG (1987b) Diets of sympatric *Planigale gilesi* and *P. tenuirostris* (Marsupialia: Dasyuridae): relationships of season and body size. *Australian Mammalogy* **10**, 11–21.

Read DG (1987c) Habitat use by *Sminthopsis crassicaudata, Planigale gilesi* and *P. tenuirostris* (Marsupialia: Dasyuridae) in semiarid New South Wales. *Australian Wildlife Research* **14**, 385–395. doi:10.1071/WR9870385

Read DG (1988) Weather and trap response of the dasyurid marsupials *Sminthopsis crassicaudata, Planigale gilesi* and *P. tenuirostris. Australian Wildlife Research* **15**, 139–148. doi:10.1071/WR9880139

Read DG (1989) Microhabitat separation and diel activity patterns of *Planigale gilesi* and *P. tenuirostris* (Marsupialia: Dasyuridae). *Australian Mammalogy* **12**, 45–53.

Read DG (2008a) Giles' Planigale *Planigale gilesi*. In *The Mammals of Australia.* 3rd edn. (Eds S Van Dyck and R Strahan) pp. 107–109. Reed New Holland, Sydney.

Read DG (2008b) Narrow-nosed Planigale *Planigale tenuirostris*. In *The Mammals of Australia.* 3rd edn. (Eds S Van Dyck and R Strahan) pp. 114–116. Reed New Holland, Sydney.

Redford KH, Eisenberg JF (1992) *Mammals of the Neotropics. Volume 2: The Southern Cone: Chile, Argentina, Uruguay, Paraguay.* University of Chicago Press, Chicago IL, USA.

Reguero MA, Gelfo JN, Lopez GM, Bond M, Abello A, Santillana SN, Marenssi SA (2014) Final Gondwana breakup: the Paleogene South

American native ungulates and the demise of the South America–Antarctica land connection. *Global and Planetary Change* **123**, 400–413. doi:10.1016/j.gloplacha.2014.07.016

Reid FA (1997) *A Field Guide to the Mammals of Central America and Southeast Mexico*. Oxford University Press, New York, USA.

Renfree MB (1983) Marsupial reproduction: the choice between placentation and lactation. *Oxford Reviews of Reproductive Biology* **5**, 1–29.

Reynolds HC (1952) Studies on reproduction in the opossum (*Didelphis virginiana virginiana*). *University of California Publications in Zoology* **52**, 223–284.

Rhind SG (1996) Habitat tree requirements and the effects of removal during logging on the marsupial Brush-tailed Phascogale (*Phascogale tapoatafa tapoatafa*) in Western Australia. *Western Australian Naturalist (Perth)* **21**, 1–22.

Rhind SG (2002) Reproductive demographics among Brush-tailed Phascogales (*Phascogale tapoatafa*) in south-western Australia. *Wildlife Research* **29**, 247–257. doi:10.1071/WR01013

Rhind SG (2003) Communal nesting in the usually solitary marsupial, *Phascogale tapoatafa*. *Journal of Zoology* **261**, 345–351. doi:10.1017/S0952836903004308

Rhind SG (2004) Direct impacts of logging and forest management on the Brush-tailed Phascogale *Phascogale tapoatafa* and other arboreal marsupials in jarrah forest of Western Australia. In *Conservation of Australia's Forest Fauna*. 2nd edn. (Ed. D Lunney) pp. 639–655. Royal Zoological Society of New South Wales, Sydney.

Rhind SG, Bradley JS (2002) The effect of drought on body size, growth and abundance of wild Brush-tailed Phascogales (*Phascogale tapoatafa*) in south-western Australia. *Wildlife Research* **29**, 235–245. doi:10.1071/WR01014

Rhind SG, Bradley JS, Cooper NK (2001) Morphometric variation and taxonomic status of Brush-tailed Phascogales, *Phascogale tapoatafa* (Meyer, 1793) (Marsupialia: Dasyuridae). *Australian Journal of Zoology* **49**, 345–368. doi:10.1071/ZO01002

Rhind S, Woinarski J, Aplin K (2008) Northern Brush-tailed Phascogale, *Phascogale pirata*. In *The Mammals of Australia*. 3rd edn. (Eds S Van Dyck and R Strahan) pp. 103–104. Reed New Holland, Sydney.

Richardson KC, Bowden TAJ, Myers P (1987) The cardiogastric gland and alimentary tract of caenolestid marsupials. *Acta Zoologica* **68**, 65–70. doi:10.1111/j.1463-6395.1987.tb00877.x

Ride WDL (1964) *Antechinus rosamondae*, a new species of dasyurid marsupial from the Pilbara District of Western Australia, with remarks on the classification of *Antechinus*. *Western Australian Naturalist (Perth)* **9**, 58–65.

Ride WDL (1970) *A Guide to the Native Mammals of Australia*. Oxford University Press, Melbourne.

Riek A, Geiser F (2014) Heterothermy in pouched mammals – a review. *Journal of Zoology* **292**, 74–85. doi:10.1111/jzo.12102

Righetti J, Fox BJ, Croft DB (2000) Behavioural mechanisms of competition in small dasyurid marsupials. *Australian Journal of Zoology* **48**, 561–576. doi:10.1071/ZO00023

Risbey DA, Calver MC, Short J, Bradley JS, Wright IW (2000) The impact of cats and foxes on the small vertebrate fauna of Heirisson Prong, Western Australia. II. A field experiment. *Wildlife Research* **27**, 223–235. doi:10.1071/WR98092

Ritchie EG, Elmhagen B, Glen AS, Letnic M, Ludwig G, McDonald RA (2012) Ecosystem restoration with teeth: what role for predators? *Trends in Ecology & Evolution* **27**, 265–271. doi:10.1016/j.tree.2012.01.001

Robinson AC (2008) Kangaroo Island Dunnart *Sminthopsis aitkeni*. In *The Mammals of Australia*. 3rd edn. (Eds S Van Dyck and R Strahan) pp. 124–126. Reed New Holland, Sydney.

Robinson AC, Armstrong DM (Eds) (1999) *A Biological Survey of Kangaroo Island, South Australia, 1989 & 1990*. Heritage and Biodiversity Section, Department for Environment, Heritage and Aboriginal Affairs, South Australia.

Robinson AC, Copley PB, Canty PD, Baker LM, Nesbitt BJ (2003) *A Biological Survey of the Anangu Pitjantjatjara Lands, South Australia*. South Australian Department for Environment and Heritage, Adelaide.

Robinson T, Gaikhorst G, Pearson D, Copley P (2008) *Sminthopsis psammophila*. In *IUCN Red List of*

Threatened Species. Version 2013.2. IUCN, Gland, Switzerland.

Rodriguez-Cabal MA, Branch LC (2011) Influence of habitat factors on the distribution and abundance of a marsupial seed disperser. *Journal of Mammalogy* **92**, 1245–1252. doi:10.1644/10-MAMM-A-315.1

Rojas AD, Körtner G, Geiser F (2014) Torpor in free-ranging antechinus: does it increase fitness? *Naturwissenschaften* **101**, 105–114. doi:10.1007/s00114-013-1136-0

Rothschild L, Dollman CG (1933) 10. On mammals collected in Dutch New Guinea by Mr. F. Shaw Mayer in 1930. *Proceedings of the Zoological Society of London* **103**, 211–219. doi:10.1111/j.1096-3642.1933.tb01587.x

Rougier GW, Wible JR, Beck RMD, Apesteguia S (2012) The Miocene mammal *Necrolestes* demonstrates the survival of a mesozoic nontherian lineage into the late Cenozoic of South America. *Proceedings of the National Academy of Sciences of the United States of America* **109**(49), 20053–20058. doi:10.1073/pnas.1212997109

Ruf T, Geiser F (2015) Daily torpor and hibernation in birds and mammals. *Biological Reviews of the Cambridge Philosophical Society* **90**, 891–926. doi:10.1111/brv.12137

Russell EM (1974) Recent ecological studies on Australian marsupials. *Australian Mammalogy* **1**, 189–211.

Russell EM (1982) Patterns of parental care and parental investment in marsupials. *Biological Reviews of the Cambridge Philosophical Society* **57**, 423–486. doi:10.1111/j.1469-185X.1982.tb00704.x

Russell BG, Banks PB (2007) Do Australian small mammals respond to native and introduced predator odours? *Austral Ecology* **32**, 277–286. doi:10.1111/j.1442-9993.2007.01685.x

Salazar DA, Fontúrbel FE (2016) Beyond habitat structure: landscape heterogeneity explains the Monito del Monte (*Dromiciops gliroides*) occurrence and behavior at habitats dominated by exotic trees. *Integrative Zoology* **11**, 413–421. doi:10.1111/1749-4877.12198

Sale MG, Ward SJ, Arnould JPY (2006) Aspects of the ecology of Swamp Antechinus (*Antechinus minimus maritimus*) on a Bass Strait island. *Wildlife Research* **33**, 215–221. doi:10.1071/WR05051

Sale MG, Arnould JPY (2012) Inflated population density of island antechinus: a case of allochthonous marine inputs leading to increased food availability? *Australian Journal of Zoology* **60**, 343–351. doi:10.1071/ZO12073

Salo P, Korpimäki E, Banks PB, Nordström M, Dickman CR (2007) Alien predators are more dangerous than native predators to prey populations. *Proceedings. Biological Sciences* **274**, 1237–1243. doi:10.1098/rspb.2006.0444

Sanchez-Villagra MR (2013) Why are there fewer marsupials than placentals? On the relevance of geography and physiology to evolutionary patterns of mammalian diversity and disparity. *Journal of Mammalian Evolution* **20**, 279–290. doi:10.1007/s10914-012-9220-3

Sanecki GM, Green K, Wood H, Lindenmayer D (2006a) The implications of snow-based recreation for small mammals in the subnivean space in south-east Australia. *Biological Conservation* **129**, 511–518. doi:10.1016/j.biocon.2005.11.018

Sanecki GM, Green K, Wood H, Lindenmayer D, Sanecki KL (2006b) The influence of snow cover on home range and activity of the Bush-rat (*Rattus fuscipes*) and the Dusky Antechinus (*Antechinus swainsonii*). *Wildlife Research* **33**, 489–496. doi:10.1071/WR05012

Sanson GD (1985) Functional dental morphology and diet selection in dasyurids (Marsupialia: Dasyuridae). *Australian Mammalogy* **8**, 239–247.

Santori RT, Astúa de Moraes D, Cerqueira R (1995) Diet composition of *Metachirus nudicaudatus* and *Didelphis aurita* (Marsupialia, Didelphoidea) in southeastern Brazil. *Mammalia* **59**, 511–516. doi:10.1515/mamm.1995.59.4.511

Santori RT, Astúa de Moraes D, Grelle CEV, Cerqueira R (1997) Natural diet at a restinga forest and laboratory food preferences of the opossum *Philander frenata* in Brazil. *Studies on Neotropical Fauna and Environment* **32**, 12–16. doi:10.1076/snfe.32.1.12.13460

Santori RT, Astúa de Moraes D, Cerqueira R (2004) Comparative gross morphology of the digestive tract in ten Didelphidae marsupial species. *Mammalia* **68**, 27–36. doi:10.1515/mamm.2004.004

Sato C (2006) Life on the edge: risk-sensitive foraging behaviour in Australian desert dunnarts. BSc (Hons) thesis. University of Sydney, Sydney.

Saunders GR, Gentle MN, Dickman CR (2010) The impacts and management of foxes *Vulpes vulpes* in Australia. *Mammal Review* **40**, 181–211. doi:10.1111/j.1365-2907.2010.00159.x

Scarff FR, Bradley JS (2006) Invertebrate prey of the bark-foraging insectivore *Phascogale tapoatafa*: distribution of biomass amongst alternative foraging substrates within south-western Australian woodlands. *Australian Journal of Zoology* **54**, 335–341. doi:10.1071/ZO05051

Scarff FR, Rhind SG, Bradley JS (1998) Diet and foraging behaviour of Brush-tailed Phascogales (*Phascogale tapoatafa*) in the jarrah forest of south-western Australia. *Wildlife Research* **25**, 511–526. doi:10.1071/WR98007

Schlegel H (1866) *Phascogale longicaudata. Nederlande Tijdchrift your de Dierkunde* **3**, 356–357.

Schmitt LH, Bradley AJ, Kemper CM, Kitchener DJ, Humphreys WF, How RA (1989) Ecology and physiology of the Northern Quoll, *Dasyurus hallucatus* (Marsupialia, Dasyuridae), at Mitchell Plateau, Kimberley, Western Australia. *Journal of Zoology* **217**, 539–558. doi:10.1111/j.1469-7998.1989.tb02510.x

Schwimmer H, Haim A (2009) Physiological adaptations of small mammals to desert ecosystems. *Integrative Zoology* **4**, 357–366. doi:10.1111/j.1749-4877.2009.00176.x

Selwood L (1982) A review of maturation and fertilization in marsupials with special reference to the dasyurid: *Antechinus stuartii*. In *Carnivorous Marsupials.* (Ed. M Archer) pp. 65–76. Royal Zoological Society of New South Wales, Sydney.

Selwood L (1983) Factors influencing pre-natal fertility in the Brown Marsupial Mouse, *Antechinus stuartii. Journal of Reproduction and Fertility* **68**, 317–324. doi:10.1530/jrf.0.0680317

Selwood L, McCallum F (1987) Relationship between longevity of spermatozoa after insemination and the precentage of normal embryos in Brown Marsupial Mice (*Antechinus stuartii*). *Journal of Reproduction and Fertility* **79**, 495–503. doi:10.1530/jrf.0.0790495

Serena M, Soderquist TR (1988) Growth and development of pouch young of wild and captive *Dasyurus geoffroii. Australian Journal of Zoology* **36**, 533–543. doi:10.1071/ZO9880533

Serena M, Soderquist TR (1989) Spatial organization of a riparian population of the carnivorous marsupial *Dasyurus geoffroii. Journal of Zoology* **219**, 373–383. doi:10.1111/j.1469-7998.1989.tb02586.x

Serena M, Soderquist TR (2008) Western Quoll, *Dasyurus geoffroii*. In *The Mammals of Australia.* 3rd edn. (Eds S Van Dyck and R Strahan) pp. 54–56. Reed New Holland, Sydney.

Shaw G (2006) Reproduction. In *Marsupials.* (Eds PJ Armati, CR Dickman and ID Hume) pp. 83–107. Cambridge University Press, Cambridge, UK.

Short J (2004) Mammal decline in southern Western Australia – perspectives from Shortridge's collections of mammals in 1904-07. *Australian Zoologist* **32**(4), 605–628. doi:10.7882/AZ.2004.006

Short J, Hide A (2012) Distribution and status of the Red-tailed Phascogale (*Phascogale calura*). *Australian Mammalogy* **34**, 88–99. doi:10.1071/AM11017

Short J, Hide A, Stone M (2011) Habitat requirements of the endangered Red-tailed Phascogale, *Phascogale calura. Wildlife Research* **38**, 359–369. doi:10.1071/WR10220

Shortridge GC (1909) An account of the geographical distribution of macropods of south-west Australia, having special reference to the specimens collected during the Balston expedition of 1904-1907. *Proceedings of the Zoological Society of London* **1909**, 803–848.

Shortridge GC (1936) Field notes (hitherto unpublished) on Western Australian mammals south of the tropic of Capricorn (exclusive of Marsupialia and Monotremata), and records of specimens collected during the Balston expeditions (November 1904 to June 1907). *Proceedings of the Zoological Society of London* **1936**, 743–749.

Sillitoe P (2013a) *Managing Animals in New Guinea: Preying the Game in the Highlands.* Routledge, Melbourne.

Sillitoe P (2013b) *A Place Against Time: Land and Environment in the Papua New Guinea Highlands.* Routledge, Abingdon, UK.

Sinclair RG, Bird PL (1984) The reaction of *Sminthopsis crassicaudata* to meat baits containing 1080: implications for assessing risk to non-target species. *Australian Wildlife Research* **11**, 501–507. doi:10.1071/WR9840501

Smith CS, Howes AL, Price B, McAlpine CA (2007) Using a Bayesian belief network to predict suitable habitat of an endangered mammal – the Julia Creek Dunnart (*Sminthopsis douglasi*). *Biological Conservation* **139**, 333–347. doi:10.1016/j.biocon.2007.06.025

Smith GC (1984) The biology of the Yellow-footed Antechinus, *Antechinus flavipes* (Marsupialia: Dasyuridae), in a swamp forest of Kinaba Island, Cooloola, Queensland. *Australian Wildlife Research* **11**, 465–480. doi:10.1071/WR9840465

Smith GC, Means K, Churchill S (2017) Aspects of the ecology of the Atherton Antechinus (*Antechinus godmani*) living in sympatry with the Rusty Antechinus (*A. adustus*) in the Wet Tropics, Queensland – a trapping and radio-tracking study. *Australian Mammalogy*, **40**, 16–25. doi:10.1071/AM16050

Smithsonian Institution. Porto Rico, 1911–1912: Field diaries, 13 December 1911–15 September 1912, Four volumes. SIA RU007006 Alexander Wetmore Papers, circa 1848–1979 and undated.

Smithsonian Institution. Correspondence, manuscripts and related material concerning general mammalogy, 1908–1924, 1927–1930. 1933. Archives: SIA RU007173 Gerrit Smith Miller, Jr Papers 1908–1940.

Soderquist TR (1993a) Maternal strategies of *Phascogale tapoatafa* (Marsupialia: Dasyuridae). I. Breeding seasonality and maternal investment. *Australian Journal of Zoology* **41**, 549–566. doi:10.1071/ZO9930549

Soderquist TR (1993b) Maternal strategies of *Phascogale tapoatafa* (Marsupialia: Dasyuridae). II. Juvenile thermoregulation and maternal attendance. *Australian Journal of Zoology* **41**, 567–576. doi:10.1071/ZO9930567

Soderquist TR (1994) Anti-predator behaviour of the Brush-tailed Phascogale (*Phascogale tapoatafa*). *Victorian Naturalist* **111**, 22–24.

Soderquist TR (1995) Spatial organization of the arboreal carnivorous marsupial *Phascogale tapoatafa*. *Journal of Zoology* **237**, 385–398. doi:10.1111/j.1469-7998.1995.tb02769.x

Soderquist TR, Ealey L (1994) Social interactions and mating strategies of a solitary carnivorous marsupial, *Phascogale tapoatafa*, in the wild. *Wildlife Research* **21**, 527–542. doi:10.1071/WR9940527

Soderquist T, Lill A (1995) Natal dispersal and philopatry in the carnivorous marsupial *Phascogale tapoatafa* (Dasyuridae). *Ethology* **99**, 297–312. doi:10.1111/j.1439-0310.1995.tb00904.x

Soderquist T, Rhind S (2008) Brush-tailed Phascogale, *Phascogale tapoatafa*. In *The Mammals of Australia.* 3rd edn. (Eds S Van Dyck and R Strahan) pp. 105–107. Reed New Holland, Sydney.

Soderquist TR, Serena M (1994) Dietary niche of the Western Quoll, *Dasyurus geoffroii*, in the jarrah forest of Western Australia. *Australian Mammalogy* **17**, 133–136.

Soderquist TR, Serena M (2000) Juvenile behaviour and dispersal of Chuditch (*Dasyurus geoffroii*) (Marsupialia:Dasyuridae). *Australian Journal of Zoology* **48**, 551–560. doi:10.1071/ZO00029

Solari S (2003) Diversity and distribution of *Thylamys* (Didelphidae) in South America, with emphasis on species from the western side of the Andes. In *Predators with Pouches: the Biology of Carnivorous Marsupials.* (Eds M Jones, C Dickman and M Archer) pp. 82–101. CSIRO Publishing, Melbourne.

Solari S (2004) A new species of *Monodelphis* (Didelphimorphia: Didelphidae) from southeastern Peru. *Mammalian Biology* **69**, 145–152. doi:10.1078/1616-5047-00129

Solari S (2007) New species of *Monodelphis* (Didelphimorphia: Didelphidae) from Peru, with notes on *M. adusta* (Thomas, 1897). *Journal of Mammalogy* **88**, 319–329. doi:10.1644/06-MAMM-A-075R1.1

Southgate RI (2015) Family Thylacomyidae (Greater Bilby). In *Handbook of the Mammals of the World. Volume 5. Monotremes and Marsupials.* (Eds DE Wilson and RA Mittermeier) pp. 350–361. Lynx Edicions, Barcelona, Spain.

Southwood TRE (1977) Habitat, the templet for ecological strategies? *Journal of Animal Ecology* **46**, 337–365. doi:10.2307/3817

Spencer B (1885a) Preliminary notice of two new species of marsupials from central Australia. *Proceedings of the Royal Society of Victoria* **7**, 222–224.

Spencer B (1885b) The Horn Expedition to central Australia. *Nature* **51**, 221–222.

Spencer B (1894) Horn Expedition to central Australia 1894 diary. Notebook. Museum Victoria Collection, Melbourne.

Spencer B (1896) Mammalia. In *Report on the Work of the Horn Scientific Expedition to Central Australia. Part 2. Zoology*. Pp. 1–52. Melville, Mullen and Slade, Melbourne.

Spencer EE, Crowther MS, Dickman CR (2014) Diet and prey selectivity of three species of sympatric mammalian predators in central Australia. *Journal of Mammalogy* **95**, 1278–1288. doi:10.1644/13-MAMM-A-300

Spencer PBS, Rhind SG, Eldridge MDB (2001) Phylogeographic structure within *Phascogale* (Marsupialia: Dasyuridae) based on partial cytochrome *b* sequence. *Australian Journal of Zoology* **49**, 369–377. doi:10.1071/ZO00080

Stannard HJ, Old JM (2010) Observation of reproductive strategies of captive Kultarrs (*Antechinomys laniger*). *Australian Mammalogy* **32**, 179–182. doi:10.1071/AM10011

Stannard HJ, Old JM (2011a) Digestibility of feeding regimes of the Red-tailed Phascogale (*Phascogale calura*) and the Kultarr (*Antechinomys laniger*) in captivity. *Australian Journal of Zoology* **59**, 257–263. doi:10.1071/ZO11069

Stannard HJ, Old JM (2011b) Rate of passage through the digestive tract of the Kultarr (*Antechinomys laniger*). *Australian Journal of Zoology* **59**, 273–276. doi:10.1071/ZO11103

Stannard HJ, Caton W, Old JM (2010) The diet of Red-tailed Phascogales in a trial translocation at Alice Springs Desert Park, Northern Territory, Australia. *Journal of Zoology* **280**, 326–331. doi:10.1111/j.1469-7998.2009.00658.x

Stannard HJ, Borthwick CR, Ong O, Old JM (2013) Longevity and breeding in captive Red-tailed Phascogales (*Phascogale calura*). *Australian Mammalogy* **35**, 217–219. doi:10.1071/AM12042

Start AN, Moro D, Adams M, Bencini R (2006) Dunnarts from Boullanger Island: new evidence and reassessment of a taxonomic issue with resource implications. *Australian Mammalogy* **28**, 51–58. doi:10.1071/AM06006

Start AN, Burbidge AA, Kendrick PG, McKenzie NL (2013) Terrestrial mammals of the south-western Little Sandy Desert, Western Australia. *Australian Mammalogy* **35**, 54–64. doi:10.1071/AM12010

Statham HL (1982) *Antechinus stuartii* (Dasyuridae, Marsupialia) diet and food availability at Petroi, northeastern New South Wales. In *Carnivorous Marsupials*. (Ed. M Archer) pp. 151–163. Royal Zoological Society of New South Wales, Sydney.

Stawski C, Körtner G, Nowack J, Geiser F (2015) The importance of mammalian torpor for survival in a post-fire landscape. *Biology Letters* **11**, 20150134. doi:10.1098/rsbl.2015.0134

Stearns SC (1977) The evolution of life history traits: a critique of the theory and a review of the data. *Annual Review of Ecology and Systematics* **8**, 145–171. doi:10.1146/annurev.es.08.110177.001045

Stein BR, Patton JL (2008a) Genus *Chironectes*. In *Mammals of South America. Volume 1. Marsupials, Xenarthrans, Shrews and Bats*. (Ed. AL Gardner) pp. 14–16. University of Chicago Press, Chicago IL, USA.

Stein BR, Patton JL (2008b) Genus *Lutreolina*. In *Mammals of South America. Volume 1. Marsupials, Xenarthrans, Shrews and Bats*. (Ed. AL Gardner) pp. 25–26. University of Chicago Press, Chicago IL, USA.

Stein GHW (1932) Einige neue beuteltiere aus Neuguinea. *Zeitschrift für Saugetierkunde* **7**, 256–257.

Stephens DW, Brown JS, Ydenberg R (2007) *Foraging: Behavior and Ecology*. Chicago University Press, Chicago IL, USA.

Stokes VL, Pech RP, Banks PB, Arthur AD (2004) Foraging behaviour and habitat use by *Antechinus flavipes* and *Sminthopsis murina* (Marsupialia: Dasyuridae) in response to predation risk in eucalypt woodland. *Biological Conservation* **117**, 331–342. doi:10.1016/j.biocon.2003.12.012

Story R, Williams MAJ, Hooper AOL, O'Ferrall RE, McAlpine JR (1969) *Lands of the Adelaide-Alligator area, Northern Territory. Land Research Series. Volume 25*. CSIRO, Canberra.

Story R, Galloway RW, McAlpine JR, Aldrick JM, Williams MAJ (1976) *Lands of the Alligator Rivers area, Northern Territory. Land Research Series. Volume 38*. CSIRO Publishing, Canberra.

Strahan R, Conder P (2007) *Dictionary of Australian and New Guinean mammals.* CSIRO Publishing, Melbourne.

Streilein KE (1982) Behavior, ecology, and distribution of South American marsupials. In *Mammalian Biology in South America.* (Eds MA Mares and HH Genoways) pp. 231–250. Pymatuning Laboratory of Ecology, University of Pittsburgh, Linesville PA, USA.

Sumner J, Dickman CR (1998) Distribution and identity of species in the *Antechinus stuartii – A. flavipes* group (Marsupialia: Dasyuridae) in south-eastern Australia. *Australian Journal of Zoology* **46**, 27–41. doi:10.1071/ZO94055

Sutherland DR, Predavec M (1999) The effects of moonlight on microhabitat use by *Antechinus agilis* (Marsupialia: Dasyuridae). *Australian Journal of Zoology* **47**, 1–17. doi:10.1071/ZO98037

Swan G, Wilson S (2012) The results of fauna recovery from a gas pipeline trench, and a comparison with previously published reports. *Australian Zoologist* **36**(2), 129–136. doi:10.7882/AZ.2012.028

Szalay FS (1994) *Evolutionary History of the Marsupials and an Analysis of Osteological Characters.* Cambridge University Press, New York, USA.

Taggart DA, Shimmin GA, Dickman CR, Breed WG (2003) Reproductive biology of carnivorous marsupials: clues to the likelihood of sperm competition. In *Predators with Pouches: the Biology of Carnivorous Marsupials.* (Eds M Jones, C Dickman and M Archer) pp. 358–375. CSIRO Publishing, Melbourne.

Taplin ME (1980) Some observations on the reproductive biology of *Sminthopsis virginiae* (Tarragon) (Marsupialia: Dasyuridae). *Australian Zoologist* **20**, 407–418.

Tasker EM, Dickman CR (2002) A review of Elliott trapping methods for small mammals in Australia. *Australian Mammalogy* **23**, 77–87. doi:10.1071/AM01077

Tate GHH (1938) New or little-known marsupials: a new species of Phascogalinae, with notes upon *Acrobates pulchellus* Rothschild. *Novitates Zoologicae* **41**, 58–60.

Tate GHH (1940) Notes on the types of certain early described species of monotremes, marsupials, muridae and bats from the Indo-Australian region. *American Museum Novitates* **1061**, 1–10.

Tate GHH (1947) On the anatomy and classification of the Dasyuridae (Marsupialia). Results of the Archbold Expeditions. No. 56. *Bulletin of the American Museum of Natural History* **88**, 97–156.

Tate GHH (1952) Results of the Archbold Expeditions. No. 66. Mammals of Cape York Peninsula. *Bulletin of the American Museum of Natural History* **98**, 567–616.

Tate GHH, Archbold R (1936) Results of the Archbold Expeditions. No. 8. Four apparently new polyprotodont marsupials from New Guinea. *American Museum Novitates* **823**, 1–4.

Tate GHH, Archbold R (1937) Results of the Archbold Expeditions. No. 16. Some marsupials of New Guinea and Celebes. *Bulletin of the American Museum of Natural History* **73**, 331–476.

Tate GHH, Archbold R (1941) Results of the Archbold Expeditions. No. 31. New rodents and marsupials from New Guinea. *American Museum Novitates* **1101**, 1–9.

Taylor JM (1989) Collection and preservation of mammals. In *Fauna of Australia. Volume 1B. Mammalia.* (Eds DW Walton and BJ Richardson) pp. 1–15. Australian Government Publishing Service, Canberra.

Taylor JM, Horner BE (1970) Gonadal activity in the marsupial mouse *Antechinus bellus*, with notes on other species of the genus (Marsupialia: Dasyuridae). *Journal of Mammalogy* **51**, 659–668. doi:10.2307/1378292

Taylor JM, Calaby JH, Redhead TD (1982) Breeding in wild populations of the marsupial-mouse *Planigale maculata sinualis* (Dasyuridae, Marsupialia). In *Carnivorous Marsupials.* (Ed. M Archer) pp. 83–87. Royal Zoological Society of New South Wales, Sydney.

Taylor R, Woinarski J, Charlie A, Dixon R, Pracy D, Rhind S (2004) 'Report on mammal survey of the Pellew Islands'. Northern Territory Department of Infrastructure, Planning and Environment, Darwin.

Teta P, de la Sancha N (2008) *Chacodelphys formosa.* In *IUCN Red List of Threatened Species. Version 2016.2.* IUCN, Gland, Switzerland.

Teta P, D'elia G, Fores D, de la Sancha N (2009) Diversity and distribution of the mouse opossums of the genus *Thylamys* (Didelphimorphia, Didelphidae) in northeastern and central Argentina. *Gayana (Concepción)* **73**, 180–199.

Thomas O (1888) *Catalogue of the Marsupialia and Monotremata in the Collection of the British Museum (Natural History)*. British Museum (Natural History), London, UK.

Thomas O (1904a) On a collection of mammals made by Mr. J.T. Tunney in Arnhem Land, Northern Territory of South Australia. *Novitates Zoologicae* **11**, 222–229. doi:10.5962/bhl.part.26833

Thomas O (1904b) Exhibition of specimens and descriptions of new species of mammals from New Guinea. *Annals & Magazine of Natural History* **14**, 402.

Thomas O (1906) On mammals from northern Australia presented to the National Museum by Sir Wm. Ingram, Bt., and the Hon. John Forrest. *Proceedings of the Zoological Society of London* **37**, 536–543.

Thomas O (1906) On mammals from northern Australia presented to the National Museum by Sir Wm. Ingram, Bt., and the Hon. John Forrest. *Proceedings of the Zoological Society of London* **37**, 536–543.

Thomas O (1906c) List of further collections of mammals from Western Australia, including a series from Bernier Island, obtained for Mr W. E. Balston; with field notes by the collector, Mr. G. C. Shortridge. *Proceedings of the Zoological Society of London* **1906**, 763–777.

Thomas O (1909) Some mammals from N.E. Kimberley, northern Australia. *Annals & Magazine of Natural History* **8**, 149–152. doi:10.1080/0022293 0908692558

Thomas MO (1910) *Lutreolina*. *Annals & Magazine of Natural History* **8**, 247.

Thomas MO (1912) Notes on *Phascogale* and *Chaetocercus*. *Annals & Magazine of Natural History* **9**, 91–92. doi:10.1080/00222931208693109

Thomas MO (1913) Two new Australian mammals. *Annals & Magazine of Natural History* **11**, 79–80. doi:10.1080/00222931308693294

Thomas MO (1921) A new genus of opossum from Southern Patagonia. *Annals & Magazine of Natural History* **8**, 136–139. doi:10.1080/00222932108632567

Thomas O (1923) The Godman Exploration Fund: list of mammals from north Queensland collected by Mr. T. V. Sherrin. *Annals & Magazine of Natural History* **9**(11), 170–178. doi:10.1080/00222932308632835

Thomas O (1924) A new pouched mouse (*Phascogale*) from northern New South Wales. *Annals & Magazine of Natural History* **9**(14), 528. doi:10.1080/00222932408633152

Thomas O (1926) The local races of *Dasyurus hallucatus*. *Annals & Magazine of Natural History* **9**(18), 543–544. doi:10.1080/00222932608633550

Thompson GG, Thompson SA (2008) Abundance and spatial distribution of five small mammals at a local scale. *Australian Mammalogy* **30**, 65–70.

Timm RM, Patterson BD (2008) Genus *Caenolestes*. In *Mammals of South America. Volume 1. Marsupials, Xenarthrans, Shrews and Bats*. (Ed. AL Gardner) pp. 120–123. University of Chicago Press, Chicago IL, USA.

Toftegaard CL, Bradley AJ (2003) Chemical communication in dasyurid marsupials. In *Predators with Pouches: the Biology of Carnivorous Marsupials*. (Eds M Jones, C Dickman and M Archer) pp. 347–357. CSIRO Publishing, Melbourne.

Toftegaard CL, Moore C, Bradley AJ (1999) Chemical characterization of urinary pheromones in Brown Antechinus, *Antechinus stuartii*. *Journal of Chemical Ecology* **25**, 527–535. doi:10.1023/A:1020901820044

Tomlinson S, Withers PC, Maloney SK (2012) Comparative thermoregulatory physiology of two dunnarts, *Sminthopsis macroura* and *Sminthopsis ooldea* (Marsupialia: Dasyuridae). *Australian Journal of Zoology* **60**, 54–63. doi:10.1071/ZO12034

Traill BJ, Coates TD (1993) Field observations on the Brush-tailed Phascogale *Phascogale tapoatafa* (Marsupialia: Dasyuridae). *Australian Mammalogy* **16**, 61–65.

Troughton E (1932) A new species of fat-tailed marsupial mouse, and the status of *Antechinus froggatti* Ramsay. *Records of the Australian Museum* **18**, 349–353. doi:10.3853/j.0067-1975.18.1932.738

Troughton E (1965) A review of the marsupial genus *Sminthopsis* (Phascogalinae) and diagnoses of new forms. *Proceedings of the Linnean Society of New South Wales* **89**, 309–321.

Tyndale-Biscoe H (1997) Marsupials and modern biology – breaking the paradigms. In *Marsupial Biology: Recent Research, New Perspectives.* (Eds NR Saunders and LA Hinds) pp. 1–19. University of New South Wales Press, Sydney.

Tyndale-Biscoe H (2005) *Life of Marsupials.* CSIRO Publishing, Melbourne.

Tyndale-Biscoe H, Renfree M (1987) *Reproductive Physiology of Marsupials.* Cambridge University Press, Cambridge, UK.

Ujvari B, Oakwood M, Madsen T (2013) Queensland Northern Quolls are not immune to Cane Toad toxin. *Wildlife Research* **40**, 228–231. doi:10.1071/WR13011

Urban MC (2015) Accelerating extinction risk from climate change. *Science* **348**, 571–573. doi:10.1126/science.aaa4984

Valente A (1984) Reproductive biology of *Antechinomys laniger* (*spenceri* form) (Marsupialia: Dasyuridae). PhD thesis. La Trobe University, Melbourne.

Valente A (2008) Kultarr *Antechinomys laniger.* In *The Mammals of Australia.* 3rd edn. (Eds S Van Dyck and R Strahan) pp. 122–124. Reed New Holland, Sydney.

van der Ree R, Soderquist TR, Bennett AF (2001) Home-range use by the Brush-tailed Phascogale (*Phascogale tapoatafa*) (Marsupialia) in high-quality, spatially limited habitat. *Wildlife Research* **28**, 517–525. doi:10.1071/WR00051

van der Ree R, Bennett AF, Soderquist TR (2006) Nest-tree selection by the threatened Brush-tailed Phascogale (*Phascogale tapoatafa*) (Marsupialia: Dasyuridae) in a highly fragmented agricultural landscape. *Wildlife Research* **33**, 113–119. doi:10.1071/WR05007

Van Dyck S (1979) Behaviour in captive individuals of the dasyurid marsupial *Planigale maculata* (Gould 1851). *Memoirs of the Queensland Museum* **19**, 413–429.

Van Dyck S (1980) The Cinnamon Antechinus, *Antechinus leo* (Marsupialia: Dasyuridae), a new species from the vine-forests of Cape York Peninsula. *Australian Mammalogy* **3**, 5–17.

Van Dyck S (1982a) The relationships of *Antechinus stuartii* and *A. flavipes* (Dasyuridae, Marsupialia) with special reference to Queensland. In *Carnivorous Marsupials.* (Ed. M Archer) pp. 723–766. Royal Zoological Society of New South Wales, Sydney.

Van Dyck S (1982b) The status and relationships of the Atherton Antechinus, *Antechinus godmani* (Marsupialia: Dasyuridae). *Australian Mammalogy* **5**, 195–210.

Van Dyck S (1985) *Sminthopsis leucopus* (Marsupialia: Dasyuridae) in north Queensland rainforest. *Australian Mammalogy* **8**, 53–60.

Van Dyck S (1986) The Chestnut Dunnart, *Sminthopsis archeri* (Marsupialia: Dasyuridae) a new species from the savannas of Papuan New Guinea and Cape York Peninsula, Australia. *Australian Mammalogy* **9**, 111–124.

Van Dyck S (1987) The Bronze Quoll, *Dasyurus spartacus* (Marsupialia: Dasyuridae), a new species from the savannahs of Papua New Guinea. *Australian Mammalogy* **11**, 145–156.

Van Dyck S (1997) Revisionary studies in *Murexia* and *Antechinus* (Marsupialia: Dasyuridae). PhD thesis. University of New South Wales, Sydney.

Van Dyck S (2002) Morphology-based revision of *Murexia* and *Antechinus* (Marsupialia: Dasyuridae). *Memoirs of the Queensland Museum* **48**, 239–330.

Van Dyck SM, Crowther MS (2000) Reassessment of northern representatives of the *Antechinus stuartii* complex (Marsupialia: Dasyuridae): *A. subtropicus* sp. nov. and *A. adustus* new status. *Memoirs of the Queensland Museum* **45**(2), 611–635.

Van Dyck S, Gibbons D (1980) Tuan predation by Powerful Owls. *Victorian Naturalist* **97**, 58–63.

Van Dyck S, Ogilvie P (1977) *Antechinus swainsonii* (Waterhouse, 1840), the Dusky Marsupial Mouse, an addition to the mammal fauna of Queensland. *Memoirs of the Queensland Museum* **18**(1), 69–73.

Van Dyck S, Woinarski JCZ, Press AJ (1994) The Kakadu Dunnart *Sminthopsis bindi* (Marsupialia: Dasyuridae), a new species from the stony woodlands of the Northern Territory. *Memoirs of the Queensland Museum* **37**, 311–323.

van Eeden LM, Crowther MS, Dickman CR, Macdonald DW, Ripple WJ, Ritchie EG, *et al.* (2018) A global review of livestock – large carnivore conflict management. *Conservation Biology*, **32**, 26–34. doi:10.1111/cobi.12959

Van Oosterzee P (2006) Drawing the Wallace Line. In. *Evolution and Biogeography of Australasian Vertebrates*. (Eds JR Merrick, M Archer, GM Hickey and MSY Lee) pp. 95–110. Auscipub, Sydney.

van Riper C, III, van Riper SG, Goff ML, Laird M (1986) The epizootiology and ecological significance of malaria in Hawaiian land birds. *Ecological Monographs* **56**, 327–344. doi:10.2307/1942550

van Weenen J (2008) *Sminthopsis aitkeni*. In *IUCN Red List of Threatened Species. Version 2013.2*. IUCN, Gland, Switzerland.

Vieira EM, Astúa de Moraes D (2003) Carnivory and insectivory in Neotropical marsupials. In *Predators with Pouches: the Biology of Carnivorous Marsupials*. (Eds M Jones, C Dickman and M Archer) pp. 271–284. CSIRO Publishing, Melbourne.

Vieira EM, Monteiro-Filho ELA (2003) Vertical stratification of small mammals in the Atlantic rain forest of south-eastern Brazil. *Journal of Tropical Ecology* **19**, 501–507. doi:10.1017/S0266467403003559

Voss RS, Jansa SA (2003) Phylogenetic studies on didelphid marsupials II. Nonmolecular data and new IRBP sequences: separate and combined analyses of didelphine relationships with denser taxon sampling. *Bulletin of the American Museum of Natural History* **276**, 1–82. doi:10.1206/0003-0090(2003)276<0001:PSODMI>2.0.CO;2

Voss RS, Jansa SA (2009) Phylogenetic relationships and classification of didelphid marsupials, an extant radiation of New World metatherian mammals. *Bulletin of the American Museum of Natural History* **322**, 1–177. doi:10.1206/322.1

Voss RS, Jansa SA (2012) Snake-venom resistance as a mammalian trophic adaptation: lessons from didelphid marsupials. *Biological Reviews of the Cambridge Philosophical Society* **87**, 822–837. doi:10.1111/j.1469-185X.2012.00222.x

Voss RS, Lunde DP, Simmons NB (2001) The mammals of Paracou, French Guiana: a Neotropical lowland rainforest fauna, part 2 – nonvolant species. *Bulletin of the American Museum of Natural History* **263**, 1–236. doi:10.1206/0003-0090(2001)263<0003:TMOPFG>2.0.CO;2

Voss RS, Gardner AL, Jansa SA (2004) On the relationships of "*Marmosa*" *formosa* Shamel, 1930 (Marsupialia: Didelphidae), a phylogenetic puzzle from the Chaco of northern Argentina. *American Museum Novitates* **3442**, 1–18. doi:10.1206/0003-0082(2004)442<0001:OTROMF>2.0.CO;2

Voss RS, Myers P, Catzeflis F, Carmignotto AP, Barreiro J (2009) Chapter 11. The six opossums of Félix de Azara: identification, taxonomic history, neotype designations, and nomenclatural recommendations. *Bulletin of the American Museum of Natural History* **331**(1), 406–433. doi:10.1206/582-11.1

Wainer JW (1976) Studies of an island population of *Antechinus minimus* (Marsupialia, Dasyuridae). *Australian Journal of Zoology* **19**, 1–7.

Wainwright M (2002) *The Natural History of Costa Rican Mammals*. Zona Tropical, Miami FL, USA.

Waithman J (1979) A report on a collection of mammals from southwest Papua. 1972-1973. *Australian Zoologist* **20**, 313–326.

Wakefield NA, Warneke RM (1963) Some revision in *Antechinus* (Marsupialia) 1. *Victorian Naturalist* **80**, 192–219.

Wakefield NA, Warneke RM (1967) Some revision in *Antechinus* (Marsupialia) 2. *Victorian Naturalist* **84**, 69–99.

Walsberg GE (2000) Small mammals in hot deserts: some generalizations revisited. *Bioscience* **50**, 109–120. doi:10.1641/0006-3568(2000)050[0109:SMIHDS]2.3.CO;2

Walters JR, Derrickson SR, Fry DM, Haig SM, Marzluff JM, Wunderle JM, Jr (2010) Status of the California Condor (*Gymnogyps californianus*) and efforts to achieve its recovery. *The Auk* **127**, 969–1001. doi:10.1525/auk.2010.127.4.969

Warburton N, Wood C, Lloyd C, Song S, Withers P (2003) The 3-dimensional anatomy of the North-western Marsupial Mole (*Notoryctes caurinus* Thomas 1920) using computed tomography, X-ray and magnetic resonance imaging. *Records of the Western Australian Museum* **22**, 1–7. doi:10.18195/issn.0312-3162.22(1).2003.001-007

Ward SJ (2000) Possible predation on Feathertail Gliders (*Acrobates pygmaeus*) by Agile Antechinus (*Antechinus agilis*). *Australian Mammalogy* **21**, 173–176.

Wardell-Johnson G (1986) Use of nest boxes by Mardos, *Antechinus flavipes leucogaster*, in regenerating Karri forest in south western Australia. *Australian Wildlife Research* **13**, 407–417. doi:10.1071/WR9860407

Warnecke L, Geiser F (2009) Basking behaviour and torpor use in free-ranging *Planigale gilesi. Australian Journal of Zoology* **57**, 373–375. doi:10.1071/ZO09097

Warnecke L, Geiser F (2010) The energetics of basking behaviour and torpor in a small marsupial exposed to simulated natural conditions. *Journal of Comparative Physiology. B, Biochemical, Systemic, and Environmental Physiology* **180**, 437–445. doi:10.1007/s00360-009-0417-6

Warnecke L, Turner JM, Geiser F (2008) Torpor and basking in a small arid zone marsupial. *Naturwissenschaften* **95**, 73–78. doi:10.1007/s00114-007-0293-4

Warnecke L, Körtner G, Burwell CJ, Turner JM, Geiser F (2012) Short-term movement patterns and diet of small dasyurid marsupials in semiarid Australia. *Australian Mammalogy* **34**, 49–54. doi:10.1071/AM10052

Waterhouse GR (1840) Description of a new marsupial mammal belonging to the genus *Phascogale. Magazine of Natural History* **2**(4), 299.

Waterhouse GR (1842) On two new species of marsupial animals from South Australia. *Proceedings of the Zoological Society of London* **10**, 47–48.

Waterhouse GR (1846) *A Natural History of Mammalia. Volume 1. Containing the Order Marsupiata, or Pouched Animals*. Bailliere, London, UK.

Watson ML, Calaby JH (2008) Fawn Antechinus *Antechinus bellus*. In *The Mammals of Australia*. 3rd edn. (Eds S Van Dyck and R Strahan) pp. 85–86. Reed New Holland, Sydney.

Watt AM (1992) The ecology of three species of *Antechinus* (Marsupialia: Dasyuridae) in upland rainforests of north-east Queensland. PhD thesis. James Cook University, Townsville.

Waudby HP, Petit S (2017) Thermoregulatory value of cracking-clay soil shelters for small vertebrates during extreme desert conditions. *Integrative Zoology* **12**, 237–249. doi:10.1111/1749-4877.12225

Webb LJ (1978) A general classification of Australian rainforest. *Australian Plants* **9**, 349–363.

Webb JK, Brown GP, Child T, Greenlees MJ, Phillips BL, Shine R (2008) A native dasyurid predator (Common Planigale, *Planigale maculata*) rapidly learns to avoid a toxic invader. *Austral Ecology* **33**, 821–829. doi:10.1111/j.1442-9993.2008.01847.x

Webb JK, Pearson D, Shine R (2011) A small dasyurid predator (*Sminthopsis virginiae*) rapidly learns to avoid a toxic invader. *Wildlife Research* **38**, 726–731. doi:10.1071/WR10206

Westerman M, Young J, Donnellan S, Woolley PA, Krajewski C (2006) Molecular relationships of New Guinean three-striped dasyures (*Myoictis*, Marsupialia: Dasyuridae). *Journal of Mammalian Evolution* **13**, 211–222. doi:10.1007/s10914-006-9023-5

Westerman M, Young J, Krajewski C (2008) Molecular relationships of species of *Pseudantechinus, Parantechinus* and *Dasykaluta* (Marsupialia: Dasyuridae). *Australian Mammalogy* **29**, 201–212.

Westerman M, Kear BP, Aplin K, Meredith RW, Emerling C, Springer MS (2012) Phylogenetic relationships of living and recently extinct bandicoots based on nuclear and mitochondrial DNA sequences. *Molecular Phylogenetics and Evolution* **62**(1), 97–108. doi:10.1016/j.ympev.2011.09.009

Westerman M, Blacket MJ, Hintz A, Armstrong K, Woolley PA, Krajewski C (2016) A plethora of planigales: genetic variability and cryptic species in a genus of dasyurid marsupials from northern Australia. *Australian Journal of Zoology* **64**, 303–311. doi:10.1071/ZO16052

Western Australian Museum. Department of Zoology: Mammals Section Correspondence. Archives: ACC 835A: G.C.Shortridge Letters 1904–1906.

White ME (2006) Environments of the geological past. In *Evolution and Biogeography of Australasian Vertebrates*. (Eds JR Merrick, M Archer, GM Hickey and MSY Lee) pp. 17–50. Auscipub, Sydney.

Whitford D, Fanning FD, White AW (1982) Some information on reproduction, growth and development in *Planigale gilesi* (Dasyuridae, Marsupialia). In *Carnivorous Marsupials*. (Ed. M Archer) pp. 77–81. Royal Zoological Society of New South Wales, Sydney.

Whitley GP, Rutledge M (1974) Krefft, Johann Ludwig Gerard (Louis). In *Australian Dictionary of*

Biography. Volume 5. Pp. 42–44. Melbourne University Publishing, Melbourne.

Wilder SM, Yuan L, Roumilhac C, LeGal A, Raubenheimer D, Simpson SJ, *et al.* (2016) Three-dimensional diet regulation and the consequences of choice for weight and activity level of a marsupial carnivore. *Journal of Mammalogy* **97**, 1645–1651. doi:10.1093/jmammal/gyw130

Williams D, Wüster W, Fry BG (2006) The good, the bad and the ugly: Australian snake taxonomists and a history of the taxonomy of Australia's venomous snakes. *Toxicon* **48**(7), 919–930. doi:10.1016/j.toxicon.2006.07.016

Williams R, Williams A (1982) The life cycle of *Antechinus swainsonii* (Dasyuridae, Marsupialia). In *Carnivorous Marsupials.* (Ed. M Archer) pp. 89–95. Royal Zoological Society of New South Wales, Sydney.

Williams SE, Pearson RG, Walsh PJ (1996) Distributions and biodiversity of the terrestrial vertebrates of Australia's Wet Tropics: a review of current knowledge. *Pacific Conservation Biology* **2**, 327–362. doi:10.1071/PC960327

Willmott WF, Moore T (2006) *Rocks and Landscapes of the National Parks of Central Queensland*. Geological Society of Australia, Queensland Division, Brisbane.

Wilson BA (1986) Reproduction in the female dasyurid *Antechinus minimus maritimus* (Marsupialia, Dasyuridae). *Australian Journal of Zoology* **34**, 189–197. doi:10.1071/ZO9860189

Wilson BA, Aberton JG (2006) Effects of landscape, habitat and fire on the distribution of the White-footed Dunnart (*Sminthopsis leucopus*) (Marsupialia: Dasyuridae) in the Eastern Otways. *Australian Mammalogy* **28**(1), 27–38. doi:10.1071/AM06004

Wilson BA, Bachmann MR (2008) Swamp Antechinus *Antechinus minimus*. In *The Mammals of Australia.* 3rd edn. (Eds S Van Dyck and R Strahan)pp. 93–94. Reed New Holland, Sydney.

Wilson BA, Bourne AR, Jessop RE (1986) Ecology of small mammals in coastal heathland at Anglesea, Victoria. *Australian Wildlife Research* **13**, 397–406. doi:10.1071/WR9860397

Wilson BA, Aberton J, Reichl T (2001) Effects of fragmented habitat and fire on the distribution and ecology of the Swamp Antechinus (*Antechinus minimus maritimus*) in the Eastern Otways, Victoria. *Wildlife Research* **28**, 527–536. doi:10.1071/WR00016

Wilson BA, Dickman CR, Fletcher TP (2003) Dasyurid dilemmas: problems and solutions for conserving Australia's small carnivorous marsupials. In *Predators with Pouches: the Biology of Carnivorous Marsupials.* (Eds M Jones, C Dickman and M Archer) pp. 407–421. CSIRO Publishing, Melbourne.

Withers PC, Thompson GG, Seymour RS (2000) Metabolic physiology of the North-western Marsupial Mole, *Notoryctes caurinus* (Marsupialia: Notoryctidae). *Australian Journal of Zoology* **48**, 241–258. doi:10.1071/ZO99073

Withers PC, Cooper CE, Nespolo RF (2012) Evaporative water loss, relative water economy and evaporative partitioning of a heterothermic marsupial, the Monito del Monte (*Dromiciops gliroides*). *The Journal of Experimental Biology* **215**, 2806–2813. doi:10.1242/jeb.070433

Woinarski JCZ (1992) Habitat relationships for two poorly known mammal species *Pseudomys calabyi* and *Sminthopsis* sp. from the wet-dry tropics of the Northern Territory. *Australian Mammalogy* **15**, 47–54.

Woinarski JCZ (2004a) *National Multi-species Recovery Plan for the Carpentarian Antechinus* Pseudantechinus mimulus*, Butler's Dunnart* Sminthopsis butleri *and Northern Hopping-mouse* Notomys aquilo*, 2004–2009.* Northern Territory Department of Infrastructure Planning and Environment, Darwin.

Woinarski JCZ (2004b) The forest fauna of the Northern Territory. In *Conservation of Australia's Forest Fauna*. (Ed. D. Lunney) pp. 36–55. Royal Zoological Society of New South Wales, Sydney.

Woinarski J (2006) Butler's Dunnart: *Sminthopsis butleri*. Threatened Species of the Northern Territory. Information sheet. Northern Territory Government, Darwin.

Woinarski J (2008a) Kakadu Dunnart *Sminthopsis bindi*. In *The Mammals of Australia.* 3rd edn. (Eds S Van Dyck and R Strahan) pp. 129–130. Reed New Holland, Sydney.

Woinarski J (2008b) *Pseudantechinus bilarni*. In *IUCN Red List of Threatened Species. Version 2013.1*. IUCN, Gland, Switzerland.

Woinarski J (2008c) *Sminthopsis bindi*. In *IUCN Red List of Threatened Species. Version 2013.2*. IUCN, Gland, Switzerland.

Woinarski J, Dickman C (2008a) *Pseudantechinus mimulus*. In *IUCN Red List of Threatened Species. Version 2013.1*. IUCN, Gland, Switzerland.

Woinarski J, Dickman C (2008b) *Sminthopsis macroura*. In *IUCN Red List of Threatened Species. Version 2013.2*. IUCN, Gland, Switzerland.

Woinarski JCZ, Woolley PA, Van Dyck S (1996) The distribution of the dunnart *Sminthopsis butleri*. *Australian Mammalogy* **19**, 27–29.

Woinarski J, Brennan K, Hempel C, Armstrong M, Milne D, Chatto R (2003) *Biodiversity Conservation on the Tiwi Islands, Northern Territory. Part 2: Fauna*. Northern Territory Department of Infrastructure, Planning and Environment, Darwin.

Woinarski J, van Weenen J, Burbidge A (2008a) *Ningaui ridei*. In *IUCN Red List of Threatened Species. Version 2013.1*. IUCN, Gland, Switzerland.

Woinarski J, Rhind S, Oakwood M (2008b) *Phascogale pirata*. In *IUCN Red List of Threatened Species. Version 2013.1*. IUCN, Gland, Switzerland.

Woinarski J, Burbidge A, van Weenen J, Dickman C (2008c) *Pseudantechinus macdonnellensis*. In *IUCN Red List of Threatened Species. Version 2013.1*. IUCN, Gland, Switzerland.

Woinarski JCZ, Armstrong M, Brennan K, Fisher A, Griffiths AD, Hill B, Milne DJ, *et al.* (2010) Monitoring indicates rapid and severe decline of native small mammals in Kakadu National Park, northern Australia. *Wildlife Research* **37**, 116–126. doi:10.1071/WR09125

Woinarski JCZ, Legge S, Fitzsimons JA, Traill BJ, Burbidge AA, Fisher A, *et al.* (2011a) The disappearing mammal fauna of northern Australia: context, cause, and response. *Conservation Letters* **4**, 192–201. doi:10.1111/j.1755-263X.2011.00164.x

Woinarski JCZ, Ward S, Mahney T, Bradley J, Brennan K, Ziembicki M, Fisher A (2011b) The mammal fauna of the Sir Edward Pellew Island group, Northern Territory, Australia: refuge and death-trap. *Wildlife Research* **38**, 307–322. doi:10.1071/WR10184

Woinarski JCZ, Burbidge AA, Harrison PL (2014) *The Action Plan for Australian Mammals 2012*. CSIRO Publishing, Melbourne.

Woinarski JCZ, Burbidge AA, Harrison PL (2015) Ongoing unravelling of a continental fauna: decline and extinction of Australian mammals since European settlement. *Proceedings of the National Academy of Sciences of the United States of America* **112**, 4531–4540. doi:10.1073/pnas.1417301112

Woinarski JCZ, Garnett ST, Legge SM, Lindenmayer DB (2017) The contribution of policy, law, management, research, and advocacy failings to the recent extinctions of three Australian vertebrate species. *Conservation Biology* **31**, 13–23. doi:10.1111/cobi.12852

Wollaston AFR (1912) *Pygmies & Papuans: The Stone Age To-Day in Dutch New Guinea*. Reprint 2013. Forgotten Books, London, UK.

Wollaston AFR (2013) Two New Guinea expeditions. In *Letters and Diaries of A. F. R. Wollaston*. (Ed. M Wollaston) pp. 97–155. Cambridge University Press, New York, USA.

Wood DH (1970) An ecological study of *Antechinus stuartii* (Marsupialia) in a south-east Queensland rain forest. *Australian Journal of Zoology* **18**, 185–207. doi:10.1071/ZO9700185

Wood Jones F (1923) *The Mammals of South Australia. Part I. The Monotremes and the Carnivorous Marsupials*. Government Printer, Adelaide.

Woolley PA (1966) Reproduction in *Antechinus* spp. and other dasyurid marsupials. *Symposia of the Zoological Society of London* **15**, 281–294.

Woolley P (1971a) Maintenance and breeding of laboratory colonies of *Dasyuroides byrnei* and *Dasycercus cristicauda*. *International Zoo Yearbook* **11**, 351–354. doi:10.1111/j.1748-1090.1971.tb01932.x

Woolley PA (1971b) Observations on the reproductive biology of the Dibbler, *Antechinus apicalis* (Marsupialia: Dasyuridae). *Journal of the Royal Society of Western Australia* **54**, 99–102.

Woolley PA (1977) In search of the Dibbler, *Antechinus apicalis* (Marsupialia: Dasyuridae). *Journal of the Royal Society of Western Australia* **59**(4), 111–117.

Woolley PA (1980) Further searches for the Dibbler, *Antechinus apicalis* (Marsupialia: Dasyuridae). *Journal of the Royal Society of Western Australia* **63**(2), 47–52.

Woolley PA (1981) *Antechinus bellus*, another dasyurid marsupial with post-mating mortality in males. *Journal of Mammalogy* **62**, 381–382. doi:10.2307/1380717

Woolley PA (1982) Phallic morphology of the Australian species of *Antechinus* (Dasyuridae, Marsupialia): a new taxonomic tool? In *Carnivorous Marsupials*. (Ed. M. Archer) pp. 767–781. Royal Zoological Society of NSW, Sydney.

Woolley PA (1984a) Phallic morphology of the New Guinean species of *Antechinus. Bulletin of the Australian Mammal Society* **8**, 182.

Woolley PA (1984b) Reproduction in *Antechinomys laniger* (*'spenceri'* form) (Marsupialia: Dasyuridae): field and laboratory observation. *Australian Wildlife Research* **11**, 481–489. doi:10.1071/WR9840481

Woolley PA (1988) Reproduction in the Ningbing Antechinus (Marsupialia: Dasyuridae): field and laboratory observations. *Australian Wildlife Research* **15**, 149–156. doi:10.1071/WR9880149

Woolley PA (1989) Nest location by spool-and-line tracking of dasyurid marsupials in New Guinea. *Journal of Zoology* **218**, 689–700. doi:10.1111/j.1469-7998.1989.tb05010.x

Woolley PA (1990a) Mulgaras, *Dasycercus cristicauda* (Marsupialia: Dasyuridae); their burrows and records of attempts to collect live animals between 1966 and 1979. *Australian Mammalogy* **13**, 61–64.

Woolley PA (1990b) Reproduction in *Sminthopsis macroura* (Marsupialia: Dasyuridae). Parts I and II. *Australian Journal of Zoology* **38**, 187–217. doi:10.1071/ZO9900187

Woolley PA (1991a) Reproduction in *Dasykaluta rosamondae* (Marsupialia: Dasyuridae): field and laboratory observations. *Australian Journal of Zoology* **39**, 549–568. doi:10.1071/ZO9910549

Woolley PA (1991b) Reproduction in *Pseudantechinus macdonnellensis* (Marsupialia: Dasyuridae): field and laboratory observations. *Wildlife Research* **18**, 13–25. doi:10.1071/WR9910013

Woolley PA (1991c) Reproductive pattern of captive Boullanger Island Dibblers, *Parantechinus apicalis* (Marsupialia: Dasyuridae). *Wildlife Research* **18**, 157–163. doi:10.1071/WR9910157

Woolley PA (1992) New records of the Julia Creek Dunnart, *Sminthopsis douglasi* (Marsupialia: Dasyuridae). *Wildlife Research* **19**, 779–783. doi:10.1071/WR9920779

Woolley PA (1994) The dasyurid marsupials of New Guinea: use of museum specimens to assess seasonality of breeding. *Science in New Guinea* **20**, 49–55.

Woolley PA (1995) Observations on reproduction in captive *Parantechinus bilarni* (Marsupialia: Dasyuridae). *Australian Mammalogy* **18**, 83–85.

Woolley PA (2001) Observations on the reproductive biology of *Myoictis wallacei, Neophascogale lorentzi, Dasyurus albopunctatus* and *Dasyurus spartacus*, dasyurid marsupials endemic to New Guinea. *Australian Mammalogy* **23**, 63–66. doi:10.1071/AM01063

Woolley PA (2003) Reproductive biology of some dasyurid marsupials of New Guinea. In *Predators with Pouches: the Biology of Carnivorous Marsupials.* (Eds M Jones, C Dickman and M Archer) pp. 169–182. CSIRO Publishing, Melbourne.

Woolley PA (2005a) Revision of the three-striped dasyures, genus *Myoictis* (Marsupialia: Dasyuridae), of New Guinea, with description of a new species. *Records of the Australian Museum* **57**, 321–340. doi:10.3853/j.0067-1975.57.2005.1450

Woolley PA (2005b) The species of *Dasycercus* Peters, 1875 (Marsupialia: Dasyuridae). *Memoirs of the Museum of Victoria* **62**, 213–221. doi:10.24199/j.mmv.2005.62.8

Woolley PA (2006) Studies on the Crest-tailed Mulgara *Dasycercus cristicauda* and the Brush-tailed Mulgara *Dasycercus blythi* (Marsupialia: Dasyuridae). *Australian Mammalogy* **28**, 117–120. doi:10.1071/AM06018

Woolley PA (2007) Observations on the oestrous cycle and breeding season of the Chestnut Dunnart *Sminthopsis archeri*, a little known dasyurid marsupial. *Australian Mammalogy* **29**, 89–96. doi:10.1071/AM07011

Woolley PA (2008a) Brush-tailed Mulgara *Dasycercus blythi*. In *The Mammals of Australia*. 3rd edn. (Eds S Van Dyck and R Strahan) pp. 47–48. Reed New Holland, Sydney.

Woolley PA (2008b) Butler's Dunnart *Sminthopsis butleri*. In *The Mammals of Australia*. 3rd edn. (Eds S Van Dyck and R Strahan) pp. 130–131. Reed New Holland, Sydney.

Woolley PA (2008c) Dibbler *Parantechinus apicalis*. In *The Mammals of Australia*. 3rd edn. (Eds S Van Dyck and R Strahan) pp. 65-66. Reed New Holland, Sydney.

Woolley PA (2008d) Family Dasyuridae. In *The Mammals of Australia*. 3rd edn. (Eds S Van Dyck and R Strahan) pp. 45–46. Reed New Holland, Sydney.

Woolley PA (2008e) Fat-tailed Pseudantechinus *Pseudantechinus macdonnellensis*. In *The Mammals of Australia*. 3rd edn. (Eds S Van Dyck and R Strahan) pp. 69–70. Reed New Holland, Sydney.

Woolley PA (2008f) Julia Creek Dunnart, *Sminthopsis douglasi*. In *The Mammals of Australia*. 3rd edn. (Eds S Van Dyck and R Strahan) pp. 136–137. Reed New Holland, Sydney.

Woolley PA (2008g) Kaluta *Dasykaluta rosamondae*. In *The Mammals of Australia*. 3rd edn. (Eds S Van Dyck and R Strahan) pp. 51–52. Reed New Holland, Sydney.

Woolley PA (2008h) *Myoictis leucura*. In *IUCN Red List of Threatened Species. Version 2013.1*. IUCN, Gland, Switzerland.

Woolley PA (2008i) *Myoictis wavicus*. In *IUCN Red List of Threatened Species. Version 2013.1*. IUCN, Gland, Switzerland.

Woolley PA (2008j) Ningbing Pseudantechinus *Pseudantechinus ningbing*. In *The Mammals of Australia*. 3rd edn. (Eds S Van Dyck and R Strahan) pp. 73–74. Reed New Holland, Sydney.

Woolley PA (2008k) Red-cheeked Dunnart, *Sminthopsis virginiae*. In *The Mammals of Australia*. 3rd edn. (Eds S Van Dyck and R Strahan) pp. 158–159. Reed New Holland, Sydney.

Woolley PA (2008l) Sandstone Pseudantechinus *Pseudantechinus bilarni*. In *The Mammals of Australia*. 3rd edn. (Eds S Van Dyck and R Strahan) pp. 67–68. Reed New Holland, Sydney.

Woolley PA (2008m) Woolley's Pseudantechinus *Pseudantechinus woolleyae*. In *The Mammals of Australia*. 3rd edn. (Eds S Van Dyck and R Strahan) pp. 76–77. Reed New Holland, Sydney.

Woolley PA (2011) *Pseudantechinus mimulus*: a little known dasyurid marsupial. *Australian Mammalogy* **33**, 57–67. doi:10.1071/AM10043

Woolley PA, Ahern LD (1983) Observations on the ecology and reproductive biology of *Sminthopsis leucopus* (Marsupialia: Dasyuridae). *Proceedings of the Royal Society of Victoria* **95**, 169–180.

Woolley PA, Elliott C (2013) Breeding in wild populations of a small dasyurid marsupial, *Planigale ingrami*, in north-western Queensland using a novel method for collection of specimens. *Australian Mammalogy* **36**, 81–89.

Woolley PA, Gilfillan SL (1990) Confirmation of polyoestry in captive White-footed Dunnarts, *Sminthopsis leucopus*. *Australian Mammalogy* **14**, 137–138.

Woolley PA, Valente A (1982) The Dibbler, *Parantechinus apicalis* (Marsupialia: Dasyuridae): failure to locate populations in four regions in the south of Western Australia. *Australian Mammalogy* **5**, 241–245.

Woolley PA, Valente A (1986) Reproduction in *Sminthopsis longicaudata* (Marsupialia: Dasyuridae): laboratory observations. *Australian Wildlife Research* **13**, 7–12. doi:10.1071/WR9860007

Woolley PA, Watson MR (1984) Observations on a captive outdoor breeding colony of a small dasyurid marsupial, *Sminthopsis crassicaudata*. *Australian Wildlife Research* **11**, 249–254. doi:10.1071/WR9840249

Woolley PA, Raftopoulos SA, Coleman GJ, Armstrong SM (1991) A comparative study of the circadian activity patterns of two New Guinean dasyurid marsupials, *Phascolosorex dorsalis* and *Antechinus habbema*. *Australian Journal of Zoology* **39**, 661–671. doi:10.1071/ZO9910661

Woolley PA, Patterson MF, Stephenson GM, Stephenson DG (2002) The ilio-marsupialis muscle in the dasyurid marsupial *Sminthopsis douglasi*: form, function and fibre type profiles in females with and without suckling young. *The Journal of Experimental Biology* **205**, 3775–3781.

Woolley PA, Westerman M, Krajewski C (2007) Interspecific affinities within the genus *Sminthopsis* (Dasyuromorphia: Dasyuridae) based on morphology of the penis: congruence with other anatomical and molecular data. *Journal of Mammalogy* **88**(6), 1381–1392. doi:10.1644/06-MAMM-A-443R.1

Woolley P, Leary T, Seri L, Flannery T, Wright D, Hamilton S, *et al.* (2008) *Dasyurus albopunctatus*. In *IUCN Red List of Threatened Species. Version 2013.1.* IUCN, Gland, Switzerland.

Woolley P, Leary T, Seri L, Wright D, Hamilton S, Helgen K, *et al.* (2008) *Murexia longicaudata*. In *IUCN Red List of Threatened Species. Version 2013.2.* IUCN, Gland, Switzerland.

Woolley P, Leary T, Wright D, Hamilton S, Helgen K, Singadan R, *et al.* (2008) *Neophascogale lorentzii*. In *IUCN Red List of Threatened Species. Version 2013.1.* IUCN, Gland, Switzerland.

Woolley P, Leary T, Seri L, Wright D, Hamilton S, Helgen K, *et al.* (2008) *Phascolosorex doriae*. In *IUCN Red List of Threatened Species. Version 2013.1.* IUCN, Gland, Switzerland.

Woolley P, Leary T, Seri L, Flannery T, Wright D, Hamilton S, Helgen K, *et al.* (2008) *Phascolosorex dorsalis*. In *IUCN Red List of Threatened Species. Version 2013.1.* IUCN, Gland, Switzerland.

Woolley P, Helgen K, Dickman C, Lunde D, Winter J, Burnett S (2008) *Sminthopsis archeri*. In *IUCN Red List of Threatened Species. Version 2013.2.* IUCN, Gland, Switzerland.

Woolley PA, Haslem A, Westerman M (2013) Past and present distribution of *Dasycercus*: toward a better understanding of the identity of specimens in cave deposits and the conservation status of the currently recognised species *D. blythi* and *D. cristicauda* (Marsupialia : Dasyuridae). *Australian Journal of Zoology* **61**(4), 281–290. doi:10.1071/ZO13034

Woolnough AP, Carthew SM (1994) Notes on the feeding behaviour of *Ningaui yvonneae* in captivity. *Australian Mammalogy* **17**, 121–122.

Woolnough AP, Carthew SM (1996) Selection of prey by size in *Ningaui yvonneae. Australian Journal of Zoology* **44**, 319–326. doi:10.1071/ZO9960319

Wroe S (1997) A re-examination of proposed morphology-based synapomorphies for the families of Dasyuromorphia (Marsupialia). I. Dasyuridae. *Journal of Mammalian Evolution* **4**(1), 19–52. doi:10.1023/A:1027379610196

Wroe S, Archer M (2006) Origins and early radiations of marsupials. In *Evolution and Biogeography of Australasian Vertebrates.* (Eds JR Merrick, M Archer, GM Hickey and MSY Lee) pp. 551–574. Auscipub, Sydney.

Wroe S, Myers TJ, Wells RT, Gillespie A (1999) Estimating the weight of the Pleistocene Marsupial Lion, *Thylacoleo carnifex* (Thylacoleonidae: Marsupialia): implications for the ecomorphology of a marsupial super-predator and hypotheses of impoverishment of Australian marsupial carnivore faunas. *Australian Journal of Zoology* **47**, 489–498. doi:10.1071/ZO99006

Wroe S, Argot C, Dickman C (2004) On the rarity of big fierce carnivores and primacy of isolation and area: tracking large mammalian carnivore diversity on two isolated continents. *Proceedings. Biological Sciences* **271**, 1203–1211. doi:10.1098/rspb.2004.2694

Wroe S, McHenry C, Thomason J (2005) Bite club: comparative bite force in big biting mammals and the prediction of predatory behaviour in fossil taxa. *Proceedings. Biological Sciences* **272**, 619–625. doi:10.1098/rspb.2004.2986

Wroe S, Chamoli U, Parr WCH, Clausen P, Ridgely R, Witmer L (2013a) Comparative biomechanical modeling of metatherian and placental sabertooths: a different kind of bite for an extreme pouched predator. *PLoS One* **8**(6), e66888. doi:10.1371/journal.pone.0066888

Wroe S, Field JH, Archer M, Grayson DK, Price GJ, Louys J, *et al.* (2013b) Perspective: climate change frames debate over the extinction of megafauna in Sahul (Pleistocene Australia-New Guinea). *Proceedings of the National Academy of Sciences of the United States of America* **110**(22), 8777–8781. doi:10.1073/pnas.1302698110

Zachos FE (2015) Letter to the editor. Taxonomic inflation, the phylogenetic species concept and lineages in the tree of life – a cautionary comment on species splitting. *Journal of Zoological Systematics and Evolutionary Research* **53**(2), 180–184. doi:10.1111/jzs.12088

Zapata SC, Procopio D, Travaini A, Rodríguez A (2013) Summer food habits of the Patagonian Opossum, *Lestodelphys halli* (Thomas,1921), in southern arid

Patagonian shrub-steppes. *Gayana (Concepción)* **77**, 64–67. doi:10.4067/S0717-65382013000100009

Zarza H, Ceballos G, Steele MA (2003) *Marmosa canescens*. *Mammalian Species* **725**, 1–4. doi:10.1644/1545-1410(2003)725<0001:MC>2.0.CO;2

Ziegler AC (1977) Evolution of New Guinea's marsupial fauna in response to a forested environment. In *The Biology of Marsupials*. (Eds B Stonehouse and B Gilmore) pp. 117–138. Macmillan Press, London, UK.

Ziembicki MR, Woinarski JCZ, Mackey B (2013) Evaluating the status of species using Indigenous knowledge: Novel evidence for major native mammal declines in northern Australia. *Biological Conservation* **157**, 78–92. doi:10.1016/j.biocon.2012.07.004

Zimicz N (2014) Avoiding competition: the ecological history of Late Cenozoic metatherian carnivores in South America. *Journal of Mammalian Evolution* **21**, 383–393. doi:10.1007/s10914-014-9255-8

Index

www.ingramcontent.com/pod-product-compliance
Ingram Content Group UK Ltd.
Pitfield, Milton Keynes, MK11 3LW, UK
UKHW061542051026
15625UKWH00011B/573
* 9 7 8 1 4 8 6 3 0 5 1 4 8 *